T0271679

Creating and Marketing New Products and Services

Creating and Marketing New Products and Services

Rosanna Garcia

CRC Press
Taylor & Francis Group
Boca Raton   London   New York

CRC Press is an imprint of the
Taylor & Francis Group, an **informa** business

AN AUERBACH BOOK

CRC Press
Taylor & Francis Group
6000 Broken Sound Parkway NW, Suite 300
Boca Raton, FL 33487-2742

© 2014 by Taylor & Francis Group, LLC
CRC Press is an imprint of Taylor & Francis Group, an Informa business

Printed on acid-free paper
Version Date: 20130910

International Standard Book Number-13: 978-1-4822-0360-8 (Hardback)

Library of Congress Cataloging-in-Publication Data

Garcia, Rosanna.
    Creating and marketing new products and services / Rosanna Garcia.
        pages cm
    Includes bibliographical references and index.
    ISBN 978-1-4822-0360-8
    1. New products. 2. Marketing research. 3. Product design. I. Title.

HF5415.153.G36 2014
658.5'75--dc23   2013035804

Visit the Taylor & Francis Web site at
http://www.taylorandfrancis.com

and the CRC Press Web site at
http://www.crcpress.com

For their endless support always,
A.J. and Tatiana Garcia Rummel and
Paul Rummel

Contents

LIST OF FIGURES   xv

FOREWORD   xxi

ACKNOWLEDGMENTS   xxiii

ABOUT THE AUTHOR   xxv

CHAPTER 1   THE PROACTIVE NEW PRODUCT DEVELOPMENT PROCESS   1

Learning Objectives   1

Introduction   2

New Product Development Is Risky   3

Service Economy   5

Stage-Gate: A Systematic, Sequential, Iterative Process   6

    The Fuzzy Front End: Discovery through Scoping   9

    Design Phase: Building the Business Case through Development   10

    Testing and Validation   12

    Launch and Postlaunch Review   12

Criticisms of Sequential Processing   13

What Is a New Product Exactly? How Can They Be Classified?   16

Types of New Products and Customizing the Development Process   16

Why Innovation Type Matters   17

Avoiding Failures   18

Marketing's Involvement in the Stage-Gate Process   20

The New Product Manager   21

Goals of the Text   22

Chapter Summary   22

Glossary   22

Review Questions   25

Assignment Questions   25

Endnotes   26

CHAPTER 2   NEW PRODUCT INNOVATION STRATEGY   29

Learning Objectives   29

Introduction   30

Corporate Strategy Dictates Innovation Strategy 31
    Innovation Strategy 33
    Innovation Strategy Dictates the New Product Development Strategy 38
    Reactive versus Proactive Innovation Development Process 39
    Which Innovation Strategy to Use? 42
    New Product Portfolio Management 43
    Tools for Portfolio Management 44
    Portfolio Maps 45
    Portfolio Review Process 48
    Formal Process versus Reality 49
Chapter Summary 50
Glossary 50
Review Questions 52
Assignment Questions 52
Endnotes 53

CHAPTER 3   OPPORTUNITY IDENTIFICATION AND IDEA GENERATION:
THE FUZZY FRONT END 55
Learning Objectives 55
Introduction 57
Opportunity Identification 57
Step 1: Generating Product Ideas 58
    Identifying Lead Users 63
Ideation Methods 63
    Scenario Generation 64
    Problem Analysis 65
    Crowdsourcing for New Product Ideas 66
    Brainstorming 67
    Inventive Templates 68
    Individual Creativity 69
    Group Creativity 71
Step 2: Aligning Opportunities with NPD Strategy 71
    Portfolio Alignment 73
Step 3: Market Identification 73
    Growth Potential 74
    Economies of Scale 75
    Competitive Attractiveness 76
    Investment 76
    Reward 77
    Risk 77
Step 4: Market Selection 78
    Substitution 78
    Selecting the Best Opportunities 80
Chapter Summary 81
Glossary 81
Review Questions 83
Assignment Questions 83
Endnotes 84
Creativity Resources 87

CHAPTER 4   OUT OF THE FUZZY FRONT END INTO THE DESIGN PHASE 89
Learning Objectives 89
Introduction 91

Go/No Go Decision Making   91
Idea Screening Gate   94
    Idea Selection Process   94
    Number of Ideas   95
Scoping: The First Stage   96
    Scoring Models   99
    Voice of the Customer Analysis   99
    Experiential Interviews   102
    Empathic Design and User Observation   103
    Elicitation Techniques   105
    Benefit Chains   105
    Web-Based "Eavesdropping"   107
    Evaluating the Data   109
Building the Business Case   111
    Situational Analysis   112
    Product Definition   113
    Project Justification   114
    Project Plan   114
Chapter Summary   115
Glossary   116
Review Questions   118
Assignment Questions   118
Endnotes   120

CHAPTER 5   THE CONCEPT TEST   123
Learning Objectives   123
Introduction   124
What Is a Concept Test?   125
Conducting the Concept Test   127
    Step One: Determine Goal of Concept Test   127
    Step Two: Select a Survey Population   128
    Step Three: Select Most Appropriate Survey Format 129
    Step Four: Prepare the Concept Statement   130
    Step Five: Develop the Questionnaire and Conduct the Survey 134
    Step Six: Interpret and Report the Results   137
    Sales Forecasts Based on Purchase Intent   139
    Forecasting New Product Sales from Likelihood of Purchase Ratings 140
Creating a Positioning Statement   141
Concerns with Concept Tests   142
Chapter Summary   143
Glossary   144
Review Questions   145
Assignment Questions   145
Appendix: Concept Description Example for TH!NK Electric Vehicle   146
Endnotes   148

CHAPTER 6   PERCEPTUAL MAPS   151
Learning Objectives   151
Introduction   153
Customers Buy Based on Perceptions   153
    Benefits and Value   154
Perceptual Maps   155
    Types of Perceptual Maps   158

Factor Analysis Summary 166
Identifying a New Dimension (Factor) 167
Overall Similarity Gap Maps and Other Mapping Techniques ... 168
Overall Similarity Gap Map 168
Value Maps and Customer Priorities 169
Chapter Summary 172
Glossary 173
Review Questions 175
Assignment Questions 175
Endnotes 176

CHAPTER 7   **ESTIMATING SALES POTENTIAL** 179
Learning Objectives 179
Introduction 181
Forecasting Techniques 182
Judgment Techniques 183
Quantitative Techniques 184
Customer/Market Research Techniques 186
New Product Forecasting Strategy 186
Forecasting Using Purchase Intention 188
Repeat Purchasing 192
ATAR with Cannibalization 194
Probability Scales 195
Forecast Prediction 196
Diffusion of Innovation 196
Estimating p and q 198
Regression to Estimate Purchase Probabilities 202
Estimation of Parameters 203
Managerial Use of the Model 204
Chapter Summary 205
Glossary 205
Review Questions 208
Assignment Questions 209
Appendices 210
Appendix A 210
Additional Forecasting References 210
Sources for Estimates of Bass Model p and q 210
Appendix B 211
Endnotes 212

CHAPTER 8   **PROACTIVE NEW PRODUCT DEVELOPMENT PROCESS** ... 217
Learning Objectives 217
Introduction 218
Design 219
Voice of the Engineer Blending with Voice of the Customer ... 222
Generate Product Designs 226
Product Architecture and Platform in Product Design 227
Product Platform 228
Product Architecture 230
Technology Roadmapping 231
Design Thinking and the NPD Process 235
Problem-Solving Approach or Process 236
Process and Methods 236

Methods   239
Role of Marketing in Design   240
Chapter Summary   243
Glossary   244
Review Questions   247
Assignment Questions   247
Appendix   248
  Additional References for Design Thinking   248
  Critiques   249
Endnotes   249

CHAPTER 9 PRODUCT/MARKET TESTING   253
Learning Objectives   253
Introduction   255
Reducing Risk   257
Product Use Testing   258
  Preuse Reactions   259
  Alpha Testing   260
  Beta Testing   261
  Conducting Product Use Tests   262
  Issues in Product Use Tests   263
  Summary of Product Use Testing Procedures   264
Market Components Testing   264
  Testing Advertising   265
  Testing Price   268
  Testing Distribution Options   270
  Summary of Marketing Components Testing   272
Premarket Testing   272
  Pseudosale   272
  Trial/Repeat Measurement   275
  Controlled Sales   276
  Summary of Premarket Testing   277
Market Testing   277
  Test Markets   277
  Rollouts   279
  Market Testing for Durable Consumer Goods and Industrial Products/
  Services   280
  Summary of Market Testing   281
Chapter Summary   281
Glossary   281
Review Questions   284
Assignment Questions   285
Endnotes   285

CHAPTER 10 INTO THE MARKET: LAUNCH   289
Learning Objectives   289
Introduction   291
Prelaunch Strategizing and Tactics   292
Strategic Launch   293
  Product Concerns   293
  Organizational Concerns   295
  Industry Concerns   298
Tactical Launch Planning   299

Product Name and Branding Strategies 301
Checklist of Criteria for a Good Product Name 303
Checklist of Things to Avoid 304
     Price 307
     Launch Timing 310
Launch Management 313
     Monitoring Launch 313
     Postlaunch Analysis 314
Product Life Cycle Management 315
     Product Failure 315
Chapter Summary 318
Glossary 318
Review Questions 321
Assignment Questions 321
Appendix: Pricing Resources 322
Endnotes 322

CHAPTER 11   GLOBAL NEW PRODUCT DEVELOPMENT 327

GLORIA BARCZAK AND ROSANNA GARCIA
Learning Objectives 327
Organizing for New Product Development 329
     Physical Proximity of NPD Teams 329
Open Innovation and Global Markets 332
Innovation in Emerging Markets 335
     Reverse Innovation 335
     Bottom-of-the-Pyramid 336
Launching Global New Products 338
     Global New Product Launches 338
Global Brands 338
     Branding Strategies 338
     Standardization or Adaptation 341
Packaging 341
Consumer Perceptions of Global Brands 342
     Protecting Your Global Brand 343
Chapter Summary 344
Glossary 344
Review Questions 345
Assignment Questions 346
Endnotes 346

CHAPTER 12   SUSTAINABILITY IN INNOVATION 351

MARIUS CLAUDY AND ROSANNA GARCIA
Learning Objectives 351
Introduction 353
     Finite Resources 355
     Stakeholder Pressure and Growing Transparency 355
The Business Case for Sustainability 355
     Costs Reduction 356
     Complying with Regulation 356
     Reputation and Brand Value 357
     Differentiation 357
     Attract and Retain Employees 357

Attract Capital Investment   357
Developing Sustainability Strategies   358
   Principles Underlying Sustainability Product Design   358
   Four Paradigms for Sustainable New Product Development   362
Product Improvement and Redesign   362
Functional and System Innovation   364
System Innovation and the Role of Services   368
   Product-Oriented Services   368
   User-Oriented Services   369
   Results-Focused Services   369
Marketing Sustainable Products   370
   Eco Labels   371
Chapter Summary   374
Glossary   375
Review Questions   377
Assignment Questions   378
Appendix   378
   Online Video Presentations   378
   H P Bulmers Ltd. Case Study   379
Endnotes   380

INDEX   385

List of Figures

Figure 1.1 Stage-gate product innovation process. 7

Figure 1.2 Flow diagram of the activities involved in the development and commercialization of a new drug. 7

Figure 1.3 Braun Syncro shaver design. 8

Figure 1.4 Prototyping at Dyson. 11

Figure 1.5 Evolution of Dyson vacuum designs. 13

Figure 1.6 Boundary box "playing field." 14

Figure 1.7 Looping NPD process. 15

Figure 1.8 Marketing's role in new product development. 20

Figure 2.1 Integration of strategies. 31

Figure 2.2 Components of the innovation strategy. 33

Figure 2.3 General Electric mission statement. 38

Figure 2.4 Portfolio map. 46

Figure 2.5 Strategic buckets. 47

Figure 2.6 Integrated stage-gate portfolio management process. 49

Figure 3.1 Five steps of idea generation and opportunity identification. 58

Figure 3.2 Sources of innovations. 58

Figure 3.3   Global innovation grid.   61

Figure 3.4   Lead users at Lego®.   62

Figure 3.5   A mosaic by R. Elbert using Post-it® notes.   64

Figure 3.6   Example of an ideation game.   65

Figure 3.7   Swiffer® carpet flick.   66

Figure 3.8   Crowdsourcing by a coffee shop.   67

Figure 3.9   Polo Harlequin evolving from attribute dependency-inclusion brainstorming.   68

Figure 3.10   Operators underlying the five inventive templates.   69

Figure 3.11   Market entry opportunities and the existing portfolio.   73

Figure 3.12   Experience curve for Ford automobiles.   75

Figure 4.1   Inside the Segway.   90

Figure 4.2   Probability of success/failure of ideas.   96

Figure 4.3   Criteria used to rank projects.   100

Figure 4.4   Incorporating voice of the customer in the design process.   101

Figure 4.5   VOC studies inform product design and business strategies.   101

Figure 4.6   Number of experiential interviews required.   104

Figure 4.7   Benefit chain structure for voice-over-the-internet telephone.   106

Figure 4.8   Means–end chain.   106

Figure 4.9   Laddering example for a mobile phone.   108

Figure 4.10   "Pulse" on the Internet for iPhones and 3G mobile banking: Discussions for January 18–July 17, 2009.   109

Figure 4.11   Factors to consider in developing the business case.   115

Figure A.001   Gravesite sculpture using Internet traces of the individual.   119

Figure 5.1   Mobile printer concept test description.   126

Figure 5.2   Visual presentation of concept description.   131

Figure 5.3   Example of virtual reality in auto promotion.   132

Figure 5.4   Purchase intention questions.   134

Figure 5.5   Frequency questions for the concept test.   135

Figure 5.6   Purchase intent relationships.   139

Figure 5.7   Quadrant analysis map 1.   142

Figure 5.8   Perceptual map for digital waiter.   142

Figure 6.1   Perceptual map for smartphones.   153

Figure 6.2   Perceptual map of pain relievers.   156

Figure 6.3   Perceptual map of transportation options.   157

Figure 6.4   Snake plot of perceptions on smartphone technology-based features. 160

Figure 6.5   SPSS output for factor analysis.   161

Figure 6.6   Scree plot.   162

Figure 6.7   Factor *loading* matrix.   164

Figure 6.8   Factor *score* matrix.   165

Figure 6.9   Descriptive statistics (means).   166

Figure 6.10   Similarity map of business schools.   169

Figure 6.11   Value map for pain relievers.   171

Figure 7.1   New product forecasting techniques.   183

Figure 7.2   Contingent forecasting strategies.   187

Figure 7.3   Purchase intention questionnaire from concept test on mobile
printers.   189

Figure 7.4   Alternative tools and devices for motivating distributors.   192

Figure 7.5   Two-state switching model to derive "repeat" purchase.   194

Figure 7.6   Bass diffusion curve for different p and q.   198

Figure 8.1   Product specification process.   221

Figure 8.2   The House of Quality (HOQ).   223

Figure 8.3   Functional decomposition of TiVo.   227

Figure 8.4   Black and Decker portable tools platform.   228

Figure 8.5   Skype's instant messaging platform.   229

Figure 8.6   Chocomize's mass customization strategy.   230

Figure 8.7   Product architecture's chunks.   231

Figure 8.8   Trade-off between distinctiveness and commonality.   232

Figure 8.9   Different types of technology roadmaps.   233

Figure 8.10   Generalize technology roadmap architecture.   235

Figure 8.11   Design thinking process.   238

Figure 8.12   d.school mindsets.   238

Figure 9.1   Role of testing in new product development.   256

Figure 9.2   Role of risk reduction in testing.   257

Figure 9.3   Preference survey for podcasts.   260

Figure 9.4   Google's example of beta sites.   261

Figure 9.5   General Motor's concept car: Equinox fuel cell vehicle.   262

Figure 9.6   A model for the consumer response hierarchy to advertising copy.   265

Figure 9.7   Average trial of 40 new brands versus believability and meaningfulness.   267

Figure 9.8   Example of conjoint including price.   269

Figure 9.9   Example of contingent valuation study for price.   270

Figure 9.10   Distribution alternatives for new products.   271

Figure 9.11   ASSESSOR capabilities by M/A/R/C research.   274

Figure 9.12   Second life as a simulated market test platform.   275

Figure 9.13   Summary of best uses for pseudo and controlled sales.   278

Figure 9.14   Competing coffee-flavored colas.   278

Figure 10.1   Launch process and postlaunch monitoring.   292

Figure 10.2   Launch considerations.   293

Figure 10.3   Morphing of zaarly.com from buyer-driven requests to small storefront offerings.   294

Figure 10.4   Launch activities.   300

Figure 10.5   The importance of product naming Bigbelly versus Seahorse Power.   301

Figure 10.6   Naming short-list criteria.   302

Figure 10.7   Brand associations. Nestlé Toll House is a registered trademark of Nestlé.   304

Figure 10.8   Best global brands and their valuations in $millions.   305

Figure 10.9   Brand hierarchy.   306

Figure 10.10   Brand depth and breadth.   307

Figure 10.11   Launch timing.   311

Figure 10.12   Product life cycle from cradle to grave.   316

Figure 10.13   A stepwise product deletion process.   316

Figure 11.1   Crowdsourcing.   334

Figure 11.2   Tata Nano.   336

Figure 11.3   Vscan-pocket-sized ultrasound from GE Healthcare.   336

Figure 11.4   Base of the pyramid.   337

Figure 11.5   Co-brand credit card landscape in Europe.   340

Figure 11.6   Classic bottle shapes.   341

Figure 12.1   The triple bottom line and the new product development process.   354

Figure 12.2   Waves of innovation.   354

Figure 12.3   Sources of inputs for a laptop computer.   356

Figure 12.4   Pepsi's eco-bottle.   359

Figure 12.5   Professional wet cleaning to minimize toxic materials.   360

Figure 12.6   Four levels of innovation in the design path towards sustainability. 362

Figure 12.7   Rebound effect.   363

Figure 12.8   Product life cycles. (a) Product life cycle of the typical innovation; (b) cradle-to-cradle product life cycle.   365

Figure 12.9   The nutrient cycles.   366

Figure 12.10   Preserve's Gimme 5 C2C approach to recycling #5 plastics.   367

Figure 12.11   Popular labels. (a) U.S. Energy Star label and "Chasing arrows" recycling label; (b) EU flower label.   375

Foreword

In 1980 the world of product development was changing rapidly with new research on the diffusion of innovations, opportunity identification, idea generation, perceptual mapping, conjoint analysis, pretest marketing, lifecycle management, and organizational structures. We sought to bring this research together to help a new generation of product developers design new products that served customer needs and were profitable for the firm. In 1993, we felt a need to revise our text because many of these methods had become mainstream. New ideas included more recent research on designing and managing over the lifetime of a product line. Alas, as time passed other demands on our time made it difficult to find the time necessary to again update *Design and Marketing of New Products* to include all of the exciting new developments. In the past twenty years, theory and practice have advanced. There are now better ways to understand customers, communicate with customers, manage product development, and launch new products. The growth of the Internet, social media, and mobile communications has further revolutionized product development. Today's product development manager must be adept at many tasks to seek the right core benefit proposition for their customers and their firm.

Rosanna has stepped up to the challenge of synthesizing new ideas on product development. She has widened the strategic framework (particularly in the upfront development phases) and made the material accessible to both managers and students. Her text is carefully focused. Rosanna identifies exemplar methods to be used in each phase of development. She has also added new ways for students to learn with a keen eye for the processes that reinforce the central lessons.

Creating and Marketing New Products and Services is an important resource for brand managers, product development teams, and marketing scientists who need to understand the analytic methods to designing new products. Thank you, Rosanna, for this effective and up-to-date new product development text. We recommend it to students, managers, and analysts interested in successfully developing new products.

John Hauser
Kirin Professor of Marketing
MIT Sloan School of Management

Glen Urban
David Austin Professor of Marketing
MIT Sloan School of Management Dean Emeritus
Chairman, MIT Center for Digital Business

Acknowledgments

I would like to thank a number of reviewers for careful reading of my manuscript and for providing helpful comments. All errors are my own. Marco Bonilla (M/A-COM Technology), Kwong Chan (Northeastern University, Boston, MA), Scott Dacko (Warwick Business School, UK), Ken Kahn (Virginia Commonwealth University & Da Vinci Center for Innovation), Fred Kinch (Northeastern University), Jeff Sieloff (Northeastern University), Rebecca Slotegraaf (Indiana University), Richard Wargo (Snap-On Tools). Samriti Bedi, Gillian Hurst, Dominik Reichel, AJ Rummel, and Selene Sizar, provided excellent administrative assistance for which I am also grateful. Special thanks go to my New Product classes of 2009–2013 at Northeastern University for working with previous versions of the manuscript. Their input was invaluable.

Thank you to Robert G. Cooper, author of *Winning at New Products*, for the use of his stage-gate process model, which I have used in most chapters. A special thanks to my past students, Adam Liebman, Michael Melo, Sean Reilly, and Gunnar Schramm for allowing me to use their Determinant Gap Map of different modes of transportation around Boston in Chapter 6. For those unfamiliar with Boston public transportation, the Charlie Pass and the Charlie Card are subway payment options. Hubway is a Boston-based bicycle rental service. The Longboard refers to a type of skateboard. The Sustainable Design Eco-system figure on the opening page of Chapter 12 on sustainability in innovation is courtesy of SolidWorks Corp., copyright 1997–2012 Dassault Systèmes SolidWorks Corp.

Rosanna Garcia
Gloucester, Massachusetts

About the Author

Rosanna Garcia, PhD, is a professor of marketing and innovation at North Carolina State University. Her undergraduate degree in chemical engineering and an MBA with a marketing focus provided her with a background that she utilized in technology-driven companies to develop and market new products and services. After more than 10 years in industry, she moved to academia to research topics, such as the diffusion of resistant innovations, the role of environmental sustainability in the innovation process, and the changing role of technology in the marketplace. Dr. Garcia is published in numerous academic journals including *Sloan Management Review*. She continually updates her knowledge on the innovation process through consulting at companies worldwide.

The Proactive New Product Development Process

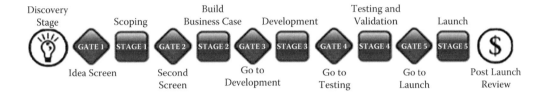

Learning Objectives

In this chapter, we will address the following questions:

1. Why study new products?
2. What is an innovation? What are the different types of innovations, and why should managers care?
3. What types of systemic processes are followed in developing new products?
4. What is the role of marketing in the innovating organization?
5. What is the role of the new product development manager in the innovating organization?

New Product Development at Whirlpool

In mid-1999, housing construction and sales of Whirlpool appliances were booming. Despite strong demand, the prices of Whirlpool appliances were falling at an average rate of 3.4% a year. Now retired, Chairman and CEO David R. Whitwam remembers those days like this: "I go into an appliance store. Now, I have pretty good eyes. I stand 40 feet away from a line of washers, and I can't pick ours out. They all look alike. They all have decent quality. They all have the same price point. It's a sea of white." The appliance maker had never paid much attention to innovation. During most of its 95-year history, it excelled at operating plants and distribution channels efficiently and at turning out washers and dryers that were solid and long-lasting. Believing that brilliant ideas were buried in the corporate hierarchy, after that visit, Whitwam invited each of the company's 61,000 employees to unleash their creativity: "Everybody everywhere," he exhorted, "go out and innovate!" This liberation, however, resulted in innovations that were useless, impractical, and poorly suited to Whirlpool's strengths. In addition to Internet bike racing, employees proposed the Unattended Box—a doorstep appliance to keep food deliveries hot or cold—and a membership club for people who wanted home repair services.

Whirlpool learned the hard way that real innovation requires a lot more than simply urging thousands of employees around the world to tap into their inner designer and then expect great ideas to roll in. It requires hard work, structure, and unwavering discipline. After its inauspicious start, the company retreated from the all-out effort to democratize innovation and moved to a more traditional centralized model of product development. New ideas came pouring in. One quartet of engineers from Whirlpool's oven factory in Oxford, Mississippi, invented a combination gas grill/refrigerator/oven/boom box for tailgate parties. Whitwam set aside $45 million from the capital budget for innovation in 2000 and doubled that amount in 2001. Revenues from products that fit the company's definition of "innovative" zoomed up from $10 million in 2001 to $760 million in 2005, or 5% of the Benton Harbor (Michigan, U.S.A.) company's record $14.3 billion in total 2005 revenue. In 2006, the company merged with Maytag. In 2008, the company reported more than $19 billion in sales.

Introduction

As the Whirlpool example shows, success in the new product development process does not occur by chance, but is a result of careful planning and execution. The intent of this book is to provide you with the skills needed to bring successful innovations to life in your organization. In this chapter, we lay the foundation for understanding the role of new product development in the innovating organization from the original ideation stage to the launch phase. We also will emphasize the important role of marketing in this process.

New products are crucial to growth and increased profits in many organizations. The development of new products is rewarding and necessary to maintain a healthy organization. Recent examples show how a company can succeed with innovation. Apple built on its Macintosh computer franchise with major innovation in personal music devices (the iPod), intelligent phones (the iPhone), and notebook computers (the iPad). This resulted in sales volumes greater than $50 billion and, in 2010, a market cap greater than Microsoft (itself a major software innovator).

Likewise, by 2010, Google had built a company of over $25 billion in sales and $150 billion market cap by innovating in information search. But Google's innovation was not just in developing a search algorithm. Google is an information company that makes its money in advertising by building a full range of advertising planning and evaluation tools to help advertisers select the best words, track conversion, and to experiment in real time to improve their ad effectiveness.

These are specific examples of success, but the importance of new products to sales and profit growth is more general. In a survey of 700 firms (60% industrial, 20% consumer durables, and 20% consumer nondurables), the consulting firm of Booz Allen Hamilton Inc.[1] found that over a five-year period new products accounted for 28% of these companies' growth. In a survey sponsored by the Marketing Science Institute, it was found that 25% of current sales were from products introduced in the previous three years.[2] It is not surprising that some of the most successful companies today, such as 3M, Procter & Gamble, Microsoft, and Mercedes Benz, are also known for their new product development strategies.

New Product Development Is Risky

Apple has had an enviable string of successes, but the company failed in the 1990s with the Newton (a small PC personal digital assistant where you interacted with handwriting instead of typing). Microsoft failed in 2010 with a social networking phone called Kin (it only stayed on the market for 48 days). Pfizer, the no. 1 pharmaceutical company worldwide, has been a successful innovator in drugs, but experienced a $2.8 billion loss in 2007 in the development and launch of Exubera—an inhalable insulin for diabetics. Exubera was meant to replace the inconvenient and often painful injections that type 1 and type 2 diabetics are subjected to daily for medication. But the drug failed miserably (internal forecasts were for billions in sales, but Exubera only sold $12 million worth) due to poor acceptance by doctors and patients. Although it eliminated the pain of a shot, the inhaling device was both large and awkward. It was based on crushing a tablet and the inhalation of the powder was difficult to calibrate (called dosing) and, for some users, required intensive inhaling. Doctors had reservations on having patients inhaling any powder into their lungs and, additionally, the patient had to be trained in the complex usage regime. How could a big, successful company miss so many obvious things? Pfizer failed to complete adequate patient and doctor testing and ignored the negative warning signs that surfaced in early tests. They were determined to be the first on the market with inhalable insulin.

Another failure was Eclipse Aviation, a VLJ (very light jet) aircraft manufacturer founded by an ex-Microsoft executive; the company spent more than $1 billion in R&D expenses on a small carbon fiber jet designed for the air taxi market and corporate use. Manufacturing problems, technical difficulties (lightning strikes posed a danger to carbon fiber so that a redesign with aluminum had to be used in some parts), slow market penetration, and gasoline price fluctuations caused the firm to file for bankruptcy in early 2009. Similarly in 2011, Solyndra, a solar panel company backed by the U.S. Department of Energy, filed for bankruptcy after raising nearly $1 billion in private equity financing.

These are not special cases; in general, companies have not always been successful in their innovation strategies. Polaroid, the innovator in instant photography in the 1970s, no longer exists. Nokia was the leader in cell phones in the early 2000s, but has suffered in the marketplace as other phone manufacturers out-innovated them (e.g., flip phone and touch screen smartphones). Innovating is risky and new product failures are common. Studies by the Product Development Management Association (PDMA[3]) show that failure rates average about 40% for products that make it to the market.[4] Another study found that 20 to 25% of industrial products and 30 to 35% of consumer products fail after market launch.[5] A study by the Association of National Advertisers found that 27% of product line extensions failed, 31% of new brands introduced into existing categories failed, and 46% of new products introduced into new categories failed.[6] Studies have shown that only one of seven new product ideas is carried to the commercialization phase.[7] The overall success rate for a project that has made it through scoping to final launch in the market is only 15%.[8]

Innovation is a high-risk activity and getting riskier as the life of successful new products becomes shorter and as technology renders products obsolete at faster rates. Cell (mobile/handy) phones are an excellent example of how technology quickly dates recent introductions. Slim phones were replaced by camera phones, which were replaced by Web-accessible phones, which have been replaced with smart, social networking-capable phones. Companies cannot afford to fail needlessly with their new products if they wish to remain competitive in a rapidly changing marketplace.

The losses that result from new products are not only due to low sales and profits, but also can result from costly research and development (R&D). The return on investment (ROI) on new product effort is also at risk due to the large developmental expenses a company accrues in the process. Large investments in R&D engineering, marketing research, manufacturing and logistics system creation, and marketing development and testing are made before the product is introduced. In 2011, firms spent over $603 billion on R&D alone.[9] This is more than double the 1997 amount. The growth has been steady and, even in the recession year of 2008, firms increased their R&D expenditures.[10] This may be because (1) innovation has become a core component of corporate strategy, (2) many companies saw the recession as an opportunity to build their advantage in the market, and (3) firms could focus spending on product development and engineering (only 20% on basic research and advanced development) to prioritize new product launches.[11] See Table 1.1 for an idea of how much is invested by the top 10 R&D spending companies in different industries. Spending varies by industry with the highest R&D evident in the top three spenders of computing and electronics at 28% of total, followed by healthcare at 21%, and automotive at 16%. The lowest spenders on R&D are telecom at 2% and consumer products at 3.0%.[12]

Because many products do not make it from the ideation phase to market, large investments are made on products that never return revenue. This means that the successful product must not only return its own development cost, but also cover the costs of other products that started in the NPD (new product development) process but never made it to market. High failure rates and high costs clearly make new product development risky.

Table 1.1   R&D spending by top 10 R&D spenders worldwide (2012)

COMPANY	R&D SPENDING ($BILLIONS)	R&D AS % OF SALES	INDUSTRY
Toyota	9.9	4.2	Autos
Novartis	9.6	16.4	Health
Roche Holding AG	9.4	19.5	Health
Pfizer	9.1	13.5	Health
Microsoft	9.0	12.9	Software/Internet
Samsung	9.0	6.0	Computing & Electronics
Merck	8.5	17.7	Health
Intel	8.4	15.6	Computing & Electronics
General Motors	8.1	5.4	Automotive
Nokia	7.8	14.5	Computing & Electronics

Source: Endnote 13.

Service Economy

R&D spending is generally attributed to manufacturing firms and the NPD process is typically approached from a products perspective. Services have grown steadily and now (banking, healthcare, entertainment, transportation, e-commerce, etc.) account for 55% of the U.S. economic activity.[14] The current list of Fortune 500 companies contains more service companies and fewer manufacturers than in previous decades. The United States is also experiencing the servitization of products where products today have a higher service component than in previous decades. There is now less of a distinction between product and service, which has been replaced by a service–product continuum. For example, IBM sells "hardware, software, and managed support."[15] Although it still manufactures computers, its Global Business Services and Global Technology Services accounted for 55% of its revenue in 2008. With this growing service economy in the United States and worldwide, service innovations play a significant role in many firms' new product/service portfolio.

It has been found that critical success factors for services differ compared to manufacturers.[16] While manufacturers focus primarily on product innovation advantage and quality, service providers focus on innovativeness in their human resource strategy. It was found that successful service firms must place greater emphasis on the selection, development, and management of employees who work directly with the customer. The "frontline personnel" are the face of the company. Employees' close contact and potentially long-term relationships with customers make them an important source of new ideas in the firm's new service development process. Because frontline employees can significantly impact the success or failure of new service launches, the human resource strategy must be executed well.

In services, innovations can spawn off current technologies. For example, in early 2009, the British Broadcasting Corporation (BBC) was reproached for being a "me-too broadcaster with a serial record of imitation." It was reported: "Pirate radio stations spawned first Radio 2 and then Radio 1. Sky News brought forth BBC News

24. ITV and Channel 4's success with reality TV and phone voting saw the BBC hurrying to catch up. The BBC is often a parasite on others' ideas."[17] The BBC appears to have made a successful new product strategy from imitation. BBC competitors will need to be even more innovative in their new service to keep the BBC from spinning off "me-too" services. In mid-2011, Hewlett-Packard, the computer hardware company, announced it was abandoning the PC market[18] in favor of the tablet and smartphone markets. Technology changes have forced HP to change its innovation strategy toward being a service-oriented approach.

Throughout this book when we speak of product innovation, this is interchangeable with service innovation. The same processes that are undertaken with product innovations, but also can be conducted for service, with the caveat that employees who are providing the service should be a team member on the integrated new product development team throughout the stage-gate process.

Stage-Gate: A Systematic, Sequential, Iterative Process

Firms need new products to grow and be profitable, but the bad news is that new products are risky and costly. The good news is that new product development can be managed so that the risks are minimized and profits are maximized. One answer to risk management is a sequential new product development process that eliminates failure early so new products don't use up scarce resources and/or fail in the marketplace.

Innovative companies employ systemic, structured processes for developing new products. For over 40 years, companies have managed risk by a go/no go set of new product development stages. One popular process, the **stage-gate innovation process**, is summarized in Figure 1.1.[19] In between each stage are evaluation tasks, or gates, which are used to determine if the project should move on to the next phase. Many companies follow this type of sequential process because of its simplicity and its ability to break down a very complex process into steps that can be easily conducted and evaluated. The challenge is to build a process that balances risk and ROI for the firm in its technology, market, and competitive environments.

In this book, we will present the new-product development decision process as a sequential set of seven activities starting with Discovery, followed by five Stage-Gates, and ending with Post Launch Review, as shown in Figure 1.1. This allows us to cover each aspect of the process in depth by providing the managerial concepts and analytical techniques necessary to minimize risks and maximize creativity.

The "gates" are the "go/no go" decision point to determine whether to advance to the next stage in the process. **Gates** are the point where the project's viability in the ever-changing marketplace is reassessed to determine its congruency with the firm's goals and the resources required to move the project to the next step. The go/no go step can be viewed as a funnel for projects that no longer meet the risk–reward objectives of

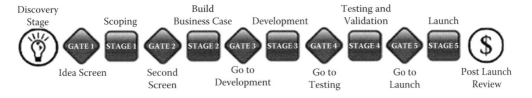

Figure 1.1 Stage-gate product innovation process. (Endnote 19.)

the firm. Often one of the seven steps will be reiterated if the project doesn't meet the criteria set by the firm for advancing to the next step.

In practice, the process is less rigid than this chart suggests and successful organizations customize it to their needs and capabilities. The process is typically not sequential. As new information is obtained (new technologies, new competitors, improved customer research, etc.), key steps may be iterated or sometimes even skipped. For example, a banking service may pass from concept description to brochures and testimonials back to testing and validation of the services with selected customers. Industrial goods may start from concept to prototype to pilot production output, followed with launch.

Figure 1.2 is an example of the stage-gate process for a new pharmaceutical drug.[20] This is different from the process in Figure 1.1, but reflects the go/no go decision process. Likewise, as shown in Figure 1.3, Braun, the shaving manufacturer, followed a slightly different process.[21] Comparing these two processes, we see that both have a discovery/ideation stage, and both end with go-to-market decisions. Both processes have checkpoints, or gates, in which the project can be stopped or revised. However, the stages along the way vary based on the development style of each company.

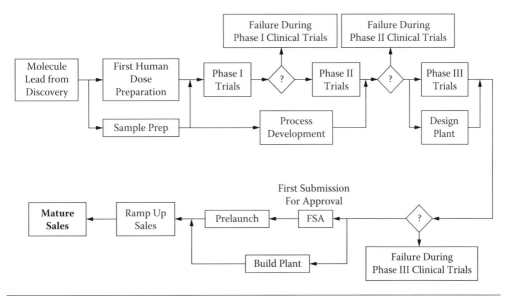

Figure 1.2 Flow diagram of the activities involved in the development and commercialization of a new drug. (Endnote 22.)

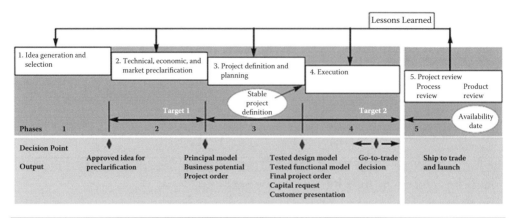

Figure 1.3   Braun Syncro shaver design. (From Endnote 21.)

Below we summarize the seven steps,[23] which we will be using as the framework for the remainder of this book. The process starts with **Discovery,** where new ideas are generated. In **Idea Screening**, the creative ideas are culled into a few great ones to take into the **Scoping** stage where the target group of customers is carefully defined and the firm's capability to meet the needs of this group is evaluated. The **Second Screen** looks at the best opportunity for the firm from the short list of ideas from scoping. **Building the Business Case** develops the business/marketing plan for the selected idea. Gate 3, **Go to Development**, is the go/no go decision gate whether the project should move into the expensive and resource heavy **Development** stage. Gate 4, **Go to Testing**, determines whether the product as designed during development should move to **Testing and Validation**, where market tests and technology evaluations prior to launch are conducted. **Go to Launch**, gate 5, is the last step to keep a potentially failing product from reaching market and ensuring a successful product is sufficiently supported to reach success. **Launch** and **Postlaunch Review** phases monitor the delivery of the product so that it fulfills customer needs and delivers on the core benefit proposition as set out in the business plan. This multiphased process assists the manager in identifying the customer's priorities and designing the product to deliver priority benefits at a price that provides high value. Techniques, such as perceptual maps, value maps, and preference analyses, are tools that help the new product manager throughout the process (these will be discussed in later chapters).

To illustrate this sequential, iterative process in detail, we consider the design process of the Dyson centrifugal bagless vacuum cleaner (www.dyson.com). James Dyson invented this type of vacuum cleaner in Great Britain in 1979. Its design was rejected by numerous vacuum companies at that time, including Hoover. Today, Dyson is one of the best-selling vacuum cleaners worldwide, much to the consternation of these competing companies, who continually try to design around the Dyson patents.

The Fuzzy Front End: Discovery through Scoping

Discovery and Scoping together are often called the *fuzzy front end;*[24] it involves the early efforts in discovering and uncovering opportunities, and in generating ideas. Perhaps the most common mistake that students and managers make is to become too quickly overzealous about a particular new product idea. Everyone has his own favorite idea about what is needed. However, today's markets are becoming more and more complex, the risks of failure are greater, and the consequences more costly. Disciplined new product design and techniques identify failures at a much lower cost to the firm while increasing the ultimate profit from successes.

In 1978, James Dyson noticed how the air filter in one of his factories' spray-finishing room, was constantly clogging with powder particles (just like a vacuum cleaner bag clogs with dust). He started thinking that there should be a better way to clean the air. This problem, of cleaner air, was the Discovery stage. "In his usual style of seeking solutions from unexpected sources, Dyson thought of how a nearby sawmill used a cyclone—a 30-foot-high cone that spun dust out of the air by centrifugal force—to expel waste. He reasoned that a vacuum cleaner that could separate dust by cyclonic action and spinning it out of the airstream would eliminate the need for both bag and filter."[25]

By understanding the sources for ideas and creative group processes, the organization can generate ideas that integrate specific engineering, R&D, production, and marketing inputs. Ideas evolve into high potential concepts that may ultimately become successful products; this is done through Scoping.

Scoping involves screening an idea with a quick, preliminary investigation of the project; mainly conducting feasibility studies from a marketing, engineering, and financial perspective. In the Scoping stage, Dyson set about making a miniature version of a cyclone for his vacuum cleaner out of an empty breakfast cereal box. He jury-rigged his old Hoover with the cardboard cyclone and tested it at home. He was amazed to discover that it worked, maintaining full cleaning power continuously, with no degradation of suction, which was common in vacuum bags. After years of experimentation and thousands of prototypes, he perfected the dual-cyclone technology that powers his products today. Dyson took this technology to several vacuum manufacturers, who kindly turned him down. Although his technology was viable, few thought the marketing position was. He instead chose to move forward on his own. The vacuum cleaner design was over 200 years old, and although no one else saw it as being vulnerable, Dyson saw a market ripe for innovation.

Questions that may be answered in Scoping include:

- Who is the target market and will targeted customers benefit from the product?
- Do we have the capabilities to manufacture the product? Can we attain the skills if we don't?

- Will the customer even buy the product as proposed (remember the product is not real at this point)? Will the product be profitable when manufactured and delivered to the customer at the target price?
- Do we have the technology, or can we acquire it in order to meet the needs and wants of our customers?

During the Scoping stage, growing, profitable, or vulnerable markets are identified. This requires forecasting global demand and identifying availability of viable technology so that technology capabilities are matched to their market opportunities. Also, during this stage, a concept test conducted by the marketing department might be conducted to determine viability of the product in the marketplace. Marketing should be an active participant in the fuzzy front end because they can act as representatives for the voice of the customer, providing insights into customer needs and wants in new product designs.

Design Phase: Building the Business Case through Development

Design includes two major components: building the business case and product development.

Building the business case involves detailed investigation with primary research (both market and technical) leading to a business case, product and project definition, project justification, and project plan. The primary deliverable is a well-conceived business plan. Typical components include:

- Engineering efforts/resources requirements
- Marketing efforts/resources requirements
- Manufacturing efforts/resources requirements
- Company synergies with existing product platforms
- Profitability and breakeven point

Because of the marketing component of the business plan, the marketing department is actively involved to help prepare the business case and, in some companies, may be responsible for putting together the business case.

Product development involves actual detailed design and physical development of the new product, and the design of the operations or production process. The primary deliverable from R&D is to produce a physical prototype or mock-up of the proposed product. The actual number of iterations in the development process depends on the complexity of the product. For example, a new automobile may move from a concept to prototype to a preproduction model in a consistent fashion. Each iteration provides better information, refines the product concept, and moves closer to a marketable product. A new toothpaste formulation may go from concept right to preproduction without a prototype.

Rapid Prototyping   Initial Prototype Modeling

Figure 1.4   Prototyping at Dyson. (From Endnote 26. With permission.)

Engineering issues addressed in the product development phase may include:

- Resources required
- Engineering operations planning
- Department scheduling
- Supplier collaboration
- Logistics planning
- Program review and monitoring
- Finalization of product attributes and features

In the product development stage, Dyson developed and built 5,127 dual cyclone prototype cleaners between 1979 and 1984. Starting first with handmade cardboard models, thousands of prototypes were then built to refine improvements one at a time (see Figure 1.4 for how this process is conducted). The first prototype vacuum cleaner, the G-Force, was built in 1983 and was followed by three more years of extensive testing before launching in Japan in 1986. Fully functional prototypes were submitted for laboratory testing only after the prototypes showed promise.

Although, it might seem that the marketing department should take a back seat during the development process, during this stage, marketing issues addressed include:

- Who is the target market and who is the decision maker in the purchasing process?
- How will consumers react to the product?
- What product features are required to make the product a success?
- What will it cost to produce? For what price should the product/service be sold?
- How will the product be distributed (place)?
- How will the product be promoted?

The **voice of the customer** is also considered in the design stage. The organization needs to hear what customers want and their willingness to trade one function for another before the product is too far along to change the design without significant costs. Marketing also considers procedures to position the product vis-à-vis competition and to target the product to the appropriate segments of customers. Sales forecasting also is conducted based on product concepts, prototypes, etc., as the product moves closer to launch.

Testing and Validation

Testing involves tests or trials in the marketplace, lab, and plant to verify and validate the proposed new product and its marketing and production/operations. Possible tasks include:

- Test the product and its packaging in typical usage situations.
- Conduct focus group customer interviews or introduce at trade shows.
- Premarket testing using concept tests and conjoint analyses (both techniques are discussed in later chapters).
- Trial runs of the marketing mix: product, price, place, promotion (the 4Ps).

Dyson boasts that every cleaner is subjected to a battery of endurance tests to ensure that it's as durable as a Dyson should be. At test facilities in Wiltshire, England, "Dysons are pushed, pulled, dropped, frozen, baked, and shaken." There are five vacuum cleaner assault courses where the most demanding test of all human use puts machines through their paces, with 28,000 hours of punishment doled out every month.[26]

Launch and Postlaunch Review

Launch includes production, marketing, distribution, and selling the new product. Tasks may include:

- Prelaunch strategic and tactical planning
- Arrange, activate, and announce product/service launch
- Postlaunch planning
- Product life-cycle management

In 1986, a production version of Dyson's G-Force was first sold in Japan. The G-Force was a new-to-the-world innovation, which was an extremely risky innovation since it took over five years to develop and was only accepted slowly by the Japanese, who are considered innovative consumers. Today, Dyson has more than 25 different designs that continually include new technology, such as root cyclone and ball technology on its DC25 model as seen in Figure 1.5. But, Dyson continues to innovate. They have introduced vacuum cleaners for pet owners, canister models, and handheld vacuum cleaners as well as a line of innovative fans.

Postlaunch Review involves the ongoing evaluation of the success of product and subsequent feedback into related projects. Marketing metrics (ROI, market share, sales) are monitored to determine the success of the new product. As Dyson vacuums

Dyson GForce circa 1993 compared to DC25 circa 2007

Figure 1.5   Evolution of Dyson vacuum designs. (From Endnote 27. With permission.)

are well accepted by consumers in over 44 countries, there is little risk involved in launching or purchasing this incremental innovation. Today Dyson vacuums can be found at numerous stores, including the ubiquitous United States-based Target stores. Dyson's postlaunch success will facilitate their future success. The Dyson vacuum cleaner provides a good example of how a planned sequential process can lead to successful innovative new products.

Part of the postlaunch review stage is to determine when existing products/service should be phased out. The G-Force was dropped to make room for the DC25, and continually, over the life of the company, products will be added and deleted from the product line. Managing these product life cycles is equally as important to the success of the firm.

Criticisms of Sequential Processing

Some companies are critical of the stage-gate process arguing that it slows the innovation process because:

- all steps are sequential/synchronized;
- time management is not considered;
- it becomes difficult to compress development cycles (speed to market);
- phased systems fight against moving on to the next step with only "partial" information; and
- phased systems with gates encourage queues/gushes (feast or famine with new information) when information becomes available at a gate; immediate responses are needed to move to the next stage.

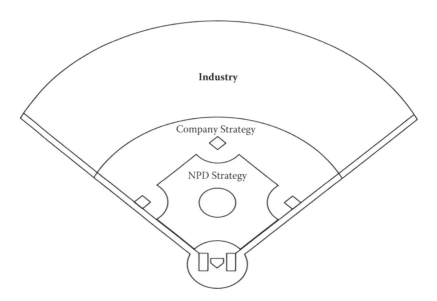

Figure 1.6 Boundary box "playing field."

Having a rigid sequential process reduces risk and will lead to longer development times. However, in fact, the process is typically not sequential, because there often is considerable overlap between the different phases. To reduce the time that the NPD process takes, many flexible companies complete several steps at the same time using concurrent engineering or speed to market strategies. This overlap of phases helps reduce time to market and is usually implemented with Pert or Gantt chart[28] project planning tools. The process also is iterative, meaning that if a project doesn't make it through one of the go/no go steps (gates), it will go back to the last phase for reassessment, thus, further increasing the development time. Yet, it is these go/no go steps that keep the project on track for ensuring success. If stages are skipped, it is wise not to completely eliminate gates and checks-and-balances should be combined at other gates.

In response to the above noted criticisms, other new product development processes are used by some companies. One method is called "**bounding boxes**" and another "**looping**." Bounding boxes set up a boundary, which the new product team and management agree to work within to keep the project on track. As long as a program is "in bounds" or within the zone, the team makes day-to-day decisions and adjustments to the NPD program without management intervention.[29] Outside the boundaries, an NPD team will escalate issues or decisions to management. Figure 1.6 demonstrates how new product development managers must abide by the company's strategic directive within a set industry. Just as the game of baseball must be played within a boundary, so too must the NPD process operate within a certain framework. For example, a directive to find a solution for improving online education with no other restrictions

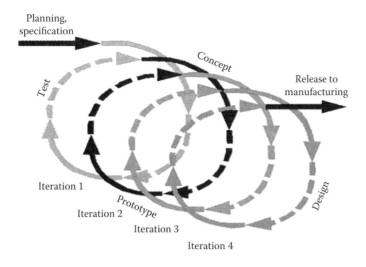

Figure 1.7   Looping NPD process. (From Endnote 30. With permission.)

may be given to an NPD team. This encourages the team to be creative in finding unique solutions.

In the case of the Westinghouse tailgate appliance (refrigerator, grill, plus) development, the "playing field" for new ideas was limited by the company's strategy for innovative, durable home appliances. If a team designed outside that space, they needed to go back to the corporation to get approval. This bounding box method gives the team flexibility to cycle back and forth through need identification, concept development, and prototypes testing activities without being limited by a sequential process. Risk is limited by the bounds of the strategic boundaries of the "planning field" as management must approve departure from the bounding box.

Looping, as shown in Figure 1.7, is another NPD process commonly used in software design where versions are developed, tested, and then looped back into the design phase. Looping differs from the sequential-iterative processing of the stage-gate in that there are no structured "gates" to stop the design process to evaluate whether the project is on mission or has drifted from the goal of the NPD strategy. This can be advantageous in the organic organization when speed to market is crucial. Many software companies, including Google, take this strategy and introduce beta versions regularly. Google Maps, Google Scholar, and Gmail were all introduced to the market in beta phases, which allows Google to make subsequent design changes based on user feedback as opposed to waiting for extensive testing prior to release.

This looping saves time by overlapping phases and rapid recycling through the development activities. The challenge is to take the product, market, technology, competition, and corporate culture into account in the organic development process in order to result in sales and profit growth, along with a good ROI.

What Is a New Product Exactly? How Can They Be Classified?

How to customize the new product process depends in part on what kind of new product you are dealing with. So, first let's define innovation and then look at their differences.

An 1991 Organization for Economic Cooperation and Development (OECD)[31] study on technological innovations best captures the essence of innovations from an overall perspective: "Innovation" is an iterative process initiated by the perception of a new market and/or new service opportunity for a technology-based invention that leads to development, production, and marketing tasks, which strives for the commercial success of the invention. This definition addresses two important distinctions: (1) the "innovation" process comprises the technological development of an invention *combined* with the market introduction of that invention to end users through adoption and diffusion, and (2) the innovation process is *iterative* in nature and, thus, automatically includes the first introduction of a new innovation and the reintroduction of an improved innovation. New products are the outcome of this process.

A new product is many things to different people. We define it as an innovation brought to the marketplace that is new to the market or new to the firm. When Apple introduced the iPod to the market, MP3 players were not new to the market but were new to Apple, who had primarily focused on personal computers. Each time it develops a new model of the iPod, even with small changes such as the U2 (the music group) co-branded iPod, it is introducing a new product. However, different types of new products carry varying risks, uncertainties, and rewards for the company, and need to be managed differently in the NPD process. Next, we examine these differences.

Types of New Products and Customizing the Development Process

Different classification schemas exist for categorizing the different types of new products. One of the most widely used classification schemas is *incremental innovation*, *really new innovation*, and *radical innovation*.[32] **Incremental innovations** are products that provide new features, benefits, or improvements to existing technologies used in existing markets. **Really new innovations** are products that introduce either a new marketing or technological innovation. Really new innovations evolve into new product lines (e.g., Sony Walkman), product line extensions with new technology (e.g., Canon LaserJet), or new markets with existing technologies (e.g., early fax machines). **Radical innovations** introduce *both* market and technological discontinuities to the marketplace. Examples include cell phones, personal computers, and Internet telephony.

Another popular categorization for innovations originates from Booz Allen Hamilton Inc.:[33]

- *New-to-the-world products*: Inventions that create a whole new market, e.g., Toyota's Prius hybrid automobile, Sony Walkman, P&G's Febreze. This category accounts for 10% of new products.

- *New-to-the-firm products*: Products that take the firm in a new direction. These products are not new to the world, but are new to the firm, e.g., Apple iPod in 2001, AT&T's Universal credit card, or Swatch's Smart car.[34] This category accounts for about 20% of new products.
- *Additions to existing product lines*: Product line extensions, flankers, or brand extensions, e.g., Crest Tartar Control toothpaste, iPod shuffle, or Dyson's hand-held vacuum cleaners. This category accounts for about 26% of new products.
- *Improvement or revisions to existing products*: Minor changes to improve existing products, e.g., MICHELIN® X One® tires, P&G's Swaddlers Sensitive New Baby diapers, or Gillette's 5-blade Fusion razor. This category accounts for about 26% of all new products.
- *Repositioning*: Products that take on new uses, e.g., Levi Jeans from work pants to fashion statement, aspirin for heart attacks. This category accounts for about 7% of new products.
- *Cost reductions*: Products that replace existing products by providing similar performance at a lower cost, e.g., many HDTVs or 8 GB iPhone. This category represents about 10% of new products.

Why Innovation Type Matters

There are several reasons why it is important to identify the different types of new products. First, as the innovativeness of the innovation increases, so does the development time. Incremental innovations typically take 6 to 10 months from idea to launch, really new innovations take 12 to 24 months from idea to launch, and radical innovations can take 5 to 10 years from idea to launch. Pharmaceutical innovations typically take more than 10 years because of the need for clinical trials. This time difference occurs because fewer stages are required when developing incremental innovations compared to really new innovations. For example, the iPod nano (incremental innovation) probably went right from ideation into development, skipping scoping and business case development, whereas the iPhone likely went through all the stages. A second reason identification matters is that radical innovations are significantly more expensive to develop compared to incremental innovations because of the resources required to develop and launch radical innovations. Firms need to know what resources are required to develop new innovations to help guarantee successful development.

A third reason for identifying different types of products is that product uncertainties also differ across the different categories, and thus risks of failure increase with increasing innovativeness of new products. Consumers are typically more willing to incrementally accept new products, such as improved toothpaste. Consumers' degree of uncertainty regarding new products must be taken into consideration in planning. Because of the technological advances in radical innovations, they are most likely to require consumer learning. Often consumers don't want to change from the status quo, which makes it difficult for radical innovations to be accepted in the marketplace. Another reason to identify innovation type is that, as technology becomes more complicated with increasing innovativeness, the possibility of failure from within the firm also increases.

Avoiding Failures

With an understanding of the underlying nature of our innovation, let's now look at the specifics of minimizing risk in the stage-gate development process. As noted earlier, not even the most respected large companies are exempt from experiencing product failures. Ford had its Edsel, the Coca-Cola Company had its New Coke, Apple Computers had its Newton personal digital assistant, Pzifer had Exubera, and the list continues. Success in the new product development process does not occur by chance but is a result of careful planning and execution.

Therefore, what can be done to reduce the risk of new product failures? To avoid such financial and marketing disasters, first, it is important for new product managers to understand why new products fail. Based on research studies[35] and the experience of colleagues, we have identified 14 main reasons for new product failures. Table 1.2 outlines these reasons and makes suggestions on how the proactive new product process reduces the risk and cost of failures by providing disciplined checkpoints at each stage of development.

In the market definition stage of opportunity identification, the new product manager must check the market for sales and volume potential, thus, avoiding the trap of *market too small*. The opportunity identification phase is also the place to systematically assess the *strategic fit* between company capabilities and product requirements. In the design phase, careful consideration of potential customer benefits, technological expertise, and the identification of a unique competitive positioning avoids the pitfalls of *not new/not different*, *no real benefit*, and *poor positioning*. This was one of the downfalls for Exubera. The product ended up having little benefit to diabetics because the inhalers were extremely complicated to use and dosing was ambiguous. More thorough prototype testing during development would have revealed the difficulties for consumers in using the product. An unbiased testing phase could have brought this to light as well. Market size also should be reassessed at the testing phase. This was one of the downfalls of Eclipse Aviation; by the time R&D was nearing completion, the potential air taxi market had failed to develop due to economic considerations. In these cases, escalation of commitment[36] became a problem because so much money had already been invested; killing the product that has significant sunk investments is rarely a consideration. A systematic new product process with vigilant go/no go gate evaluations avoids *poor timing*. Good launch planning and control minimizes the damages from *competitive response* and *changes in customer tastes*.

Table 1.2 Reasons for New Product Failures

FAILURE REASON	ELABORATION	SUGGESTED SAFEGUARD
1. Not new/not different, low value added	A poor idea that offers nothing new; no unique benefits or superior value	Creative and systematic idea generation; do upfront homework—preliminary market & technology assessments
2. Market too small	Insufficient demand for this type of product	Know the market with strong market orientation and estimate rough forecast in opportunity identification and concept test phase
3. Poor strategic fit	Company capabilities do not match product requirements	Opportunities are matched to company's capabilities and strategic plans before development begins
4. No real benefit	Product does not offer better performance	In the design stage and test stage, test perceived benefits of concepts as well as benefits from actual product use
5. Poor positioning/ misunderstanding of consumer needs	Perceived attributes of the product are not unique or superior	Use of perceptual mapping and preference analysis to create well-positioned products
6. Inadequate support from channel	Product fails to generate expected channel support	Assessment of trade response in pretest–market phase.
7. Forecasting error	Overestimation of sales	Use of systematic methods in design, pretest, and test phase to forecast consumer acceptance
8. Competitive response	Quick and effective copying by competitors	Good design and strong positioning to preempt competition; quick diagnosis of, and response to competitive moves
9. Changes in consumers' tastes	Substantial shift in consumer preference before product is successful	Frequent monitoring of consumers' perceptions and preferences during development and after introduction; use a market-driven and customer-focused NPD process
10. Changes in environmental constraints	Drastic change in key environmental factor	Incorporation of environmental factors in opportunity analysis and design phases; adaptive strategy
11. Insufficient return on investment	Poor profit margins and high costs	Careful selection of markets, forecasting of sales and costs, and market response analysis to maximize profits
12. Organizational problems	Intraorganizational conflicts and poor management practices	Multifunctional approach to new product development to facilitate intraorganizational communication; recommendations for a sound formal and informal organizational design
13. Local success/global failure	Product taken to other markets are not successful	Develop global products that look at global factors in addition to local needs and wants
14. Product R&D expenses over budget	Too many resources allocated to project at expense of other opportunities	Use portfolio management

Source: Endnote 37.

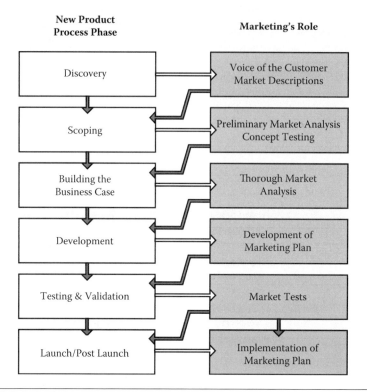

Figure 1.8   Marketing's role in new product development.

Marketing's Involvement in the Stage-Gate Process

To close this section, we examine the role of marketing in the development process. R&D, engineering, and manufacturing are often recognized as critical skills in development, but marketing is often underutilized. Figure 1.8 is a flowchart of the new product process with the role of the marketing manager indicated at each step of the process. Marketing managers should be involved at the beginning of the process—at Opportunity Identification. Many companies do not involve the marketing manager until Building the Business Case stage or Testing and Validation, which is one reason so many new products fail; the voice of the customer is not considered until after product development is complete. Robert G. Cooper, credited with developing the stage-gate process discussed earlier, argues that involving marketing managers early in the process helps considerably to reduce failure rate.[38] In the remainder of this book, we will investigate thoroughly how marketing is influential in the success of a new product and how marketing inputs function in the multidisciplinary skill set necessary to be successful in new product development.

The New Product Manager

So, what is the role of the new product manager in the stage-gate process? For a firm to successfully innovate, dedicated personnel are required to drive the innovation process. A new product (NP) manager is concerned with the development of new products within the organization and usually, but not always, the commercialization of that product. The NP manager typically takes on three major roles within the organization:[39]

1. *Team leader/project management*: The NP manager leads the new product team comprised of team members with diverse skills who also may have part-time commitments to other parts of the organization. In this role, the primary responsibility of the manager is to the specific project and the team. The primary goal of the new product manager is to make sure an individual project moves ahead smoothly and successfully.

2. *Functional new product management*: The NP team integrates functional skills from marketing, manufacturing, R&D, and finance departments. The primary focus of each team member is to their functional area and not to the new product team. The NP manager may fill one of the functional roles him/herself (i.e., marketing representative), but the overall responsibility is for overseeing all functional needs of the project with a set of team members assuring that they work together effectively.

3. *New product program management*: In this role, the NP manager oversees many new product projects and helps the project managers implement excellent new product processes. The NP manager's primary responsibility is to the organization, the process of development, and the NPD portfolio. A program manager may oversee multiple projects sequentially or even simultaneously.

Salary reports in 2013 show the typical salary for a new product development manager in marketing in the United States is from $58,000 to $94,000, although those with a technical background can earn higher.[40] New product managers typically have advanced graduate degrees or, at minimal, an undergraduate degree with several years in a functional department within a company, such as R&D, operations, marketing, or sales. They are multidisciplinary, managed risk takers, and extremely organized even in chaotic situations. They must see how the complex process can come together from various perspectives, including R&D, marketing, and production. One of the premier organizations that focus on new product management is the Product Development Management Association (PDMA). The PDMA blog site, http://blog.pdma.org/, is a good place to go to see what issues new product managers address. There is also a glossary of terms commonly used in new product development (http://pdma.org) that is useful to review. PDMA's practitioner-oriented newsletter, *Visions*, is also a good place to read timely articles on the NPD process.

Goals of the Text

The goal of this text is to help you learn how to manage the development and marketing of new products and services. After completing this book, we expect that you will be able to:

1. Select a new product strategy that matches the needs of your organization.
2. Set up a disciplined process for new product development.
3. Define target market opportunities and search out high potential ideas.
4. Understand customer needs, structure them, and prioritize the needs to clearly define the benefits and values that your product will deliver when positioned against competition in your target market segment.
5. Integrate marketing, engineering, R&D, and production representatives to design a high-quality product that satisfies customer needs and delivers value.
6. Forecast sales before market launch based on testing of the product and the marketing plan.
7. Know the role of analytical support tools (market research and models) in new product development including: when to use them, their limitations and advantages, and the appropriate level of sophistication at each stage of the process.

We expect that the concepts and techniques discussed in the following chapters will help you become a better manager. The ideas in this book will help you improve the performance of your organization whether it is consumer or industrial, public or private, large or small, national or global, producing durable goods, frequently purchased consumer packaged goods, or services. Some of the concepts presented are generic, others must be modified for each application. Together, however, they should lead to greater profitability and reduced risk in new product development activities within your organization.

Chapter Summary

In this chapter, a foundation is laid for understanding the role of new product development in the innovating organization from the ideation stage to the launch phase. The stage-gate process was introduced as a systematic approach for strategizing for innovation. The important role of marketing in this process also was emphasized. Firms that plan accordingly will make sure that an adequate stream of new product projects is continually entering the process and the ROI for new products meets their goals.

Glossary

Additions to existing product lines: Product line extensions, flankers, or brand extensions, e.g., Crest Tartar Control toothpaste, iPod shuffle, or Dyson's handheld vacuum cleaners.

Bounding boxes: A type of new product development process that sets up a boundary, which the new product team and management agree to work within to keep the project on track. Outside the boundaries, an NPD team will escalate issues or decisions to management.

Building the business case: A process in the Design stage of the stage-gate process involving detailed investigation with primary research—both market and technical—leading to a business case, product and project definition, project justification, and project plan.

Cost reductions: Products that replace existing products by providing similar performance at a lower cost, e.g., many HDTVs or 8 GB iPhone. This category represents about 10% of new products.

Design phase: A stage in the stage-gate process where scoping, building a business case, and product development occur in order to design a product/service that meets the need of customers with focusing technology to address those needs.

Discovery: Another name for the opportunity identification stage.

Fuzzy front end: The early stages of the NPD process, when the ideas are still quite fuzzy.

Functional new product management: Product management where the new product team integrates functional skills from marketing, manufacturing, R&D, and finance departments. The primary focus of the each team member is to his/her functional area and not to the new product team.

Gates: Points in the stage-gate innovation process where the project's viability is assessed to determine its congruency with the firm's goals.

Improvement or revisions to existing products: Minor changes to improve existing products, e.g., MICHELIN® X One® tires, P&G's Swaddlers Sensitive New Baby diapers, or Gillette's 5-blade Fusion razor.

Incremental innovation: Innovations that have small incremental technology or small marketing changes to an existing product.

Innovation: An iterative process initiated by the perception of a new market and/or new service opportunity for a technology-based invention, which leads to development, production, and marketing tasks striving for the commercial success of the invention.

Launch and postlaunch life-cycle management: A stage in the stage-gate process where the product/service that is being designed is tested prior to launch. Life-cycle management focuses on the entire life of the product from launch to disposal at end-of-life.

Looping: A type of new product development process commonly used in software design where versions are developed, tested, and then looped back into the design phase. Looping has no structured "gates" to stop the design process.

New product program management: A new product development management process where the new product (NP) manager oversees many new product projects and helps the project managers implement excellent new product processes. The NP manager's primarily responsibility is to the organization, the process of development, and the NPD portfolio.

New-to-the-firm products: Products that take the firm in a new direction. These products are not new to the world, but are new to the firm, e.g., Apple iPod in 2001, AT&T's Universal credit card, or Swatch's Smart car.[41]

New-to-the-world products: Inventions that create a whole new market, e.g., Toyota's Prius hybrid automobile, Sony Walkman, P&G's Febreze.

Opportunity identification: A stage in the stage-gate process where the target group of customers is carefully defined and ideas are generated to address their needs.

Product development: A process in the Design stage of the stage-gate process involving actual detailed design and physical development of the new product, and the design of the operations or production process. The primary deliverable from the product development stage is to produce a physical prototype or mock-up of the proposed product.

Radical innovations: Innovations that introduce *both* market and technological discontinuities to the marketplace. Examples include cell phones, personal computers, and internet telephony.

Really new innovations: Innovations that introduce either a new marketing or technological innovation. Really new innovations evolve into new product lines (e.g., Sony Walkman), product line extensions with new technology (e.g., Canon LaserJet), or new markets with existing technologies (e.g., early fax machines).

Repositioning: Products that take on new uses, e.g., Levi Jeans from work pants to fashion statement, aspirin for heart attacks. This category accounts for about 7% of new products.

Scoping: A process in the Design stage of the stage-gate process where screening of an idea with a quick, preliminary investigation of the project is conducted using feasibility studies from a marketing, engineering, and financial perspective.

Stage-gate innovation process: Systematic, structured process for developing new products with gates (stopping points) that force a thorough evaluation of the product/service under design before progressing to the next stage of development; formalized by Robert G. Cooper (http://www.stage-gate.com).

Team leader/project management: The new product manager leads the new product team comprised of team members with diverse skills who also may have part-time commitments to other parts of the organization. In this role, the primary responsibility of the manager is to the specific project and the team.

Testing: A stage in the stage-gate process where tests or trials of a new product are tested in the marketplace, lab, or plant to verify its potential success in the marketplace.

Voice of the customer: Gathering data on what customers want and need from new products. Usually obtained through qualitative or quantitative research analyses.

Review Questions

1. What is a new product? What is not a new product? Why are new products important?
2. Why should firms innovate given the risky nature of innovation?
3. Why design a sequential process if it takes more time and may eliminate a product that could have been a success? Why not just get a good idea and go to market?
4. Discuss the Initiating Factors that may lead to new product ideas in the next decade for the following industries:
   a. Computers
   b. Automobiles
   c. Healthcare
   d. Education
   e. Food products
   f. Financial services
   g. Mobile telecommunication

5. Give an example from each of the new product categories:
   a. New-to-the-world products
   b. New-to-the-firm products
   c. Additions to existing product lines
   d. Repositioning
   e. Cost reductions

Assignment Questions

1. Describe the new product development process within a company in which you are familiar. If you cannot speak to a representative of this company, what type of process do you think would be most beneficial for this company?
2. Select a well-publicized and successful new product introduced within the last year. What seem to be the main factors accounting for its success? Identify any initiating factors that may have impacted the new product.
3. Describe a new product that failed or was withdrawn from the market in the last few years. What factors led to its failure? Could these have been avoided? How?
4. Select one of the following new product categories listed below. How would you think about putting together a new product development strategy?
   • Healthcare delivery

- Internet telephone
- Luxury sports car
- "Cloud" computing services
- Tailgate cooler
- Online management school

Endnotes

1. Booz Allen Hamilton, "New Product Management for the 1980s" (New York, 1982).
2. J. Wind, V. Mahajan, and B. Bayless, "The Role of New Product Models in Supporting and Improving the New Product Development Process: Some Preliminary Results" (Cambridge, MA: The Marketing Science Institute, 1990).
3. PDMA is the premier global advocate for product development and management professionals. It provides New Product Development Professional certification. If you are interested in a career as a new product manager, PDMA is a great resource.
4. See Abbie Griffin, *Journal of Product Innovation Management* 14 (1997): 429–458, for a review of past studies and survey of best practice and PDMA 2004 Best Practices Study (CPAS) available at www.pdma.org.
5. C.M. Crawford, "New Product Failure Rates—Facts and Fallacies," *Research Management* (September), 9–13 (1997).
6. "Association of National Advertisers, Prescription for New Product Success" (New York: Association of National Advertisers, Inc. (1984).
7. Booz Allen Hamilton, New Product Management.
8. PDMA. Online at: http://www.pdma.org/ (accessed July 4, 2013).
9. Barry Jaruzelski, John Loehr, and Richard Holman, "The Global Innovation 1000" (New York: Booze Allen Hamilton, Inc.) Online at: http://www.booz.com/media/file/BoozCo_The-2012-Global-Innovation-1000-Results-Summary.pdf
10. Ibid.
11. Ibid.
12. Ibid.
13. Online at: http://www.booz.com/global/home/what-we-think/global-innovation-1000/top-innovators-spenders (accessed June 28, 2013).
14. U.S. Census Bureau. Online at: http://www.census.gov/econ/www/servmenu.html (accessed February 1, 2013).
15. Online at: http://www-935.ibm.com/services/us/index.wss (accessed July 4, 2013).
16. Kwaku Atuahene-Gima, "Differential Potency of Factors Affecting Innovation Performance in Manufacturing and Services Firms in Australia," *Journal of Product Innovation Management*, 13(1) (1996): 35–52.
17. Online at: http://www.independent.co.uk/news/media/tv-radio/metoo-bbc-criticised-for-copying-others-1658044.html (accessed May 12, 2009).
18. Online at: http://www.eweek.com/c/a/Desktops-and-Notebooks/HP-Strategy-Leaves-Opening-for-Apple-Dell-Analysts-694614/ (accessed September 1, 2011).
19. Based on R. G. Cooper, *Winning at New Products: Accelerating the Process from Idea to Launch*, 3rd ed. (New York: Basic Books), which is also described in detail at http://www.stage-gate.com/ (accessed June, 28, 2013).
20. Gary E. Blau, Joseph F. Pekny, Vishal A. Varma, and Paul R. Bunch, "Managing a Portfolio of Interdependent New Product Candidates in the Pharmaceutical Industry," *Journal of Product Innovation Management* 21(4) (2004): 227–245.

21. Braun: The Syncro Shaver (A), Ivey Case Study (Boston: Design Management Institute, 2006).
22. Gary E. Blau, Joseph F. Pekny, Vishal A. Varma, and Paul R. Bunch, "Managing a Portfolio of Interdependent New Product Candidates in the Pharmaceutical Industry," *Journal of Product Innovation Management* 21(4) (2004): 227–245.
23. Urban and Hauser combine the seven stages into four: opportunity identification, design phase, testing, and launch–postlaunch; see Glen L. Urban and John R. Hauser, *Design and Marketing of New Products* (Upper Saddle River, NJ: Prentice Hall, 1993).
24. Fuzzy front end refers to early stages of the NPD process when the ideas are still quite fuzzy.
25. Online at: http://www.designtoimprovelife.dk/index.php?option=com_content_custom& view=article&id=742:dyson-vacuum-cleaner&catid=46:finalists-2005&Itemid=18 (accessed May 9, 2013).
26. Online at: http://www.dyson.com/homepage.asp (accessed July 4, 2013).
27. 1993 Dyson. Online at: http://manchestervacs.co.uk/DysonForum/index.php?topic=4.0 and 2007 (accessed July 4, 2013).
28. The Program (or Project) Evaluation and Review Technique (PERT) project management tool designed to analyze and represent the tasks involved in completing a given project. A Gantt chart is a type of bar chart that outlines a project schedule.
29. Laura Doyle, "Fast & Flexible Insights from Tektronix: Using the 'Bounding Box' to Accelerate Development." Online at: http://www.roundtable.com/research-publications/ publication/814 (accessed July 5, 2013).
30. New Product Dynamics (1999). Online at: http:www.newproductdynamics.com/newsletter99.htm#9-99. Used with permission from Preston G. Smith (accessed July 4, 2013).
31. Organization for Economic Cooperation and Development. "The Nature of Innovation and the Evolution of the Productive System," in *Technology and Productivity–The Challenge for Economic Policy* (Paris: OECD, 1991), 303–314.
32. Rosanna Garcia and Roger Calantone, "A Critical Look at Technological Innovation Typology and Innovativeness Terminology: A Literature Review," *Journal of Product Innovation Management* 19(2) (2002): 110–132.
33. Booz Allen Hamilton Inc., *New Product Development for the 1980s* (New York: Booz Allen Hamilton, Inc., 1982). The percentages are from Abbie Griffin, *Drivers of NPD Success: The 1997 PDMA Report* (Chicago: Product Development & Management Association, 1997).
34. Swatch is no longer involved in this project. Daimler AG was last to market the Smart car brand.
35. A great sources of reasons new product fails and how to overcome these failures is Robert G. Cooper, *Winning at New Products*.
36. For more on escalation of commitment, the continual support of obviously failing projects, see Jeffrey Schmidt and Roger Calantone, "Escalation of Commitment during New Product Development," *Journal of the Academy of Marketing Science* 30(2) (2002): 103–118.
37. Based in part on Robert G. Cooper, *Winning at New Products*.
38. Ibid.
39. Adapted from C. Merle Crawford and C. Anthony Di Benedetto, *New Products Management*, 9th ed. (New York: McGraw-Hill/Irwin, 2007).
40. Online at: http://www.payscale.com/af/calc.aspx?job=Product+Development+Manager%2 c+Marketing&state=&country= (accessed July 5, 2013).
41. Swatch is no longer involved in this project. Daimler AG was last to market the Smart car brand.

2

NEW PRODUCT INNOVATION STRATEGY

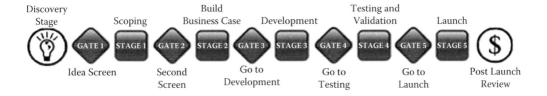

Learning Objectives

In this chapter, we will address the following questions:

1. What different types of innovation strategies can be used to drive the NPD process?
2. How are the corporate strategy, the innovation strategy, and the NPD process inter-related?
3. What is the role of portfolio management in the NPD process?
4. What are some methods of evaluating and maintaining the product portfolio?

Innovative Corporate Cultures

INNOVATING AT APPLE[1]

"Apple designers are required to come up with 10 entirely different mock-ups of any new feature," reports Michael Lopp, senior engineering manager at Apple. Ten versions give room to design without engineering or creative restrictions. These 10 are whittled down to 3, which are then thoroughly scrutinized, and then, finally, 1 strong design is chosen. To end up with the one design, each week the team has two meetings—one in which to "go crazy" brainstorming that is free of constraints; the second, a production meeting, is the other's antithesis. Here designers and engineers are required to nail everything down; to work out how this crazy idea might actually work. The best ideas from the paired design/production meetings are presented to leaders, who might just decide that some of those ideas, although crazy, are worth pursuing. This innovation process continues throughout the development of any application, though the balance from "crazy" to "producible" shifts as the application progresses.

Google has a high tolerance for chaos and ambiguity in its innovation design process. David Glazer, engineering director, explains, "We started running a bunch of experiments. We set an operational tempo: When in doubt, do something. If you have two paths and you're not sure which is right, take the fastest path. What's true in physics about objects in motion is true when you're creating a product. It's easier to keep moving and change course than when you're sitting and thinking and thinking."

Google's innovation strategy is to "throw something on the wall and see what sticks." Marissa Mayer, vice president of Search Products and User Experience at Google describes this process. "The hardest part about indoctrinating people into our culture is when engineers show me a prototype and I'm like, 'Great, let's go!' They'll say, 'Oh, no, it's not ready.' I tell them, 'The Googly thing is to launch it early on Google Labs and then to iterate, learning what the market wants—and making it great.' Google Labs is Google's technology playground."[3] The Web site claims that "Google Labs showcases a few of our favorite ideas that aren't quite ready for prime time. Your feedback can help us improve them. Please play with these prototypes and send your comments directly to the Googlers who developed them." Google also distributes new services marked as "Beta" indicating that the product isn't without bugs—Google Scholar, Google Mail, Google Maps, and others were introduced this way.

In the beginning of the twenty-first century, both Apple and Google are equally successful companies at innovating, yet, they have dramatically different innovation strategies. Apple strives for perfection prior to launch and Google strives for innovative beta versions for launch, regardless if they are bug-free. Each organization approaches new product development with an effective strategy that is likely to achieve success, while minimizing costs and risk. These companies know that potential rewards and risks from developing successful new products are high and the strengths of the company must drive its strategy. In this chapter, you will gain an understanding of why Apple's innovation strategy differs from Google's innovation strategy, and how each can be successful.

Introduction

In the last chapter, we became convinced of the importance of new products and gained an understanding of the risks in new product developments and how to deal with them in a disciplined "go/kill" decision process. In this chapter, we take a step back from the details of the process to understand the strategic framework that drives the process. With this understanding, we will be able to implement a better new product development process. As we will see in this chapter, the corporate strategy is the framework that gives an organization its overall directions and impels it to action.

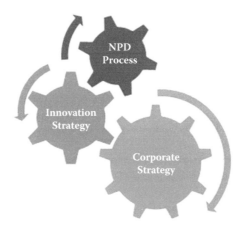

Figure 2.1 Integration of strategies.

The corporate strategy establishes the guidelines for the innovation strategy, which sets the environment that encourages discoveries, idea generation, and inventions. The innovation strategy subsequently guides the new product development strategy, which determines the proactive or reactive stance a firm will take and provides the specifics on when and what new products will be developed to meet the goals of the innovation strategy. The relationship between the three levels of strategy (corporate, innovation, new product) is shown in Figure 2.1. Corporate strategy drives the innovation strategy, which dictates the detailed new product development process. We next examine how the three types of strategies are interconnected. Throughout this chapter, we will contrast the Google and Apple strategies to emphasize how a range of strategies may be crafted to breed success.

Corporate Strategy Dictates Innovation Strategy

Corporate strategy is a framework that gives an organization its overall directions and impels it to action. The strategy formulation process begins with a systematic diagnosis of the threats and opportunities in the environment, an inventorying of the organization's strengths and weaknesses, and an understanding of the key phenomena underlying demand and competition.[4] In strategic planning, goals, programs, plans, and budgets are formulated to build on the organization's competitive advantages and market opportunities. Corporate strategies are often autocratic and structured in order to guide the entire organization on goal setting and decision making.

Corporate strategies are often summarized in a mission statement. Table 2.1 shows Google's mission statement to be focused on the user, whereas Apple's is focused on products. We see these mission statements reflected in the innovation strategy summarized in the beginning of the chapter where Apple strives for product perfection and Google strives to provide innovative applications for information seekers and advertisers. We see that corporate strategy drives these companies' innovation strategies.

Table 2.1   Google and Apple's Corporate Strategies

GOOGLE MISSION STATEMENT—ANNUAL REPORT 2008

Our mission is to organize the world's information and make it universally accessible and useful. We believe that the most effective, and ultimately the most profitable, way to accomplish our mission is to put the needs of our users first. We have found that offering a high-quality user experience leads to increased traffic and strong word-of-mouth promotion. Our dedication to putting users first is reflected in three key commitments:

- We will do our best to provide the most relevant and useful search results possible, independent of financial incentives. Our search results will be objective and we do not accept payment for search result ranking or inclusion.

- We will do our best to provide the most relevant and useful advertising. Advertisements should not be an annoying interruption. If any element on a search result page is influenced by payment to us, we will make it clear to our users.

- We will never stop working to improve our user experience, our search technology and other important areas of information organization.

We believe that our user focus is the foundation of our success to date. We also believe that this focus is critical for the creation of long-term value. We do not intend to compromise our user focus for short-term economic gain.

APPLE'S MISSION STATEMENT, 2009

Apple ignited the personal computer revolution in the 1970s with the Apple II and reinvented the personal computer in the 1980s with the Macintosh. Today, Apple continues to lead the industry in innovation with its award-winning computers, OS X operating system and iLife and professional applications. Apple is also spearheading the digital media revolution with its iPod portable music and video players and iTunes online store, and has entered the mobile phone market with its revolutionary iPhone.

Source: Endnote 5.

The innovation strategy should be synergistic, meaning that innovations should draw upon the strengths of the organization and opportunities in the marketplace in order to accomplish the goals of the company as set forth in the corporate strategy. Below are some of the questions that reflect the interface between corporate and innovation strategy. The answer to these questions will help to establish the innovation strategy.

1. What are the firm's core competences? How can we tap into these areas of expertise?
2. Where are we vulnerable from competitors? How do we fix that?
3. What is the mission statement of the organization? How does innovation manifest itself in the statement?
4. Is the company conservative or does it encourage risk-taking?
5. Why have we succeeded in the past?
6. What is the forecast for material costs and availability?
7. What technological advances can we expect?
8. What actions will the government take that will affect us?
9. What can we expect from our competitors and what are their strengths and weaknesses?
10. What opportunities exist in the marketplace that the firm can take advantage of with its innovation direction?

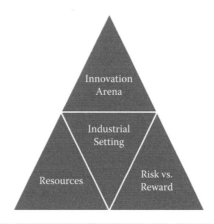

Figure 2.2 Components of the innovation strategy.

Innovation Strategy

One definition of *strategy* is "an adaptation or complex of adaptations (as of behavior, metabolism, or structure) that serves or appears to serve an important function in achieving evolutionary success."[6] Using this definition, we recognize the innovation strategy as an overarching plan within an organization that guides a firm in achieving success in an ever-changing, or evolutionary, market. It is the "gestalt"[7] that drives the firm's innovation direction. It clearly delineates the arena within which the new product development (NPD) team must operate to be successful. We can surmise that Apple's innovation strategy is to "develop revolutionary products for personal computing, portable digital music, and mobile communication experiences" and Google's innovation strategy is to "design new services that provide a high-quality user experience for information management and advertising." Four components make up the innovation strategy: innovation arena, resources, risk versus reward, and industrial setting, as shown in Figure 2.2.

Innovation Arena It is not very often that a firm will give its NPD department innovation carte blanche in which there are no constraints on the markets or technology areas in which to focus new products. Typically, boundaries are set where the innovation arena may be large, such as Apple's focus on communications technology (computers, music players, mobile phones), or more confined, such as Coca-Cola's focus on beverages in a can or bottle (soft drinks, water, juices), many of which are line extensions of its more than 2,800 different types of drinks. Google also has a large boundary as it creates services that exploit the knowledge base of the World Wide Web. Defining the innovation arena will answer questions, such as: which markets, customers, product categories, technology, competitors, and distribution channels are the primary focuses of the company? The innovation arena helps the NPD team focus on the firm's core competencies that have been identified in the corporate strategy.

Market/Customers/Product Categories   Both Apple and Google have expansive market and customer focus. Neither of these companies is geographically or demographically constrained and both companies service many industries. In their business statement, Apple indicates its customers are consumers, students, educators, businesses, and government entities. Google would have a similar target, although their target market is closely aligned to the profile of Internet users. However, many companies are market specific, such as Adidas, which has a target market of consumers who wear sport shoes, typically in the 14 to 30+ age range.

Market size and profit margins can affect the choice of development strategy. In large markets with economies of scale or experience in production, distribution, or marketing, a proactive first-in innovation may establish market dominance and give the firm an advantageous position. On the other hand, in markets that have neither the volume nor the margins, a firm may not be able to return its investment in product development, especially if there are high overhead costs. A reactive strategy might be better. The market and target customers drive the product categories that a firm will innovate within.

Technology   Technology clearly defines the innovation strategy of both Apple and Google. Both companies try to stay on the cutting edge of new technological advances in their respective industries with innovative products. For other companies, technological innovation is not as important, such as with the Coca-Cola Company whose innovation is not highly dependent upon new discoveries in the food industry. However, when food innovations do become available, Coca-Cola is quick to adapt to them as they did with aspartame, the nonsaccharide sweetener, which allowed them to offer Diet Coke in 1982 and Coke Zero in 2004. How rapidly technology changes will influence what type of strategy a company will embrace.

Competition   The intensity of the competitive environment may be critical to selecting a strategic posture. Many aggressive competitors may force firms to an imitative reactive strategy in order to be competitive. If products can be sped to market, if there are few entry costs, if the innovation is not protected by patents, and if the organization can quickly achieve economies of scale, being reactive may be appropriate. Fewer competitors also may allow the firm to be relaxed in inventing truly innovative products.

The relative size of competitors is also important. A small firm may be particularly vulnerable to competitive reaction and, thus, must be preemptive in its innovation plans. Similarly, a large firm may be proactive to protect its lead. For example, in appliances, although imitation is common, Black & Decker allocates substantial resources to design new appliances.

Channels   In many markets, middlemen serve a physical distribution, inventory, selling, or servicing function. Theses wholesalers, brokers, distributors, or retailers are independent decision makers. For example, a retail food store decides whether it will stock a new product—not the manufacturer. By making the product attractive to the retailer through an adequate margin, special deals, high quality, and consumer advertising, the firm can gain shelf space. In all cases, the desires of the channel participant must be considered to assure adequate distribution and selling effort.

In some situations, one firm in the chain of distribution may be proactive, with other firms reacting to that firm's innovation. In many industrial markets, the supplier of the materials or even the final user may develop the product. ALCOA invented the aluminum truck trailer and then sold it to the trucking industry by showing that less weight in the trailer structure meant more payloads that would pay back the higher initial investment costs. In consumer industries, the producer is the usual innovator, but powerful retailers will specify innovative products and then have other firms produce them. For example, Sears' Craftsman line of tools is well respected and commands a premium price. Whether or not a firm is innovative or imitative can depend upon the stance of other firms in the distribution channel and on its relative power within that channel.

Some firms actually gain power as well as profits by innovation in distribution. For example, Hanes Corporation was simply another apparel producer until it introduced L'eggs, a distinctively packaged panty hose, through innovative distribution in supermarkets and drugstores. It is now a dominant force in the multibillion dollar women's hosiery market.

Resources   Every NPD project requires resources in the form of monetary funding, man-hours, facilities, management guidance, outsource funding, and other firm commitments. As we saw in Chapter 1, many firms allocate a significant amount of dollars toward NPD projects. Apple's research and development expenditures totaled $1.1 billion compared to Google, which spent $2.4 billion in R&D in 2008.[8] Besides its substantial budget Google's employees, who include some of the best and brightest software programmers in the world, are one of its best resources. The resources a firm has can be a competitive advantage for them. If resources, such as employee knowledge, can't be replicated by other firms, then a company can gain an advantage in the marketplace by continually introducing more innovative products than its competition.

A focus on resources also can help a firm in deciding whether to license a new technology or to develop it internally. Technology that cannot be replicated outside the firm will need to be developed internally. Technology that utilizes no special strengths of the company may best be licensed from outside firms. Imitable innovations hold no competitive advantage for the firm, and firms can look to outsource these innovations to others so as to reserve resources for more profitable endeavors.

Following on the idea of open innovation, where innovations originate external to the organization as well as internally, resources also should be allocated for sourcing innovations out-of-house, including acquisitions. Acquisitions are formal alliances initiated by legally combining two organizations. Alliances are designed to bring together the pool of skills in technology, marketing, production, finance, and geographic experience so that the alliance members can be competitive in the market and achieve their goals. Such alliances provide opportunities to the initiating firm to gain skill at lower costs. The participants gain the opportunity to grow without bearing the full risk of market development. Alliances need not be limited to manufacturers or service providers. They can include suppliers or distributors or even customers. Explicitly partnering with customers puts the firm in close touch with their needs, and the market in general. For example, IBM partnered with USAA, a pioneering insurance company, so that it could leverage this customer's innovativeness to build better software and systems for transmitting and manipulating images.

Risk versus Reward   It is important to define in an NPD strategy how much risk is accepted in the organization. Developing and launching innovative products requires making decisions that are inherently risky because new products, services, and programs have a high potential for failure. Highly innovative programs frequently do not make it beyond the research stage and, if successfully launched into the marketplace, may take years for payback. For example, the innovative Apple Newton, the first personal digital assistant (PDA), was a marketplace failure. But taking risks in new product development often pays off, as Apple well knows.

A certain level of risk-taking by senior managers and their willingness to accept occasional failures should be a normal part of business. In the absence of openness to risks, employees are likely to be reluctant to try innovative ideas and are more likely to stay within their individual risk comfort levels. This should suggest to NPD managers that accepting risk and encouraging innovativeness in the NPD process are essential requirements for achieving high NP performance gains.[9]

> Part of risk-taking is knowing when inventions should be patented. Google has been very aggressive in defending its patents. The Mountain View, California-based company has fought and won patent suits filed against it since starting an aggressive patent defense policy in 2007. Google won a ruling that ended a lawsuit, which sought to shut down Google Earth, its Internet program that shows images and data from satellites and airplanes.[10]

Protection may be granted by the market itself instead of through patents when the first firm introduces a good product and achieves a predominant position. For example, although Burger King and Burger Chef and others have copied McDonald's food franchising operation, McDonald's is still the biggest chain, is very profitable, and continues to grow. This is because of its worldwide brand recognition. In other product categories, such as small appliances, a product pioneer can be quickly copied. The innovator may have only a short period of competitive advantage. For example, six months after the first electric knife was introduced, more than 10 brands were on the market. Thus, firms that can achieve good protection should be proactive while those that cannot may be better off in a reactive mode.

Industrial Setting   Apple is primarily a product-oriented company, whereas Google is primarily a service-oriented company. The type of industry that a firm operates in will significantly impact its innovation strategy. A consumer packaged goods manufacturer, such as Procter & Gamble, will have a different innovation strategy than a pharmaceutical company, such as Merck, and both companies will have extremely different strategies than a technology-based company, such as Dell. Dell, in fact, is a service-oriented product manufacturer. With its direct-sales model via the Internet and telephone network, it is able to deliver a unique service of individual PCs configured to customer specifications. Dell computers also are heavily supported by customer support centers located around the world. Innovation for Dell must occur at both the product level as well as the service level.

The structure of many service companies allows them to more quickly develop and test out new products, as we have seen with Google. Because of this, it is easier for Google to be a leader instead of a follower of many new Web-based innovations. In several studies on service companies,[11] it was found that employees are an important resource for service-oriented companies. Employees' close contact and potentially long-term relationships with customers can be an important source of new ideas. These employees hear on a daily basis the needs of customers, and areas where better services may be needed. These frontline employees can often make-or-break the successful launch of new services. Poorly trained employees can result in failed launches. As much time should be spent on understanding the role of employees in a service organization as is spent on understanding consumers' needs and wants.

Consumer packaged goods (CPG) companies, like Procter & Gamble, will have a major focus on incremental innovations and quickly introduce new variations of existing products to the consumer marketplace. Due to intense competition, P&G often will come out with a "me too" product, yet it also devotes considerable amount of resources to really new products, such as the Swiffer product line. CPG companies concentrate on regularly introducing new products to the marketplace. Gianni Ciserani, president in 2009 for P&G Western Europe, stated, "In laundry, we have set the trend for the future—first with Ariel Cool Clean and now with Ariel Excel Gel, a breakthrough product that delivers our best cleaning while touching three points of the sustainable product innovation cycle: less chemicals, less water, and less energy."[12] Large CPG companies must continually innovate to remain competitive even if it seems as there aren't any new trends in laundry soap products. Incremental innovations will be the typical strategy for these types of firms.

Merck, whose business is "discovering, developing, and delivering novel medicines and vaccines that can make a difference in the lives of people around the world,"[13] spent $8.4 billion in R&D in 2009.[14] Incremental innovations and "me too" strategies don't make sense for them. With the extensive testing needed for new drugs before introduction to the market and with product development times taking up to 15 years, imitative products are rare and to speed to market is not a strategy that often makes sense.

"For GE, imagination at work is more than a slogan or a tagline. It is a reason for being."

JEFF IMMELT, CHAIRMAN AND CEO

Figure 2.3 General Electric mission statement.

Another company, General Electric, is involved in a wide variety of markets including: financial services, electricity generation, lighting, industrial automation, medical imaging equipment, aircraft jet engines, aviation, as well as entertainment with 80% ownership of NBC Universal. "Imagination at work" is the company's mission (see Figure 2.3). Their Web site reads, "From jet engines to power generation, financial services to water processing, and medical imaging to media content, GE people worldwide are dedicated to turning imaginative ideas into leading products and services that help solve some of the world's toughest problems."[15] With its "imagination at work" perspective, the company has a focus on a new strategic initiative called Ecomagination. GE has committed more than $20 billion in the endeavor to build green products. The program has resulted in over 70 green products that are being brought to market, ranging from halogen lamps to biogas engines. GE strives to be one of the world's most innovative companies.

Innovation Strategy Dictates the New Product Development Strategy

Taken together the innovation arena, industrial setting, resources, and risk/rewards will culminate into an innovation strategy that the firm will use to guide decisions on what type of new product development strategy to undertake. Here are some of the questions that reflect the interface between innovation and new product strategy:

1. Where are we vulnerable from competitors? How do we fix that?
2. What is the forecast for material costs and availability?
3. What technological advances can we expect?
4. How has consumer consumption changed and how can we exploit this?
5. What products and markets must be defended to maintain the corporate image?
6. How much does your organization depend upon growth?
7. Does the organization have experience in certain channels of distribution?
8. Can synergies with existing products be exploited in terms of materials, production lines, advertising, brand equity, or promotion?

Answering these questions will help a firm determine how reactive or how proactive it may need to be in the new product development program. The final specification in the innovation strategy is the choice of reactive versus proactive innovation strategies.

Reactive versus Proactive Innovation Development Process

Today's organizations face a variety of circumstances; some call for innovation from the ground up (new markets and totally new products) while some call for a rapid defensive response that might include imitating a competitor's innovation. One strategy may not be sufficient for maintaining a competitive leg up, so managers also must understand how to put together a portfolio of products requiring flexibility in a new product strategy. As the Apple–Google example shows, companies can take on a number of different approaches to innovation. There is no "one-size-fits-all" strategy.

The organization can choose from a range of alternative strategies. One of the basic strategic decisions is whether to be **reactive** or **proactive**. Reactive product strategy is based on addressing initiating pressures as they occur. A reactive view of the competition is to wait until the competition introduces a product and copy it if it is successful. Google's search engine rose to prominence by displacing numerous browsers, such as AltaVista and Ask Jeeves, and by offering a superior service centered on page rankings, which showed popularity of a Web site. This was a reactive strategy in a developed market, but Google went on to proactively introduce many advertising innovations to their search site. A proactive strategy would be based on preempting competition by being first on the market with a product competitors would find difficult to match or improve. A proactive strategy explicitly allocates resources to achieve goals and preempt undesirable future events. Apple's iTunes online store is an example of such a proactive strategy. They were the first legal online digital music store that allowed users to seamlessly add new music to a MP3 player and to manage their music library.

Table 2.2 identifies several reactive and proactive strategies, including: imitative versus innovative, defensive versus offensive, mechanistic versus entrepreneurial, customer-focused ideation versus market-focused ideation. Each strategy is appropriate under certain conditions, which we discuss next.

Reactive Processes   An imitative strategy is based on quickly copying a new product before its maker develops exclusive market share. This imitator or "me too" strategy is common practice in the fashion and design industries for clothes, furniture, and small

Table 2.2 Proactive versus Reactive Strategies

REACTIVE	PROACTIVE
Imitative	Innovative
Defensive	Offensive
Mechanistic	Entrepreneurial
Customer Ideation	Market/Customer Innovation

appliances. For example, once Cuisinart, an entrepreneurial start-up, demonstrated that a market existed for expensive food processors, many of the major appliance companies followed with imitative products. Sunbeam and Hamilton Beach saw this innovation as an opportunity for product line expansion; with expertise in the channel, in production, and in marketing, they felt they could succeed in the marketplace with just a "me too" product.

"**Fast follower**" is another type of imitative strategy. In this strategy, a firm will quickly develop an imitation to counter the attack on its market share. In Japan, once Asahi demonstrated a market for "dry" beer, Kirin, Suntory, and others were forced to develop their own dry beers in an attempt to maintain sales. When Kirin attacked the U.S. market, Anheuser Busch responded with imitative products, Michelob Dry and then Bud Dry. Although Kirin had a frontal attack, it was never able to displace Anheuser Busch products as the top dry beer sales in the U.S.

"**Second but better**" is another imitative strategy. In this case, the firm does not just copy the competitive product, but identifies ways to improve the product and its positioning. For example, both Apple and Google have taken this strategy. Apple's iPod was not the first hard-drive MP3 player. That honor actually goes to Hewlett Packard (HP) through an acquisition of Compaq.[16] However, Apple captured primary market share by offering a player that was stylish, easy to use, and supported with strong marketing compared to the HP offering. Google's search engine followed Yahoo! Directory, Look Smart, WebCrawler, Lycos, and many others.

Another type of "second but better" strategy might not attack a new product head on, but rather identify a niche where it can provide unique benefits. For example, Microsoft's Excel spreadsheet program gained a share by providing a superior graphical user interface, flexibility, efficiency, and compatibility to Apple computers. In the early 1990s, the Lotus123 spreadsheet had the dominant share of IBM compatible machines, and Excel had the dominate share among Apple's Macintosh users. As of the late 2000s, Lotus123 is all but extinct. Lotus failed to embrace the Windows operating system, guaranteeing that Excel would take the no. 1 position. As the market for operating platforms changed, Lotus123 did not change with it. This is another reminder that firms need to continually innovate, especially as technology changes.

A **defensive strategy** protects the profitability of existing products by countering competitive new products. For example, when Datril entered the analgesic market with a position of "the same ingredients as Tylenol, but less expensive," Johnson & Johnson, the makers of Tylenol, responded with Extra Strength Tylenol to successfully keep out Datril. Some defensive strategies are primarily marketing mix responses to advertising, promotion, or price, while some strategies include counteroffensives of new flankers and new products. For example, once Tylenol countered Datril's attack, Tylenol Extra Strength established their brand among consumers who demanded a more effective pain reliever.

Customer ideation is purposively reacting to customers' requests and ideas. For example, because scientific instruments users often modify and improve the equipment they use, manufacturers can identify new opportunities (and new designs) by internalizing information flow from users. Such a strategy implies an emphasis on applications engineering and manufacturing. MITRE, a Bedford, Massachusetts-based company, provides engineering and technical services to the U.S. federal government by responding to Request for Proposals from the Department of Defense, and other government agencies. Responsive strategies also are used by manufacturers in a chain of distribution in which some other channel member is dominant. Teflon cookware was developed in response to customer requests, which in turn were encouraged by the material supplier, DuPont. A **mechanistic** strategy focuses on incremental innovations, churning out the next product in an almost mechanical way. Coca-Cola and Pepsi do this with their soft drink lines. Diet Coke was followed by Vanilla Coke, Cherry Coke, Coke Black, Coke Zero, and others. Mechanistic companies are good at production and marketing and look to take advantage of these competencies by focusing on developing product lines instead of creating new product categories.

Proactive Processes   An alternative new product strategy is for organizations to be proactive and initiate change. A proactive aerospace company does preemptive R&D and product development. Boeing has designed the innovative, fuel-efficient 787 Dreamliner, which was planned for launch in 2010, using reinforced plastic composite material instead of metals. They believe this new jet will change the direction of airplane manufacturing.

The proactive strategy, **innovative**, is based on research and development efforts to develop technically superior products. Some companies have been notably successful with this strategy. Both Google and Apple are examples of organizations that devote considerable energies to the potential of technological innovation.

Another proactive strategy, an **offensive** strategy, also focuses on innovations that are disruptive to the marketplace.[17] Many start-up companies or new ventures will take this approach by improving a product or service by leapfrogging the current technology and introducing innovations that do not meet the demands of the current marketplace, but are designed for a different set of consumers. This "leapfrogging" strategy is sometimes called **creative destruction** or **disruptive technology**. For example, Long-playing (LP) records were replaced by CDs, which were replaced by MP3 players, which are surely to be replaced by yet another upcoming technology. The disruption can occur at the marketing level or the technology level or both. Apple's iTunes introduced a new distribution method for digital music, causing a major disruption in the music industry. Disruptive technologies are particularly threatening to the leaders of an existing market, who are locked into a technology that serves an existing customer base. For more on disruptive technologies, see the work by Christensen.[18]

Another proactive form of product development is **entrepreneurial**. Both Google and Apple started out as entrepreneurial companies. Google started out in a Menlo Park, California, garage with Larry Page and Sergey Brin at the helm. Apple also started in a Silicon Valley garage with Steve Jobs, Steve Wozniak, and Ron Wayne. Wayne, not a risk-taker, sold his share of Apple Computers back to Jobs and Wozniak for $800. Some large companies have tried to instill an entrepreneurial strategy within their organization; this has been termed entrepreneurial, meaning it comes from within a company. At 3M (Minnesota Manufacturing and Mining), a separate new venture division has been established where entrepreneurs can take a leave from their regular job to work on their ventures. The Post-it note is a result of this entrepreneurial spirit.

A firm also can be proactive in identifying customer needs and developing products that provide the benefits to satisfy those needs through a **market/customer innovation strategy**. Such strategies require that the organization devote energies to understanding the input from the customer; this can include market research, talking to users, and/or rotating personnel so that they have contact with the customer. Procter & Gamble, General Foods, McDonald's, and most consumer product companies utilize this market-based philosophy. Voice of the customer is paramount in driving the direction of the innovation strategy. A sophisticated version of this strategy is user-based innovation. In this paradigm, customers are not just the source of ideas and requests (see reactive), but actually develop solutions to their needs. These consumers have needs that foreshadow the market and have developed solutions. Eric Von Hippel describes this user-centered innovation process in his book, *Democratizing Innovation*,[19] where innovation starts at the bottom (with end users). He gives the example of mountain bike riders, who were extremely influential in creating design innovations in this industry. These types of creative customers are often called "lead users" because firm collaboration with these consumers can be proactively pursued by observing, capturing, and facilitating user innovation.

Which Innovation Strategy to Use?

The question the NPD manager must ask is: "Which innovation strategy to undertake?" Often a firm will take multiple strategies based on a product line or project orientation. A reactive strategy may be necessary within one product category, whereas a proactive strategy is needed in another product category. One product line may even require being both reactive and proactive. Apple was proactive in introducing the iPhone 3G with more than 28,000 applications, yet, the competitive nature of the mobile phone market has required Apple to be reactive with incremental iPhone 3.0 OS and iPhone Pro versions launched in 2009. The Product Development Management Association (PDMA) found in a 2004 survey that 32% of firms used a first to market strategy (proactive), 37% were fast followers (proactive), 21% were niche protectors (reactive), and 7% were nonaggressive and reacted only when forced to by environmental pressures.[20]

To guide the process of determining the proper mix, firms use portfolio management techniques to set priorities and allocate resources. Not only does one product contribute to the innovation success, but a portfolio of products act to produce cumulative results to meet innovations and corporate criteria. We discuss portfolio management next.

New Product Portfolio Management

Portfolio management for new products is very similar to the concept of portfolio management in financial investments. A firm needs to have an adequate mix of new product projects that are sure winners versus a bit more risky, immediate money makers versus long-term revenue generators, incremental versus more radical innovations, recently launched versus in-decline, etc. Portfolio management is crucial for determining which group of products should receive resources, prioritization, and acceleration to market. It's about balance: the optimal mix between risk and return, radical and incremental innovation, maintenance and growth, and short- and long-term projects that support the overall company strategy. Each project must be assessed for profitability (rewards), investment requirements (resources), risks, and overall strategic fit. Once these evaluations have occurred, budget and resource allocation can be undertaken to balance the portfolio for maximum profitability. The three goals of portfolio management are value maximization, portfolio balance, and strategic alignment.[21]

Apple's portfolio mix is well defined. The company offers a range of personal computing products including desktop and portable personal computers (MacBook/Pro/Air, iMac, Mac mini, Xserve), Music Products and Services (iPod Nano/shuffle/classic/touch, iTunes), the iPhone mobile communication device, Peripheral Products (Cinema Display, Apple TV), Software and Computer Technologies (Mac OSX, iLife, iPhoto, iMovie, iDVD, Garage Band, etc.), and Internet Software and Services (Safari, QuickTime, Mobile Me, etc.). Google has an equally extensive portfolio mix. Their product lines include Google.com Services (Google Web search, image search, book search, etc.), Google Applications (Google docs, calendar, Gmail, YouTube, etc.), Google Client Services (toolbar, Chrome, Picasa, etc.), Google GEO (Google Earth and maps), Google Mobile and Android, Google Checkout, and Google Labs. Google's revenue is primarily generated by AdWords and AdSense online advertising services.

How each company decides what products to add to their portfolio will depend on their strategy. A structured approach to portfolio management manages risks, prioritizes projects, evaluates resource requirements, and determines project synergies. Cooper and colleagues[22] define portfolio management as a dynamic decision process, whereby a business's list of active new product (and R&D) projects is constantly updated and revised. In this process, the *mix* of projects is evaluated together, prioritized, and allocated resources; existing projects may be accelerated, deprioritized, or killed. A good portfolio review will terminate products instead of continuing to fund projects that no longer meet the strategic requirements of the firm. "Escalation of commitment," or the continual funding of a project because of sunk costs, is a problem for many companies.[23] Tough choices must be made and skilled managers are needed to implement these choices.

There are numerous methods for evaluating portfolios with each firm having its own preferred techniques. Some of the most popular methods are NPV/IRR (net present value/internal rate of return), the risk–reward bubble diagrams (also called portfolio maps), strategic buckets (also called the business strategy method), and scoring models. Complaints about these methods are that they are laborious to use and complicated. However, Cooper and colleagues[24] have shown that failure to use portfolio analysis often results in an overabundance of low-value projects with no strategic alignment between projects, projects fighting for limited resources, increased time to market, and higher failure rates. PDMA found that more than 50% of firms use explicit portfolio reviews as part of their new product processes.

Tools for Portfolio Management

Economic Models   Commonly used economical models in NPD include payback period, break-even analysis, return on investment (ROI), discounted cash flows (DCFs), net present value (NPV), decision tree analysis (a sensitivity analysis method), and internal rate of return (IRR). These methods are popular because they are easy to calculate, familiar to top managers who are used to making capital expenditure decisions, and easy to interpret by shareholders. The drawback to these methods is that solid, reliable financial data are not always available, especially in the early stages of the new product development process. DCFs and NPVs are highly sensitive to terminal values as well. For example, if you have a project with an undefined life, the last cash flow in your calculation essentially represents not only the cash flow for that period, but also all the future cash flows past that point. Managers can bias the results by overestimating this amount and can easily come up with a very positive NPV. Additionally, economic results are compared to a company-established hurdle rate; if the project does not meet the criteria, it is killed, if it does meet the criteria, it is added to the portfolio. Synergies or redundancies between projects cannot be considered in economic models. Cooper reports that businesses with the poorest performing portfolios rely almost exclusively on financial selection criteria. We include an example of an NPV calculation below.

Economic models also are biased against risky, radical innovations. Cooper writes,[25] "In DCF [discounted cash flow] analysis, the assumption is that the project is an all-or-nothing investment—a single and irreversible expenditure decision. In reality, however, investments in new product projects are made in increments; that is, management has a series of go/kill options along the way. As new information becomes available, the decision is made to invest more or to halt the project. The go/kill options, of course, reduce the risk of the project (versus an all-or-nothing approach). When DCF is used, this lost option value is an opportunity cost that should be incorporated when the investment is analyzed. But, traditional spreadsheets used to generate NPV do not. By contrast, options pricing theory recognizes that management can kill the projects after each incremental investment is made—that management has options along the way." Economic models are better suited to incremental product where accurate financial data is much easier to come by. These types of models are less useful for revolutionary new products where financial data may not exist.

Options theory is one method of taking into consideration the mitigated risk over time of radical or breakthrough innovations. Because coming up with unbiased and meaningful cash flows is necessary for accurate calculations, economic models are most accurate for incremental innovations, such as line extensions, where historical data are available for comparison. The real option method enables corporate decision makers to leverage uncertainty and limit downside risk.

The simplistic economic evaluators for a new product investment are break-even and payback analysis, which are very popular economic methods. Payback period looks at how much time after *launch* it takes to offset the initial expenditures. This period is sometimes referred to as "the time that it takes for an investment to pay for itself." Some payback periods are decades, such as with the microwave. The payback period has limitations because it does not properly account for the time value of money, inflation, financing, or portfolio consideration. The break-even time is how long it takes to *pay off* all the expenditures incurred in developing the new product from its original initiation. A popular method that values early over later funds flows is Net Present Value. In order to calculate NPV for a new product project, annual cash flows need to be determined and discounted to the present. To estimate these annual cash flows, five main types of information are needed: initial investments (including manufacturing, R&D, and marketing costs), annual revenues, and discount rate. The discount rate is usually the company's internal rate of return (IRR) for investments.

$$NPV = \sum_{t=1}^{T} \frac{C_t}{(1+r)^t} - C_0 \tag{2.1}$$

where C_t is the cash flow in year t, T = time horizon of the program in years, and r = the discount rate.

Example: C_0 = -400,000, C_1 = 75,000, C_2 = 200,000, C_3 = 300,000, r = 10%

$$NPV = \frac{-400,000 + 75,000}{(1+0.10)} + \frac{200,000}{(1+0.10)^2} + \frac{300,000}{(1+0.10)^3} = \$58,865.51$$

Through this example, you can see that radical innovations, which may have large negative cash flows in the early years, would be negatively biased compared to an incremental innovation, which has positive cash flows in the early years.

Portfolio Maps

Methods have been developed to compare different types of projects against one another. Portfolio maps (also called *bubble diagrams*) plot projects on two criteria on

Table 2.3   Popular Bubble Diagram Dimensions

TYPE OF CHART	X-AXIS	Y-AXIS
Risk vs. Reward	Reward: NPV, IRR, benefits after years of launch; market value	Probability of success
Newness	Technical newness	Market newness
Ease vs. Attractiveness	Technical feasibility	Market attractiveness (growth potential, consumer appeal, general, attractiveness, life cycle)
Strength vs. Attractiveness	Competitive position	Attractiveness (market growth, technical maturity, years to implementation)
Cost vs. Timing	Cost to implement	Time to impact
Strategic vs. Benefit	Strategic focus or fit	Business intent, NPV, financial fit, attractiveness
Cost vs. Benefit	Cumulative reward	Cumulative development costs

Source: Endnote 26. With permission.

an x–y map. Any two-by-two dimensional plot can be used for evaluation. Table 2.3 provides a list of bubble diagram dimensions, but you can define your own that are relevant to your strategy or develop a profile of each project and multiple dimensions. Each new product development process needs to specify the portfolio maps that are most relevant for that company. If the criteria can be rated as high or low, such as probability of success on the x-axis and risk on the y-axis, projects in the first quadrant (high risk/low probability of success) will want to be avoided. This is the traditional risk/return financial tradeoff.

Figure 2.4 shows a plot of new product projects with the y-axis "months until launch" and the x-axis "position in the stage-gate process." The *size* of the bubble indicates projected peak revenue, the *darker shade* of the bubble indicates the likelihood of success, and the *shade* of the bubble indicates the target market. This gives a

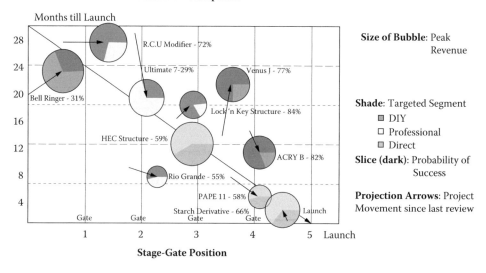

Figure 2.4   Portfolio map. (From Endnote 27. With permission.)

NPD Portfolio Strategy and Strategic Buckets

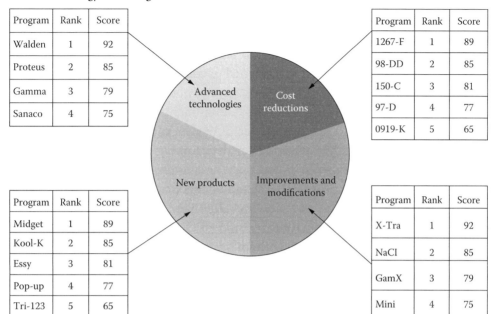

Program	Rank	Score
Walden	1	92
Proteus	2	85
Gamma	3	79
Sanaco	4	75

Program	Rank	Score
1267-F	1	89
98-DD	2	85
150-C	3	81
97-D	4	77
0919-K	5	65

Program	Rank	Score
Midget	1	89
Kool-K	2	85
Essy	3	81
Pop-up	4	77
Tri-123	5	65

Program	Rank	Score
X-Tra	1	92
NaCI	2	85
GamX	3	79
Mini	4	75

Figure 2.5   Strategic buckets. (From Endnote 28. With permission.)

good visual representation of the current portfolio, and managers can use it to assess the size, composition, and balance of the portfolio. These types of maps make it very easy to compare one project against another and to determine where gaps are in the portfolio. This map also indicates how the project has changed position since the last portfolio review. Any number of criteria can be used to represent the size of the bubble including, manhours, monetary resources needed, annual revenues, NPV, etc. In this manner, the portfolio map is based on both subjective assessments as well as economic data. The firm's objective should be to have a portfolio of projects that generate a continuing stream of innovative new products and meet financial criteria of profit and sales growth at a minimum risk.

Strategic Buckets   A strategic bucket[28] is another method for evaluating the mix of products that align with a new product strategy. In this approach, projects are allocated to buckets and then resources are assigned to each class of projects. Buckets may be product type, markets, technologies, product lines, or any category that defines the company's strategic goals. Figure 2.5 depicts a new product portfolio with strategic buckets based on four product types: advanced technologies, cost reductions, new products, and improvements and modifications. The projects that are in each bucket are listed, along with their rank and score within the bucket. A company may use the scoring model to rank projects within a single bucket, combining the two techniques.

Strategic buckets are used to ensure that the company has the appropriate mix of products within their portfolio that meets the overall corporate strategy. One of the benefits of strategic buckets is that there is no biasing of high risk/high reward projects that are often rated as unattractive by economic model analyses. Radical innovations that may be assigned to a "bucket" are considered a separate group and are judged against other innovations only in the same bucket. This method forces the company to understand how the different types of new products support their strategic goals and can assure that enough resources are assigned to risky or major innovation projects.

As noted earlier, there are three major goals of portfolio evaluation: value maximization, portfolio balance, and strategic alignment. By looking at the various portfolio evaluation methods, we can see how each method helps to reach these goals. The economic models are good for determining value maximization; subjective assessment tools also help with value maximization by weeding out projects not within the strategic guidelines; and bubble diagrams and strategic buckets assist with portfolio balance and strategic alignment. Thus, many companies use more than one portfolio evaluation method to evaluate their new product mix. Also, we have seen that different methods are more appropriate at different stages of the NPD process and with different types of products. For example, checklists are best applied at the early stages of the NPD process, and strategic buckets with scoring models are best for evaluating the place of radical innovations in the new product portfolio.

Portfolio management is crucial for determining which group of products should receive resources, prioritization, and acceleration to market. It's about balance—the optimal mix between risk and return, radical innovations and incremental innovations, market maintenance and growth, and short- and long-term projects that support the overall company strategy. Each project must be assessed for profitability (rewards), investment requirements (resources), risks, and overall strategic fit. Once these evaluations have occurred, budget and resource allocation can be undertaken to balance the portfolio for maximum profitability.

Portfolio Review Process

The portfolio review process and the stage-gate process can be integrated together as shown in Figure 2.6. The gates provide a natural place for senior management to evaluate separate projects together as a portfolio. You may recall that gates are where a new product project is evaluated for its continued alignment with strategic goals, needed resource requirements, and quality of execution. Because a gate is typically focused on a single project, all-project portfolio reviews should be conducted whenever a new project reaches a major step in the NPD process and resources are required to continue with the new project. This forces management to step back and strategically realign all projects, not just the one under review. In an all-project review, management should be checking for the right balance of projects, the right match to corporate strategy, and the reassessment of resources and priorities. Figure 2.6 shows the tasks that should occur in the portfolio review and how they integrate with the stage-gate process.

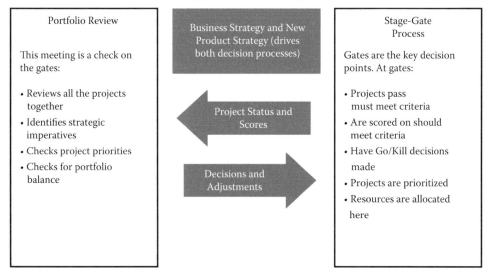

Figure 2.6 Integrated stage-gate portfolio management process. (From Endnote 29. With permission.)

Formal Process versus Reality

Although firms develop clearly defined processes, they are not always followed. In reality, firms practice one of the following strategies:

- *Who has a new idea today?* Someone with political influence comes up with an idea and pushes it through without idea screening, business evaluation, or comparison with alternatives.
- *Here is a new technology. Let's run with it*: A hot new technology becomes a "must have" criteria, even if it does not fit the firm's and/or market needs.
- *Me Too*: Hey, our competition is doing it; we have to as well even if it does not fit our capabilities and may not work.
- *Let's run it up the flag pole and see who salutes*: Here, we do a lot of trial and error with no strategy on where we are going and what core capabilities we have.
- *Rush to Market*: The Internet in the 1995–2000 time period was ripe with ideas that went to market without completing the design and testing steps. As expected, only a few of these ideas succeeded and billions of dollars were lost in the rush to get to the market.

It takes discipline to design and implement a process for NPD that matches strategy, reduces risk, and leads to a continuing stream of innovations. However, this discipline can achieve a consistent positive ROI for firms.

Chapter Summary

In this chapter, we saw how a firm's corporate strategy is the precursor for a firm's innovation strategy. Four components drive the innovation strategy: innovation arena, industrial setting, resources, and risk/reward. These factors will influence the type of innovation strategy that a firm will put together. The innovation strategy will then determine whether a firm will take a proactive or reactive stance in the product development program. There are several types of new product processes a firm can undertake: imitative versus innovative, defensive versus offensive, mechanistic versus entrepreneurial, customer ideation versus market/customer innovation. Firms will put together a portfolio of products that may use a mix of these strategies. The NPD projects that make up the portfolio will be picked based on their alignment with the company goals and their likelihood of success as determined by both qualitative and quantitative evaluation methods, such as economic models and portfolio maps. Reassessment of the portfolio mix should occur regularly at major project evaluation gates, which assess a project's overall alignment with the innovation goals of the firm, resource requirements, and probable success in advancing the company's strategic goals. With this strategic understanding, we are now ready to address each step in the development process in the next chapters.

Glossary

Creative Destruction: A term coined by Joseph Schumpeter in his work entitled *Capitalism, Socialism and Democracy* (Routledge, 1942) to denote a "process of industrial mutation that incessantly revolutionizes the economic structure from within, incessantly destroying the old one, incessantly creating a new one."[30]

Defensive strategy: An innovation strategy that protects the profitability of existing products by developing new products to counter competitors' products.

Disruptive technology: Improving a product or service by leapfrogging the current technology and introducing innovations that do not meet the demands of the current marketplace, but are designed for a different set of consumers.

Economic models: Economic approaches to evaluating financial success of an NPD project including payback period, break-even analysis, return on investment (ROI), discounted cash flows, net present value (NPV), decision tree analysis (a sensitivity analysis method), and internal rate of return (IRR).

Entrepreneurial strategy: A proactive form of product development where a firm develops a new product or service previously not available in the market.

Fast follower: A type of imitative strategy where a firm will quickly develop an imitation of an existing product in order to gain market share.

Imitative innovation strategy: An innovation strategy of quickly copying a new product before a competitor develops exclusive market share.

Innovation strategy: A firm's stated strategy on how it approaches the innovation process in the firm.

Innovative strategy: A proactive innovation strategy where the focus is on research and development (R&D) efforts to develop technically superior products.

Market/customer innovation strategy: A proactive innovation strategy where customer needs are identified and products are developed that provide the benefits to satisfy those needs.

New product development strategy: A firm's stated strategy on how it will approach the new product process whereby innovations are developed and brought to market.

Offensive strategy: A proactive innovation strategy that improves on a product or service by leapfrogging the current technology and introducing innovations that do not meet the demands of the current marketplace, but are designed for a different set of consumers.

Portfolio management: A mix of new product projects that as a combination seek to maximize a firm's profits. The mix may include sure winners versus a bit more risky, immediate money makers versus long-term revenue generators, incremental versus more radical innovations, or recently launched versus in decline.

Portfolio maps: Also called *bubble diagrams*, provide a means to compare different new product projects being undertaken by a firm to allow comparison on several criteria, typically plotted on two axes.

Proactive innovation process: Preempting competition by being first on the market with a product competitors would find difficult to match or improve.

Reactive innovation process: Addressing initiating pressures as they occur, particularly in regards to new competitive offerings.

Second but better: A type of imitative strategy where the firm does not just copy the competitive product, but identifies ways to improve the product and its positioning.

Strategic bucket: A method for evaluating a portfolio of products where the projects are allocated to buckets and then resources are assigned to each class of projects.

Strategy: "An adaptation or complex of adaptations (as of behavior, metabolism, or structure) that serves or appears to serve an important function in achieving evolutionary success."[31]

Review Questions

1. How does an innovation strategy drive a firm's new product development process?

2. Discuss the advantages and disadvantages of each of the reactive and proactive new product development strategies.

3. Consider two firms competing in the same market. Would it ever be reasonable for one to adopt a proactive strategy while another adopts a reactive strategy? If not, why not? If so, under what circumstances?

4. Mr. Hardy Cell, a manager with Widgets, Inc., has a view of the consumer that is not totally unique. Mr. Cell says, "Consumers are fickle. They don't know what they want. New products are necessary only because consumers quickly become bored with old products. Consumers should be treated as if they were children. They must be told what they should buy. Advertise anything and it will sell. If your product sales are falling, tell people the product is new and improved. That's all you need to do."
   a. Based on what you have learned in this chapter and the last, are Mr. Cell's remarks consistent with a successful NPD program?
   b. Why might Apple, Inc. support Mr. Cell's position?
   c. Why might Google support Mr. Cell's position?
   d. Could Mr. Cell's remarks be interpreted as representing a proactive new product strategy? A reactive strategy?

5. Describe how the four components of the innovation strategy (innovation arena, industrial setting, risk versus reward, resources) differ for a product-oriented company as opposed to a service-oriented company.

Assignment Questions

1. Pick a company that you admire for its innovation strategy (or use your own company). Based on the 10 questions in the section "Corporate strategy dictates innovation strategy" and the questions in the section "Innovation strategy dictates new product strategy," describe how the firm's strategies are interconnected.

2. Give an example of a company for each of the proactive and reactive strategies. If you can, stay within a single industry.

3. Take your firm's new product development portfolio and create a two-dimensional bubble diagram for it. Do you see any problems with the new product mix? Share this diagram with others in the company and see whether they are in agreement with your assessment.

4. You have just been hired as the new product manager by Electric Go!, a start-up company that is looking to introduce economical electric vehicles to India. Using the components of the innovation strategy described in this chapter (Figure 2.2), describe what type of strategy you would implement for this company.

5. Interview a new product manager from a company with which you are familiar. Create a composite criteria scorecard for two new products the company is currently designing. Is there a more appropriate evaluation method you would recommend they use? Why?

Endnotes

1. Helen Walters, "Apple's design process," The Tech Beat, *Business Week*, March 2008. Online at: http://www.businessweek.com/the_thread/techbeat/archives/2008/03/apples_design_p.html (accessed March 24, 2009).
2. Chuck Salter, "Fast Company," February 14, 2008. Online at: http://www.fastcompany.com/magazine/123/google.html? page=0%2C0 (accessed March 24, 2009).
3. As of October 2011, Google had stopped using its Google Labs as a platform for launching new products.
4. See Glen L. Urban and Steven H. Star, *Advanced Marketing Strategy* (Englewood Cliffs. NJ: Prentice Hall, 1991), for a complete review of this topic.
5. FAQ section of Apple Web site. Online at: http://www.apple.com/investor/ (accessed March 31, 2009).
6. Online at: http://www.merriam-webster.com/dictionary/strategy (accessed March 31, 2009).
7. Gestalt refers to a structure, configuration, or pattern of physical, biological, or psychological phenomena so integrated as to constitute a functional unit with properties not derivable by summation of its parts. Online at: http://www.merriam-webster.com/dictionary/gestalt (accessed March 31, 2009).
8. 2008 Annual Report. Online at: http://www.sec.gov/Archives/edgar/data/320193/000119312508224958/d10k.htm (accessed June 28, 2013).
9. For more on this topic, see Gerard J. Tellis, *Unrelenting Innovation: How to Create a Culture for Market Dominance* (San Francisco: Jossey–Bass, 2012).
10. Online at: http://www.bloomberg.com/apps/news?pid=20601109&sid=ar3V._UIg9CM&refer=home (accessed July 4, 2013).
11. Kwaku Atuahene-Gima, "Differential Potency of Factors Affecting Innovation Performance in Manufacturing and Services Firms in Australia," *Journal of Product Innovation Management* 13(1) (1996): 35–52.
12. Online at: http://www.packagingdigest.com/article/CA6647530.html?nid=3463 (accessed April 28, 2013).
13. Online at: http://www.merck.com/ (accessed June 28, 2013).
14. "2011 Global R&D Funding Forecast," *Advantage Business Media*, December 2010.
15. Online at: http://www.ge.com/ (accessed June 28, 2013).
16. Online at: http://news.cnet.com/Bragging-rights-to-the-worlds-first-MP3-player/2010-1041_3-5548180.html (accessed April 28, 2013).
17. Christensen, Clayton, "The Innovator's Dilemma: When New Technologies Cause Great Firms to Fail," *Harvard Business Press* (1997).
18. Ibid.
19. Eric Von Hippel, *Democratizing Innovation* (Cambridge, MA: MIT Press, 2006).
20. Product Development Management Association Comparative Performance Study (CPAS), 2004. Online at: http://www.pdma.org/p/cm/ld/fid=318 (accessed July 4, 2013).
21. R. G. Cooper, *Winning at New Products: Accelerating the Process from Idea to Launch*, 3rd ed. (Cambridge, MA: Perseus Books, 2001).
22. R. G. Cooper, S. J. Edgett, and E. J. Kleinschmidt, *Portfolio Management for New Products*, 2nd ed. (Cambridge, MA: Perseus Books, 2001).

23. Jeffrey B. Schmidt and Roger J. Calantone, "Escalation of Commitment during New Product Development," *Journal of the Academy of Marketing Science* 30(2) (March 2002).

24. R. G. Cooper, *Winning at New Products: Accelerating the Process from Idea to Launch*, 3rd ed. (Cambridge, MA: Perseus Books, 2001).

25. Ibid. p. 228.

26. R. G. Cooper, S. Edgett, and E. Kleinschmidt, *Portfolio Management for New Products*, p. 98.

27. Online at: http://boards.core77.com/viewtopic.php?t=17061 (accessed July 4, 2013).

28. Raul O. Chao and K. Stylianos, "A Theoretical Framework for Managing the New Product Development Portfolio: When and How to Use Strategic Buckets," *Management Science* 54(5): 907–921.

29 R. G. Cooper, S. J. Edgett, and E. J. Kleinschmidt, 2000. "New Problems, New Solutions: Making Portfolio Management More Effective," *Research Technology Management* 43(2) (2000): 18–33.

30. Online at: http://www.investopedia.com/terms/c/creativedestruction.asp (accessed July 4, 2013).

31. Online at: http://www.merriam-webster.com/dictionary/strategy (accessed March 31, 2009).

OPPORTUNITY IDENTIFICATION AND IDEA GENERATION

The Fuzzy Front End[1]

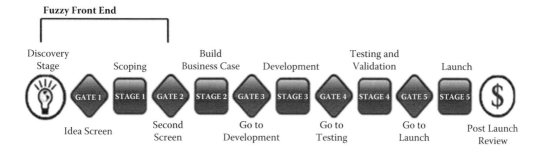

Learning Objectives

In this chapter, we will address the following questions:

1. How do firms identify new product opportunities?
2. Where do ideas come from?
3. What methods are there for generating new product ideas?
4. How do firms decide which markets to enter?

Opportunity Identification at Dell[2]

In the 1980s, IBM was the largest suppliers of personal computers (PCs) and Apple was struggling to find a place in the growing market. While a student at the University of Texas at Austin, Michael Dell recognized the market potential for a full-service PC supplier and started selling custom-designed personal computers from his dorm room. By placing an 800 number in computer magazines, customers would directly contact Dell to place their orders. The IBM and Apple brand names were not as important to experienced PC users as was the savings from eliminating the middleman distributor. Dell's initial target market was experienced computer users who valued the customizations and lower prices.

Large institutions quickly followed as buyers because they knew which features they desired in a personal computer and were happy to buy at the lower prices.

Dell explains this strategy: "So, we let our competitors introduce machines with rock-bottom prices and zero margins. We figured they could be the ones to teach consumers about PCs while we focused our efforts on more profitable segments. And then, because we're direct and see who is buying what, we noticed something interesting. The industry's average selling price to consumers was going down, but ours was going up. Consumers who were now buying their second or third machines—who wanted the most powerful machines and needed less handholding—were coming to us. And, without focusing on it in a significant way, we had a $billion consumer business that was profitable. So, we decided in 1997 that it was time to dedicate a group to serving that segment."[3] The company quickly grew to $73 million by selling PCs direct to end-users, whereby eliminating resellers' markups and the cost of carrying large inventories of finished products.

As the years progressed, Dell continued to divide customers into finer target segments as seen in the accompanying diagram. The finer the segmentation, the better Dell was able to serve the needs of these customers. Michael Dell further explains how this strategy brought the company closer to their customers. "It allows us to understand their needs in a really deep way. This closeness gives us access to information that's absolutely critical to our strategy. It helps us forecast what they're going to need and when. And, good forecasts are the key to keeping our costs down." In 2009, Dell focused on "home," "small and medium business," "public sector," and "large enterprise" segments in the United States. By observing the needs of a select market, experienced PC users who knew what they wanted in a computer, Dell was able to build a company worth several $billion.[3] The innovation for Dell was not new computer technology, but a new business model of delivering customized computers to savvy users. Opportunities lie in many areas when it comes to innovations.

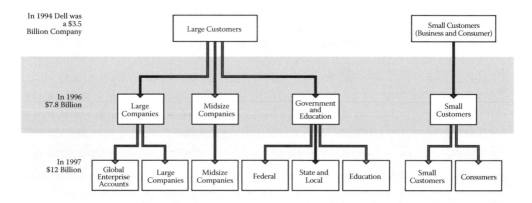

Figure Source: The power of virtual integration: An interview with Dell Computer's Michal Dell, Joan Magretta, *Harvard Business Review*, March-April 1998, pg. 73–84.

Introduction

After an organization has adopted a proactive approach to new product development, it must then identify areas of opportunity, also called *target markets* or *market segments*, a concept you should be familiar with through your marketing courses. Although a firm could just begin to serve all segments within the marketplace, a more targeted direction leads to greater success as Michael Dell explained. In launching the company, Dell decided not to initially sell products to the mass market, but instead focused on the more profitable computer savvy end user and the large institution market. In its new product strategy, Dell continually evaluates where the greatest potential lies and then refines products and services to meet the needs and wants of that target market. One of the biggest innovations that Dell provided to the PC market was the business model innovation of how computers were custom built to a consumer's requirements and then directly shipped to that consumer. Knowing where to look for market opportunities and what to look for are crucial skills for the new product manager.

Opportunity Identification

Typically, managers go through five steps in the opportunity identification stage of new product development. These five steps are often called the *fuzzy front end* of the new product development process because at this stage the product/service concepts are still quite fuzzy for the NPD manager.[4] Specific steps include:[5]

1. *Idea Generation*: Generate product ideas to tap the potential of selected markets.
2. *Opportunity Analysis*: Align these opportunities with new product development strategy.
3. *Market Identification*: Identify markets that offer the best opportunities for the organization.
4. *Market Selection*: Select the most opportunistic target markets for new product introductions and/or product line extensions.
5. *Concept Development and Refinement*: Refine and screen these ideas, taking the best to the design phase.

This five-step process can be thought of as a sieve or funnel, where a broad marketplace is narrowed down into a target market selection that makes sense for the firm based on their strengths and core competencies. Figure 3.1 demonstrates this process.

It should be noted that the steps need not be sequential or in this order. Some of the steps may occur simultaneously or in a different order. What is important is that the firm acknowledges the benefit of conducting each step in order to drive the successful development of new products. Steps 1 to 4 will be discussed in this chapter and Step 5, concept development and refinement, will be detailed in the next chapter.

Figure 3.1   Five steps of idea generation and opportunity identification.

Step 1: Generating Product Ideas

New products, by their very nature, are innovations and these innovations result from creative insight and free thinking. However, innovations are not limited to products. They may be services, concepts (AIDS prevention), business models (Dell business), process innovations (assembly line manufacturing), or ideas (recycling awareness). Figure 3.2 summarizes where innovations can come from—both in- and out-of-house. Setting up an innovation environment that encourages employees to come up

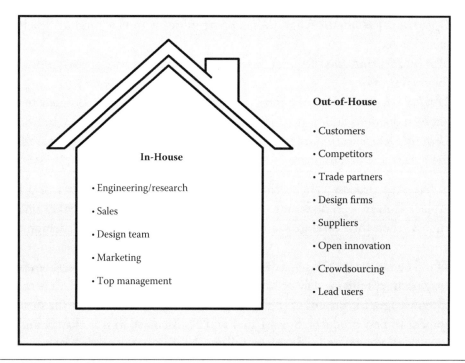

Figure 3.2   Sources of innovations.

with new product ideas can be extremely profitable. There are a variety of ways that firms can encourage in-house ideation.

Let's first look at internal ideation sources to the firm:

- **Engineering/Research:** The R&D department should be aware of new technologies that present new opportunities to meet consumer needs and fill needs that may be latent. For example, recent advances in biotechnology have created new markets as new drugs are synthesized. Genetic engineering and nanotechnologies are creating new drugs for treatment of hypertension, cancer, and AIDS. Materials science also has generated new technologies that have resulted in major new products, such as composite airplanes, carbon fiber sail boats, plastic auto bodies, and super conductors.

- **Sales:** Sales employees are the front-line contact with customers. They hear customers' needs and wants on a daily basis. Existing customers sometimes modify a product and redesign it for their particular need. These modifications can be a source of new incremental innovations for a company.

- **Production:** Production is often neglected as a source of innovation. The problem-solving skills of production engineers produce an important source of new ideas, particularly ***process innovations***. User-driven innovation is often found in process engineering at customers' sites. Products like pultrusion machines (fiber-reinforced plastic pulled through a die) and machines to insert integrated circuits on a printed circuit board are examples where production engineers pushed the products to new limits of performance.[6] Process innovations most often come from production.

- **Service Personnel:** Service and warranty experience also can be the source of new products. Service records can identify product needs and gaps in quality. Customer service or "hot lines" indicate not only short run problems, but reflect longer run unmet needs. Amazon, the online retailer, is very active in soliciting consumer feedback in order to improve its service offering. Repairs may reflect product limitations and can promote new uses for a product. Instead of voiding warranties when a user modifies a machine to run beyond its recommended performance parameters (e.g., speed or tolerances), the manufacturer may instead find new ideas for new products from user modification.

- **Marketing:** The marketing department should continually monitor the external environment for ideas. The aging of the baby boomer generation (those born between 1946 and 1965) continues to spur technological innovation and changes in the healthcare system, as Walmart, the retail conglomerate, has recognized through its in-store clinics. Deregulation of ethical pharmaceuticals has opened up opportunities for new over-the-counter (OTC) drugs, such as Advil and Nuprin, which are based on the previously prescription ingredient, ibuprofen. These are only some of the ideas that have resulted from the changes in overall marketplace social, economic, and political phenomena. Identifying these trends can be rewarding to the innovating firm.

- **Legal Teams:** Some firms may look for patent innovations to improve upon and add to their product line.
- **Top Management:** Sometimes product/service ideas come from the top down. It is said that Steve Jobs, ex-CEO, drove new product ideas at Apple, Inc.

Ideas also can come from out-of-house. For example, a competitor may introduce a new product and, thus, force your organization to innovate. While there is pressure to respond with a "me too" or second-but-better alternative, more effective ideas might come from a complete examination of market needs and recent technological developments. Some external sources include:

- **Customers:** Users of the products are important sources of product improvements. One of the bases for a proactive strategy is a customer-oriented philosophy of seeking to understand customer needs and desires. Results have shown that the financial return from customer-driven products tends to be higher. For example, women's use of eggs added to shampoos to give more body to hair was the solution adopted in the protein shampoo market. Highly involved customers, also called *lead users* (described in further detail below), are a particularly good source of new product ideas.
- **Competitors:** Knowing what leads to competitors' success and their developmental strategies are important inputs to idea generation. Even if a firm is a proactive leader in an industry, it must be ready to defend against competitors by preempting such innovation or improving on competitive firms' new products. Apple's Web browser, Safari, introduced in 2007, "borrowed" features from other innovative browsers, such as Opera, Chrome, and Firefox.
- **Trade Partners:** Noncompetitive firms in other industries also may be the source of new product ideas. Often innovation flows from one industry to another. For example, software for composition of newspapers was carried over to desktop publishing. Often, looking internationally can bring a fresh point of view to an industry. The Body Shop uses natural shampoo and cosmetic products based on native formulations from many developing parts of the world.
- **Design Firms:** While internal R&D and engineering are valuable sources of technological ideas, external sources should not be overlooked. Contacting inventors and searching for patents may present new ideas for consideration. IDEO of Palo Alto, California,[7] and Continuum,[8] headquartered in the Boston, Massachusetts area, are two popular U.S.-based firms that design innovations for their clients.
- **Suppliers:** We tend to think of manufacturers themselves as the major source of innovation, but many times the locus of innovation is elsewhere. DuPont invented Teflon cookware, which benefitted other cookware manufacturers as well as DuPont. ALCOA, supplier of aluminum materials, wanted to pioneer

the idea of aluminum truck trailers for heavy-duty hauling of fill, but manufacturers were reluctant. ALCOA had to build its own demonstration trailers and have truckers use them before truck manufacturers would buy the material.

- **Channels of Distribution:** Channel of distribution members are also a source of ideas. Discussions with the butchers in grocery stores led Mrs. Budd's Foods, a New England manufacturer of frozen meat pies, to develop a meat pie sold in the fresh meat section of the store. The concept was new and entry was much easier than trying to garner freezer space, which was already over-demanded. Channel members are becoming more powerful in many industries and they are a critical link in selling and servicing products. For example, Walmart partnered with local hospitals and healthcare providers to establish independent clinics within Walmart stores, a new service for their customers. In 2008, more than 70 Walmart stores housed in-store clinics.

- **Open Innovation:** Corporate innovation strategies used to have an aversion to "not-invented-here" innovations, where there was an unwillingness to adopt innovations invented outside company-owned R&D labs and discoveries were kept highly secretive as a competitive advantage. However, in recent years, major advances in technology, including the Internet, and global networks have facilitated the ease of dissemination of information. Innovating organizations have begun to embrace "open innovation." Open innovation refers to the concept that companies look to outsourcing innovations by buying them, licensing the technology, or collaborating on inventions with other companies. It also suggests that internal inventions should be taken outside the company through licensing, spin-offs, joint ventures, or other types of alliances. IBM and Procter & Gamble are two companies that have been known to take advantage of open innovation.[9]

The **global innovation grid**, shown in Figure 3.3, demonstrates this "promise of a simple, transparent, instant, and worldwide access to globally distributed R&D resources and services by accessing computer, databases, and experimental facilities

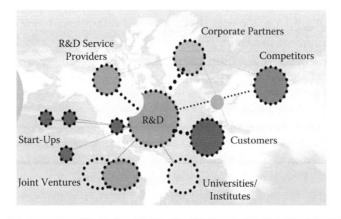

Figure 3.3 Global innovation grid. (From Endnote 10. With permission.)

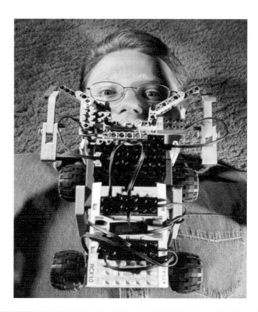

Figure 3.4   Lead users at Lego®. (From Endnote 11. With permission.)

without consideration where the data or facilities are located."[12] This type of network of firms allows innovation openness between partners.

- **Lead Users:** Sometimes the best ideas come from outside the firm and, in particular, from end users themselves. Dr. Eric Von Hippel,[13] an MIT professor, suggests that some of the best sources of insight into user needs and potential product prototypes are "lead users," customers whose strong product preferences in the present foreshadow the needs of the general marketplace in the future. These users often have such a compelling need to solve their problems that they develop their own solutions using existing products. In some cases, these users represent a very specialized niche market, but in many cases they anticipate the needs of the larger market. Von Hipple provides an example of NASCAR racers who are followed by automobile manufacturers because racing teams face new challenges and often invent new solutions that can later be applied to a more general market. Lego, the interlocking toy brick company (Figure 3.4), has been known for involving lead users in new product innovation. As reported in *Wired* magazine:

  In January [2006], Lego put out word that it's looking for 100 more citizen developers. Until recently, companies were skittish about customer innovation, fearing that outsiders might leak trade secrets or that they simply lacked the necessary technical skills [to innovate]. But Lego has warmed to the power of the open source ethos. It's clear to the Lego execs that Mindstorms NXT would be a lesser product without the MUPers' [Mindstorm Users Panel] input. Inviting customers to innovate isn't just about building better products. Opening the process engenders goodwill and creates a buzz among the zealots, a critical asset for products like Mindstorms that rely on word-of-mouth evangelism.[14]

Identifying Lead Users

Lead users are a unique group and should not be confused with a company's best customers, although it is possible they are one and the same. Lead users display two characteristics: (1) they face general needs months or years before the general marketplace encounters them, and (2) they seek these solutions because they benefit greatly from the outcome. Users who have real-world experience with an unmet need/want are in the best position to provide market researchers with new insights based on lead users' experiences. Von Hippel, an MIT professor, describes how to identify lead users, and then how to incorporate their insights into the product design process in a five-step process:[15]

1. Identify a new market trend or product opportunity (e.g., greater computer portability, zero emission vehicles, etc.).
2. Define measures of potential benefit as they relate to customer needs.
3. Select individuals who are "ahead of their time" and who will benefit the most from a good solution (e.g., power users).
4. Extract information from the "lead users" about their needs and potential solutions and then generate product concepts that embed these solutions.
5. Test the concepts with the broader market to forecast the implications of lead user needs as they apply to the market in general.

It is important to recognize that the needs of lead users foreshadow the needs of the general market in tomorrow's predicted market and may not be an indicator of today's market environment. Indeed, the literature on diffusion of innovations suggests that, in general, the early adopters of a novel product or practice differ in significant ways from the bulk of the users who follow them.[16] Therefore, the NP (new product) manager should assess how lead user ideas are perceived by the more typical users in the target market. This is done by employing traditional survey test procedures after segmenting lead and nonlead user responses.

Ideation Methods

Organizations tapping into in-house employees for new product ideas may use a variety of ideation methods. Such creativity is crucial to the successful generation of winning new product ideas. If a company wants to encourage ideation within an organization, there are many processes it can take. The Minnesota-based 3M Company is famously known for encouraging innovation by allowing employees to take 15% of their work time to experiment on personal pet projects. Much of 3M's innovation-rich culture comes from the principles introduced by William L. McKnight, former president and chairman of the board from 1949 to 1966. McKnight challenged managers to encourage employee initiative and innovation by accepting risks and possible failures in the innovation process. The ubiquitous Post-it® note[17] that has been used in unusual ways,

Figure 3.5   A mosaic by R. Elbert using Post-it® notes. (From Endnote 18. With permission.)

evolved from this philosophy (Figure 3.5). It's a great story of creativity, perseverance, and the success of market testing.

Methods for idea generation can be as simple as setting up communication channels that are sensitive to idea sources (e.g., an in-company idea box) or as sophisticated as using creative group methods and market research. A wide variety of ideation methods have been proposed including: **scenario generation**[19,20] **problem analysis** (also sometimes referred to a major revenue generators),[21,22] **crowdsourcing**,[23] **brainstorming**,[24] **inventive templates**,[25] **individual creativity**, and **group creativity**. We focus on these methods because of their popularity in organizations. Other methods include morphological analysis,[26] archival analysis (also called *theory of inventive problem solving* TRIZ),[27] open/democratizing innovation, group sessions (also called *synetics*),[28] and varied perspectives (also called *lateral thinking*).[29] Entire books have been written about the creativity process and a list of references is provided for you at the end of this chapter.

Scenario Generation

Scenario generation involves brainstorming by developing scenarios of the future—best and worst cases, as well as most likely. This process involves taking into consideration the idea sources discussed in the previous section to determine how these factors impact the world of tomorrow. Some key questions to ask include:

- What is the best future scenario?
- What is the worst possible scenario?
- What macro and micro environmental forces would impact these scenarios?

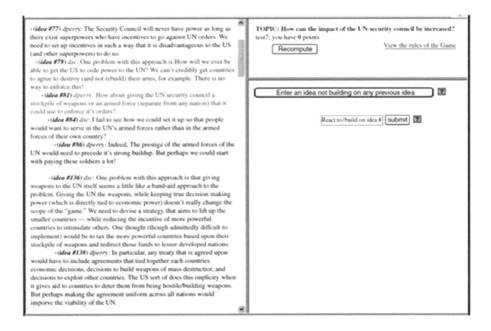

Figure 3.6   Example of an ideation game. (From Endnote 30. With permission.)

A form of structured scenario generation that can easily be conducted on the Internet is the "ideation game" created Dr. Olivier Toubia,[31] a Columbia University professor. It involves end users in a problem-solving task using a "chat room" format to encourage discussion between brainstorming participants. "Ideas are organized into 'trees' such that the 'son' of an idea appears below it in a different color and with a different indentation (one level farther to the right). The relation between a 'father-idea' and a 'son-idea' is that the son-idea builds on its father-idea."[32] The important difference between ideation games and brainstorming is that participants are provided incentives for which they are individually rewarded for their impact on the group's ideas. The advantages of this method are that (1) each participant strives for the maximum contribution, (2) brainstorming occurs in a structured environment facilitating emergence of the best ideas, and (3) participants don't need to be co-located so that a global perspective is easy to achieve. Figure 3.6 provides an example of a "chat" conducted during an ideation game.

Problem Analysis

Many companies specialize in "problem analysis" workshops where the main task is to brainstorm for new product ideas. Design firms, such as IDEO and Continuum, specialize in conducting problem analysis events. A *US News* article describes how IDEO worked with Procter & Gamble on designing the Swiffer CarpetFlick product (Figure 3.7).

The collaboration [with IDEO] combined existing P&G consumer research with IDEO'S research process, such as visiting people in their homes to see how they cleaned. Then

Figure 3.7   Swiffer® carpet flick.

came a couple of days of brainstorming, IDEO style. "So you had all these P&G managers down on the floor on their hands and knees working with tape dispensers and scrapers and things trying to pick crumbs off the floor," Godfroid recalls. "We weren't just pondering about this stuff on a whiteboard." The eureka moment came after an IDEO team member was messing with a squeegee and realized that crumbs and other small particles could be collected by pressing them down and popping them back up like tiddlywinks. That approach was quickly incorporated into rough prototypes.[33]

Some other products IDEO has worked on include:

- Technology for Prada's New York flagship store
- The Handspring Treo
- "Dilbert's Ultimate Cubicle" for Scott Adams, creator of the Dilbert comic strip
- Interiors for Amtrak's Acela high-speed train
- Leap Chair for Steelcase
- Insulin pens for Eli Lilly
- The Palm V
- Polaroid's I-Zone instant camera
- Oral-B's soft-grip kids' toothbrushes
- Crest's Neat Squeeze stand-up toothpaste tube

For more information about the company, read IDEO's general manager Tom Kelley's book, *The Art of Innovation*.[34] For more on these types of discovery methods, there is a video on Continuum's Web site that describes some of the products they have helped design.[35]

Crowdsourcing for New Product Ideas

Another form of idea generation is crowdsourcing using an online community. A few innovative companies have begun to use the Internet as a means of generating new

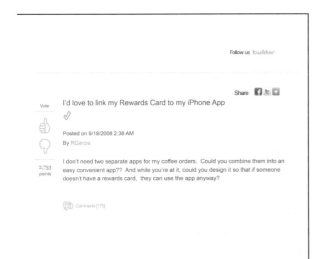

Figure 3.8   Crowdsourcing by a coffee shop.

product ideas using **crowdsourcing**, where an online community is utilized to find solutions to various types of problems. According to Jeffery Howe in *Wired*, "Crowdsourcing represents the act of a company or institution taking a function once performed by employees and outsourcing it to an undefined (and generally large) network of people in the form of an open call. This can take the form of peer-production (when the job is performed collaboratively), but also is often undertaken by sole individuals. The crucial prerequisite is the use of the open call format and the large network of potential laborers."[36] One company that acts as a broker of the process is InnoCentive,[37] who crowdsources research and development for biomedical and pharmaceutical companies. Anyone, anywhere, with interest and Internet access can become an ideation contributor. InnoCentive posts "challenges" from a "seeker" company to "solvers," who can submit their solutions to the challenge. Contributors whose solutions are selected by the sponsoring organization are compensated for their ideas. Innocentive's Open Innovation Marketplace[38] lists hundreds of product solutions that are being sought by firms.

Starbucks® is a company that has caught on to the concept of crowdsourcing. They use a blog format for customers to leave ideas. As of last viewing, more than 115,000 ideas had been posted by customers. Starbucks encourages the chat board readers to vote on the idea. More and more innovative companies are using crowdsourcing for delivering their customers the products and services they want. Figure 3.8 demonstrates a hypothetical example of an idea for a coffee shop left by a customer and then the possible implementation of that idea.

Brainstorming

One of the first group methods of creating ideas was **brainstorming**,[39] a technique developed in the 1950s and 1960s based on the belief that randomized thoughts led to greater creativity. In this approach, a group tries to generate a large number of diverse ideas. No criticism is allowed and group members are encouraged to improve on other people's ideas. It is hoped that through this method, a wide variance of ideas will result, and some will be really new. This basic meeting format has been modified for new product situations by adding structure in **attribute listing** methods. Attributes are listed for the existing products and efforts are made to adapt, modify, magnify, substitute, rearrange, reverse, or combine them into new product ideas. For example, among the attributes of the Oreo cookie is the filling. In one session, this was magnified to produce the highly successful double-thick Oreo. Reversal produces a white cookie and black filling or an Oreo cookie with a chocolate covering.

However, more recent studies have shown that groups generating ideas using traditional brainstorming techniques tend to be less successful than individuals working alone.[40] Recent studies suggest that structure and systematic approaches is the key to creativity.[41] Two important studies on this philosophy include inventive templates[42] and the theory of inventive problem solving (TRIZ),[43] which we discuss below.

Inventive Templates

It has been proposed that ideation is more effective when the NPD team focuses on well-defined schemes derived from a historical analysis of new products. Called **inventive templates**, these include *attribute dependency, component control, replacement, displacement*, and *division*. The template is a systematic change between an existing solution and a new solution and provides a method by which the NPD team can create these transitions in a series of smaller steps called **operators**, which include *exclusion, inclusion, unlinking, linking, splitting, and joining*. For example, the *attribute dependency* template operates on existing solutions by first applying the *inclusion* and then the *linking* operators.[44] An example of how a new car concept was developed by creating a dependency between color and the location of a car's parts is provided by the creators of this method.[45] Volkswagen's Polo Harlequin (Figure 3.9), which featured different colors for each panel of the car, was initially intended as an April Fools' joke, but became quite popular in Europe. Originally the idea was an advertising campaign to

Figure 3.9 Polo Harlequin evolving from attribute dependency-inclusion brainstorming. (From Endnote 46. With permission.)

Operator	Definition	Illustration
Exclusion	The exclusion operator removes an unlinked component from the configuration boundaries	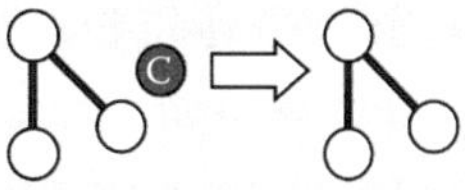
Inclusion	The inclusion operator introduces an external component into the configuration boundaries	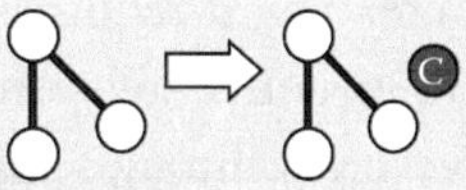
Unlinking	An unlinking operator eliminates a link	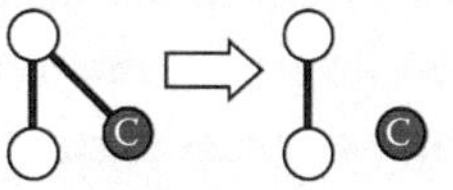
Linking	A linking operator connects two unlinked components or attributes	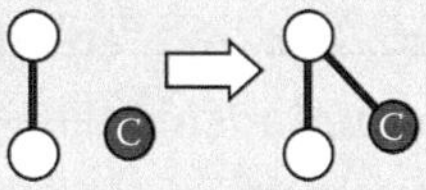
Splitting	A splitting operator removes an internal component from the link. The link maintains the original functions	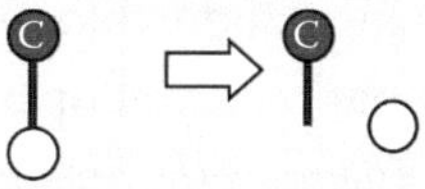
Joining	A joining operator adds a (new) component to a dangling link	

Figure 3.10 Operators underlying the five inventive templates. (From Endnote 47. With permission.)

highlight the car's components, such as engine and chassis, interior, optional equipment, and paintwork.[48] Customers were soon asking for custom colors. Since Polo Harlequin's colors really stood out in a sea of single-colored cars, Volkswagen took the opportunity to deliver this unusually colored car to customers.

Figure 3.10 demonstrates the underlying operators in the incentive templates. For example, "inclusion" could include Wi-Fi capabilities to a car's description because it is typically external to the actual automobile. OnStar, an in-vehicle communication system, was formed in 1995 as a collaboration between General Motors (GM), Electronic Data Systems (EDS), and Hughes Electronics Corporation. GM delivered the vehicle design, EDS brought information management, while Hughes contributed satellite technology and automotive electronics to the new product concept.[49] What examples can you come up with for the other types of templates shown in Figure 3.10?

Individual Creativity

Individual creativity also can play an important role in new product ideation. So, what is creativity? It has been defined by advertising guru, William Weilbacher, as "an arbitrary harmony, an expected astonishment, a habitual revelation, a familiar surprise, a generous selfishness, an unexpected certainty, a formidable stubbornness, a vital triviality, a disciplined freedom, an intoxicating steadiness, a repeated initiation, a difficult delight, a predictable gamble, an ephemeral solidity, a unifying difference, a demanding satisfier, a miraculous expectation, an accustomed amazement."[50] Creativity techniques are based on the idea that an individual has the capabilities of being creative. It requires drawing on an individual's personal knowledge and applying it to inventive solutions no matter how crazy they may seem. For example, the next creative breakthrough might be triggered by an individual's interest in astronomy, rock climbing, knitting, or even business management. Sometimes the most absurd ideas become innovative solutions. As we saw with Dyson, the centrifugal force for removing sawdust became a home vacuum cleaner.

Roger von Oech, a creativity researcher, feels that knowledge is the key input to creativity, but that "creative thinking requires an attitude that allows one to search for ideas and manipulate your knowledge and experience. … You use crazy, foolish, and impractical ideas as stepping-stones to practical new ideas. You break the rules occasionally, and explore for ideas in unusual outside places. In short, by adopting a creative outlook, you open yourself up both to new possibilities and to change."[51] His book, *Whack on the Side of the Head*,[52] describes how to free yourself from 10 mental locks that prevent individual creativity. These mental locks include:

- Looking for the one right answer
- Excessive logical thinking
- Rigidly following the rules
- Being too practical
- Not being willing to have fun and play
- Overspecializing
- Avoiding ambiguity
- Fear of looking foolish
- Unwillingness to make an error
- Believing you are not creative

Ways to create mental flexibility is through taking on different roles. von Oech outlines four creativity roles as:

- *Explorer*: Seeks new ideas from unconventional places, perhaps taking a direction no one else has thought to explore.
- *Artist*: Follows intuition to rearrange the conventional perspective, sometimes even breaking the rules to create something different.
- *Judge*: Critically weighs the evidence delivered by the Explorer and the Artist to further evolve the concept.
- *Warrior*: Strategizes how the idea can be made to succeed. If the idea is truly creative, then excuses will need to be overcome, idea killers quelled, and roadblocks dismantled.

Edward de Bono[53] proposes yet another creative approach based on **lateral thinking** that generates many new alternatives and new patterns rather than vertical thinking that restricts alternatives and concentrates on correct analysis and direct problem solution. This process encourages new ways of looking at and structuring problems and solutions so that creative alternatives may be forthcoming. Although many authors have written about creativity, there is no recipe for creative thinking. However, breaking loose and having the confidence to let our ideas out in an open response to a problem can free an amazing amount of creativity if you do not evaluate prematurely the ideas for correctness and feasibility.

Group Creativity

Brainstorming has been a popular method for new product ideas based on group interactions. Recent studies have suggested that brainstorming does not appear to provide a measurable advantage in creative output, and that it might be better for teambuilding as an enjoyable exercise. However, brainstorming is a common activity in firms because it is relatively easy to organize.

Firms are keenly aware that new ideas are crucial to their success, so many utilize formal creativity group dynamics to generate new product ideas. These types of groups encourage fruitful climates that eliminate inhibitions and unproductive structure. Group methods also provide the ability to integrate marketing, R&D, engineering, and production in these idea-generating environments. For example, marketing may have seen the need for easier means of monitoring one's health, and R&D has nanotechnology it is waiting to use on a new product, but it's not until the group starts complaining how difficult it is to make a doctor's appointment that the topic will come up on creating a running shoe that also reads blood pressure.

Whether group methods or individual efforts are used, firms that commit resources to ideation can generate a sizable number of innovative, diverse ideas. An innovative organization should set up reward structures to encourage these ideas. A willingness to accept "mistakes" in new products and a tolerance for the ambiguity of the invention process are positive cultural incentives for creativity in an organization. Another effective method is to designate idea generation as a responsibility for certain individuals and give clear organizational recognition to these people. A new product team or task force may be charged with coming up with new ideas to tap market opportunity.

Step 2: Aligning Opportunities with NPD Strategy

Clorox introduced its Ready Mop as a direct competitor to Swiffer's Wet Jet, but has not been able to gain much market share from Procter & Gamble. Organizations with competitive strengths, such as excellent financial resources or good channels of distribution or a line of complementary goods or an advantageous geographic location, still need to align opportunity with NPD strategy. Ideas come easily and developing new markets can be easily identified. However, not every idea and

Table 3.1 Dexter Corporation Core Technologies and Priority Markets

CORE TECHNOLOGIES	PRIORITY MARKETS				
	Plastic Compounding	Surface Finishing Adhesive	Composite Structure	Long Fiber/Wet Laid Paper	Biotechnology
Aerospace	⊗		⊗		
Automotive	⊗	⊗	⊗		
Electronics		⊗			
Food Packaging		⊗		⊗	
Industrial Assembly and Finishing			⊗	⊗	
Medical				⊗	⊗

⊗ Current product entries

Source: Endnote 54.

opportunity should be pursued. So, how should a new product manager determine which market is most appealing? This is done through a **market profile analysis**. In this analysis, the firm's capabilities are matched to market opportunities. Dell's capabilities are in superior supplier relationships and mass customization of products, thus, economies of scale, market size, and limited risk have resulted in the success that Dell has experienced in the past. Procter & Gamble's strengths are in developing superior products, securing strong distribution channels, and having an extensive market reach. What do you see as Clorox's strengths? Why has it been ineffective against Swiffer in the mop category?

Additionally, technological capabilities should not be overlooked when a firm has strong marketing capabilities. In technologically based companies, matching market opportunities to the organization's technological capabilities is critical for helping to focus the NPD direction. Table 3.1 shows the Dexter Corporation's (a $1 billion/year specialty material company) technology and market interactions. Even if a market had an ideal environment of good growth, few competitors, low investment requirements, high profit potential, and low risk, it would not be identified as an opportunity unless the firm had the core technologies to address the market opportunity. This matching of technology and market focus is very important in screening new product development projects and developing competitive technological advantages in focused markets.

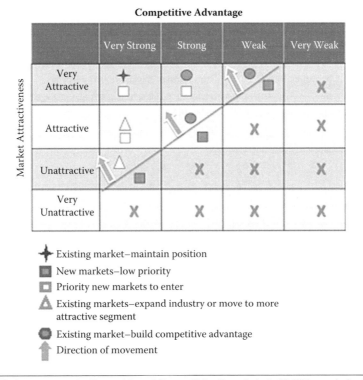

Figure 3.11   Market entry opportunities and the existing portfolio. (From Endnote 55. With permission.)

Portfolio Alignment

Another important strategic consideration is aligning new product ideas with the firm's existing product portfolio. A new product should not be added to a product line unless it adds value to the entire portfolio. For example, a new product for Campbell's Soups was the addition of Soup for One. The initial introduction included several flavors (Golden Chicken & Noodles, Old World Vegetable, etc.). The product line was later extended with additional flavors once the success of the new product was realized. Portfolio evaluation involves evaluating the current product line on a grid of market attractiveness and competitive advantage as shown in Figure 3.11. We would like to be in markets that are attractive and where we have a sustainable competitive advantage. If we are not, we can expend resources to gain competitive advantage and move to the upper left of the grid. Or, we can enter new markets that are attractive and where we have a competitive advantage.

A company may come up with many ideas through ideation exercises, but not all of these ideas align with the company's NP strategy. In **portfolio alignment**, once ideas have been vetted for portfolio fit, the firm can then examine which markets are most opportunistic.

Step 3: Market Identification

In 1984, when Michael Dell entered the personal computer marketplace, he had the foresight to recognize that this market had high sales potential, was in a nascent stage with few competitors, had good economies of scale, and required low risk small investments

Table 3.2   Desirable Characteristics of Markets

GENERAL CHARACTERISTICS	MEASURE
Growth Potential	Size of market/life cycle
Economies of Scale	Cumulative sales volume/learning opportunities
Competitive Attractiveness	Share of market potential/rivalry intensity
Investment	Investment in dollar, technology, and managerial profits
Reward	Profits/ROI
Risk	Stability/profitability of losses

Source: Endnote 56. With permission.

with large rewards—an ideal marketplace for a visionary entrepreneur. Of course, not all markets are in this highly fertile state. Firms should, however, look for markets with these types of opportunities as they begin to explore where to introduce new products. General criteria for what characteristics to look for in market opportunities are shown in Table 3.2,[57] along with some of the specific measures to look for in identifying these characteristics. These market attributes are discussed in detail below.

Growth Potential

Market potential is measured by the size of the market in sales and the growth rate of the market. For example, sales for carbonated drinks are in the billions of dollars in the United States, but growth for this domestic market is small. While size is important, growth potential is a key to identifying new opportunities. In the early 1980s, Michael Dell saw an opportunity to provide customized computers to those who didn't want to assemble their own computers, the only means at that time for obtaining a computer customized to your needs. Dell was aware of the growth opportunities, especially for large institutions. Within a year of launching his company, Dell was making over $70 million in revenues.

Growth is not limited to new untapped markets. It is possible to find a successful product opportunity in a mature market; however, sales will have to be taken directly from competitors. For example, Amazon saw the potential in providing convenience in book buying. It has forever changed the way that books are bought and it has brought on the near annihilation of bookstores within the United States.

In order to assess the growth potential of a market, a forecast and an understanding of the position of the market in its life cycle are required. Various government agencies, trade associations, and syndicated data services provide information on total market size. Standard Industrial Classification (SIC) data are commonly used in identifying industrial products' market size. In consumer goods, store audit suppliers (e.g., Information Resources Inc. and A. C. Nielsen) can be utilized to gain a picture of the current marketplace. In durable industries, numerous specialized data services exist to supply data to predict market growth rates.

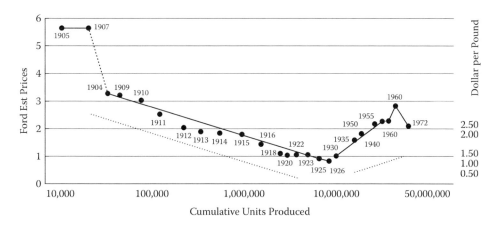

Figure 3.12 Experience curve for Ford automobiles. (From Endnote 58. With permission.)

Economies of Scale

Another place to look for opportunities is in a firm's **experience curve**. For many manufacturing industries, this curve indicates that the unit cost of producing and distributing a product declines at a constant rate for each doubling of the cumulative sales by the firm. As sales increase, the firm becomes more proficient at marketing and manufacturing a new product. As more experience is gained about producing a product, it is frequently possible to decrease production costs.[59]

Figure 3.12 below shows another experience curve, this one for Ford automobiles.[60] After the introduction of the Model T, costs dropped 15% for every doubling of sales from 1908 to 1925. This drop was associated with standardization and innovation in the production process, vertical integration, labor specialization, and better bargaining power over input material costs. This is a good example of how improvements can lead to **process innovation** as opposed to product innovation. In process innovation, firms focus on improving their manufacturing/delivery processes as opposed to coming up with new product designs. This type of curve is common in manufacturing based industries. In Figure 3.12, we see that prices begin to increase after 1925. This could be when Ford began to again focus more on product innovation than on process improvements. In this time period from 1925 to 1965, Ford introduced the V8 engine, power brakes, power steering, air conditioning, and numerous new styles, including the Thunderbird, the "Woody" station wagon, and the Mustang. Scales of economy (representing process innovation potential) and market share (representing product innovation potential) are both attractive properties of a market. Strategically, when looking for growth opportunities, firms should consider the opportunities available in process innovation as much as the opportunities available from product innovation.

Competitive Attractiveness

Although a market is growing, has few entrants, and is subject to economies of scale, it may not be a good opportunity if the competitive environment is hostile and a strong share position cannot be obtained. Some vulnerability to product improvement should be evident. For example, in 2011, contemporary feature mobile phones were vulnerable to smartphones technologies, such as those operated by Apple's iOS, Google's Android, and RIM's BlackBerry OS. In contrast, Apple has shown little vulnerability in the smartphone market due to its strong dominance with the iPhone. Yet, Nokia and Intel co-developed MeeGo, an operating system designed for hardware platforms, primarily targeted at mobile devices, such as netbooks, nettops, tablet computers, and information appliances, such as vehicle infotainment devices, and SmartTV, in order to upset Apple's popularity in the smartphone market. Rapid technological changes can quickly alter the market environment.

The service industry is also driven by competitiveness. Many consider U.S.-based Southwest Airlines' or Ireland-based Ryanair's no-frills flights as just the type of service innovation that can win in a highly competitive market. Southwest's unconventional strategy of lowering expectations about air travel while rewarding consumers with reliable service, truthful advertising, and inexpensive tickets has led to its success in a saturated marketplace.

"Sleepy" markets in which sales are stable and innovation has been absent can represent an attractive opportunity if they are penetrated effectively. Really new innovations in the mature home cleaning systems category, or, more plainly, in mops, are few. However, Procter & Gamble, along with collaboration from the design firm Continuum, has generated more than $1 billion in global sales from their Swiffer line of mops and sweeper since its introduction. Although, many thought the marketplace for mops was in a mature state, P&G found an opportunity through a really new innovation. Likewise, in the mature U.S. auto market, domestic producers have been vulnerable to higher quality and fuel efficient cars produced in Japan and Europe since the 1980s. Yet, Tesla, the electric vehicle company headquartered in California, is shaking up the auto market with its $49,900, seven-passenger Model S sedan launched in 2013. Changing technological environments and breakthroughs can significantly alter the environment in "sleepy" markets.

Investment

The more investment required by a market for entry and penetration, the less attractive it is. The term investment should be interpreted broadly to include direct financial investment as well as managerial talent, capital equipment, and laboratory resources. The investment that Procter & Gamble made in the Swiffer line was considerable, as it included fees to Continuum for the design, new manufacturing facilities, and extensive advertising. For a given level of sales volume, a larger financial investment makes a market potentially less profitable. In its 1999 annual report, Procter & Gamble noted its investments into R&D:

> With an investment of $1.7 billion this year, P&G is the 21st largest U.S.-based and 52nd largest global investor in research and development. We invest to drive clear product superiority in our core businesses and to acquire new technologies and fund entrepreneurial programs that create big, discontinuous product innovation. Ten years ago, our investment in R&D was 2.9% of net sales. Today it represents 4.5% of net sales.[61]

When billions are invested in R&D, the final stakes are often high and a new product failure can be devastating to an organization. Hewlett-Packard's 2011 discontinuance of its TouchPad tablet significantly impacted its share price. Large investment entry costs may act as barriers to discourage competitive entry and, as a result, could lead to long-term profitability.

Reward

The upside is that rewards from large investments can be big. Since 2002, Procter & Gamble has more than doubled sales in its home and fabric care category (which includes Swiffer), resulting in more than $23 million in revenue in 2008. Large investments are not unattractive if they lead to high returns on investment. For example, iRobot is a robotic company that sells robotic vacuums (www.irobot.com). They started making robots for military purposes at a very high expense. These innovations have had a significant return as they expanded from the military markets to consumer markets. Some firms reject product ideas with low investments if the return is not satisfactory. For example, one major package goods firm will not enter a market unless gross volume is greater than $10 million and payback is less than three years.[62] In determining the viability of a project, return on investment must be considered rather than sales volume or investment alone.

Risk

A final consideration of market selection is risk. Markets characterized by uncertainty are less attractive. If demand is unknown or subject to rapid and large fluctuation, the resulting risk of new product failure is high. For example, if we want to consider the market for alternative fuel vehicles (e.g., electric, ethanol, plug-in hybrid), we would be much less certain about its size and composition in the near term than if we considered the four-door sedan market for gasoline-powered cars. It is still uncertain which automotive technology will eventually replace petroleum-fueled vehicles. As the oil market has shown, uncertainty can be evident as well in the supply and subsequent prices of key raw materials. Firms in different industries will have higher or lower risk tolerance. Biotechnology firms will have much more risk tolerance than a consumer packaged goods company.

Competitive risk also is important to consider. The more certain firms are of the demand, cost, and competitive aspects of a market, the more attractive it becomes. This may have been a reason that Clorox introduced a me-too product like the Ready Mop. Based on the size of the global market for home cleaning supplies, they knew at a competitive price they could take a small market share away from Procter & Gamble and still achieve a reasonable return on their investment.

Step 4: Market Selection

After the market potential has been evaluated and the new product assessed as part of the entire product portfolio, a firm must begin to narrow down which market segments it will target.

Most companies do not have the resources necessary to adequately mass market. Even Dell in its early years mainly focused on "large customers," with less of a concentration on consumers. If the market is defined too broadly, say personal computers, then the new product development team has a difficult time in evaluating the market, selecting target customers, understanding competition, and uncovering what is necessary to succeed in the market. On the other hand, if the market is defined too narrowly, say, PC with 4 GB of memory, 16-inch screens with Vista operating systems to students, then the new product development team is certain to find the market too small and will miss opportunities. There is a balance between too narrow and too wide. It is up to the NP manager to find the target market with the greatest opportunity for the firm.

There are numerous ways to segment markets, with some of the most popular being demographics, attitudes, preference for benefits, usage, product form, and competitive products.[63] It is up to the new product team to determine which segmentation technique is appropriate, but the final decision should balance breadth with focus. The different segmentation methods can be interrelated. For example, Table 3.3 defines the PC market by attitude toward computers. But Table 3.3 also demonstrates that these attitude segments are related to demographic segments (income, years of college), competitive product segments (purchasing of laptops), benefit segments (tradeoffs among ease of use, ease of learning, speed, and power), and usage segments (PC use at home and at college).

Substitution

Observing or measuring directly which products substitute for one another is another method for market segmentation. **Substitution**, by questioning or by means of a simulated buying opportunity, means one can ask customers (1) what brands they would consider, (2) what their first choice product would be, and (3) what would their next choice be given their first choice was not available? In this manner, sets of competing products can be identified, for example, coffee segments can be based on form (ground

Table 3.3   Possible Attitude Segmentation of the Home PC Market

I hate computers:
- High school education
- $45,000 in income
- Shops at a discount store
- Dislikes change
- Politically conservative
- Comfortable with existing technology
- Wants children to go to college
- Ease of use of computer is important

Computers have to do the job with no hassle:
- 3 years of college
- $65,000 in income
- Shops online as well as at discount stores
- Adopts technology if proven
- Politically liberal
- Want to succeed
- Wants children to go to college
- Easy to learn is important

I love my laptop:
- College graduate
- $100,000 in income
- Shops online and at specialty stores
- Owns a laptop along with a PC
- Likes change
- Independent voter
- Strives for success
- Wants children to excel in college and graduate
- Power and portability is important

versus instant coffee), usage occasion (brands used in the morning versus brands used at dinner), or any other grouping that identifies distinctive usage.

The NP manager then tests to see whether the identified segments of products make sense. If, for example, we were testing a segmentation of caffeinated versus decaffeinated coffee, we would assign customers to the segment that contains their most preferred brand, and then calculate the probability that they would choose another product in that segment if they were forced to switch. If the probability of staying in the segment was higher than the market share of the segment, the hypothesized segmentation scheme would be supported. If it was low or if an alternate segmentation scheme was significantly better (e.g., ground versus instant coffee), the original segmentation scheme would be rejected.[64]

Substitutes also may be based on "use." For example, if we were examining the recreation market for new product opportunities, many consumers would indicate that TV competes with computer time or even live performances as a recreational diversion. A car may be competitive with a vacation or children's tuition if the dollars are allocated out of the same budget.[65]

Selecting the Best Opportunities

Segmentation analysis should result in an understanding of which customers are in the most attractive target segment, a description of those customers, their basic needs, their buying habits, and a knowledge of the products that now compete in that segment. So, now, how should a firm select the best opportunities for them given the viable segments? Segmentation analysis together with the market definition provides the structure in which the market profile analysis is conducted. This market profile analysis is used to screen the potential market opportunities from a large number (say 20 to 30 across various markets and/or segments) down to a relatively few (say 3 to 4). Analysis of the remaining candidate markets is done more analytically to identify the market boundaries and target consumers. This information is then used to set priorities explicitly and to select the one or two best markets. These "best" markets then pass to the "idea" generation phase.

To illustrate this, look at some of the firms in the coffee market and analyze their coverage and duplication in terms of the hierarchical market definition. Table 3.4 shows the coffee offerings of General Foods, Nestlé, Starbucks, and Hills Brothers.

General Foods has good coverage, but there appears to be a potential duplication between Brim and Sanka. General Foods may be wise to consider further research to see if Brim and Sanka compete. If this is true, dropping Brim and investing those resources in Sanka or other new products would be appropriate. The opportunities for General Foods to introduce a new brand may not be attractive because a new ground or instant regular caffeinated coffee would cannibalize their existing brands. The new product effort at General Foods might be better directed toward creating a new market branch with a major product innovation, rather than adding brands to the existing product segments. Nestlé had an opportunity to add a ground coffee and, indeed, in 1985, Nestlé acquired Hills Brothers. In 1991, Nestlé had over 13% of the U.S. ground coffee market. Starbucks is only in the ground market. It doesn't seem

Table 3.4   Coffee Products of Select Manufacturers

MANUFACTURER	GROUND	CAFFEINATED INSTANT		DECAFFEINATED INSTANT	
		REGULAR	FREEZE DRIED	REGULAR	FREEZE DRIED
General Foods	Maxwell House Brim Sanka Yuban	Maxwell House Yuban	Maxwell House	Maxwell House Sanka	Maxwell House Brim Sanka
Nestlé		Nescafé	Taster's Choice	Nescafé	Taster's Choice
Starbucks	Starbucks				
Hills Bros.	Hills Bros.	Hills Bros.			

to make sense for them to try and dominant the instant market unless they have an innovative feature that assures a high probability of becoming the dominant brand in that market.

Chapter Summary

In this chapter, we looked at opportunity identification: idea generation, opportunity analysis, and market identification. During idea generation, firms have a variety of methods they can use to find new product ideas. But, not all of these ideas should be taken to market. Opportunity analysis examines the options and determines which are best to take into development based on the firm's available resources and competencies. Profitable markets are then identified and final product ideas chosen based on the best market to target with the innovative product. Not all firms take these steps sequentially, and the opportunity identification process is typically quite organic and fluid. However, those firms that keep these guidelines in mind through this fuzzy front end of the development process are one more step closer to a successful new product. Firms must continually innovate. Dell has had its ups and downs in the past few years. Firms that continually innovate have been found to be more profitable than firms who struggle with innovation.[66]

Glossary

Artist: A creativity perspective in the book *Whack on the Side of the Head*, by Roger von Oech,[51] that follows intuition to rearrange the conventional perspective, sometimes even breaking the rules to create something different.

Attribute listing: A method for generating new product ideas where attributes are listed for existing products and efforts are made to adapt, modify, magnify, substitute, rearrange, reverse, or combine them into new products.

Brainstorming: A technique developed in the 1950s and 1960s based on the belief that randomized thoughts led to greater creativity.

Concept development and refinement: Refine and screen new product ideas, taking the best to the design phase.

Crowdsourcing: The practice of obtaining needed services, ideas, or content by soliciting contributions from a large group of people, and especially from an online community, rather than from traditional employees or suppliers.[67]

Experience curve: This curve indicates that the unit cost of producing and distributing a product declines at a constant rate for each doubling of the cumulative sales by the firm.

Explorer: A creativity perspective in the book *Whack on the Side of the Head*, by Roger von Oech,[51] that seeks new ideas from unconventional places, perhaps taking a direction no one else has thought to explore.

Global innovation grid: A map that demonstrates "worldwide access to globally distributed R&D resources and services by accessing computer, databases, and experimental facilities without consideration where the data or facilities are located,"[68] introduced by InnoCentive, the crowdsourcing company.

Idea generation: Generate product ideas to tap the potential of selected markets.

Individual creativity: The process where an individual, as opposed to a group, identifies new product ideas.

Inventive templates: A systematic change between an existing solution and a new solution and provides a method by which the NPD team can make these changes in a series of smaller steps called "operators": exclusion, inclusion, unlinking, linking, splitting, and joining.

Judge: A creativity perspective in the book *Whack on the Side of the Head*, by Roger von Oech,[51] that critically weighs the evidence delivered by the Explorer and the Artist to further evolve the concept.

Lateral thinking: A creative approach, introduced by Edward de Bono,[69] that generates many new alternatives and new patterns rather than "vertical thinking," which restricts alternatives and concentrates on correct analysis and direct problem solution.

Lead users: Customers whose strong product preferences in the present foreshadow the needs of the general marketplace in the future.

Market identification: Identify markets that offer the best opportunities for the organization.

Market profile analysis: Procedure for matching a firm's capabilities to market opportunities.

Market selection: Select the most opportunistic target markets for new product introductions and/or product line extensions.

Open innovation: The idea that innovative firms look to outside inventors to buy or license inventions.

Operators: A series of smaller steps, which include exclusion, inclusion, unlinking, linking, splitting, and joining, in the "inventive template" technique for sourcing new product ideas.

Opportunity analysis: Align opportunities with new product development strategy.

Portfolio alignment: Balancing the firm's new product portfolio to align with the firm's new product strategy to maximize return on investment in new product development.

Process innovation: Innovation in the way a product or service is manufactured, created, or distributed.

Scenario generation: Brainstorming by developing scenarios of the future—best and worst cases, as well as most likely.

"Sleepy" markets: Markets in which sales are stable and innovation has been absent can represent an attractive opportunity if they are penetrated effectively.

Substitution: A market segmentation technique identifies segments by observing or measuring directly which products substitute for one another.

Warrior: A creativity perspective in the book *Whack on the Side of the Head*, by Roger von Oech,[51] that strategizes how the idea can be made to succeed. If the idea is truly creative, then excuses will need to be overcome, idea killers quelled, and roadblocks dismantled.

***Whack on the Side of the Head*:** A popular book on creativity by Roger von Oech.[51]

Review Questions

1. Why should managers start new product development projects by studying consumers rather than engineering designs?

2. Suppose a very strong "experience curve" effect is observed in a particular market. Under what conditions would this effect encourage market entry? Hinder market entry?

3. Why is market definition important? How could proper market definition affect the actions of the multinational company?

4. Perk-a-Cup Coffee Company is thinking of launching a new single-serving-size coffee. Mr. Kava, the company's chief marketing executive, believes a great potential exists for this new product based on the size of the beverage market. How could proper use of market segmentation help focus on this brand's ultimate market?

5. How might a hospital supply firm encourage users to innovate? How might the firm use lead users in generating new product ideas?

6. Banks are interested in delivering more online services to their customers. What sources internal and external to a banking institution might be used to generate new service ideas?

7. What strategies does a firm use to determine which market to enter?

Assignment Questions

1. Investigate two recent and possibly significant technological advances. For each generate three new product ideas for three different consumer needs.

2. Select three of the methods of generating new product ideas that were discussed in the chapter. For each method:

    Discuss the major players who would be involved. Marketing? Engineering? Customers?

    State which industries would find the method most valuable and why.

    Describe the generation of a hypothetical product when employing the method.

3. Describe how a company you are familiar with might use open innovation or crowdsourcing to generate a new product idea.

4. Use the desirable characteristics of markets discussed in this chapter to provide an analysis of market segments best suited for a new electric vehicle.

5. Identify segments for the following new products/services:

   A netbook, a laptop computer designed for wireless communication and access to the Internet to use cloud-computing capabilities

   McCafe, McDonald's coffee shops

   Fairtrade certified coffee

   Brain-health snack bars with DHA Omega–3

   Stevia-flavored fruit juices as branded by Coca-Cola or, instead, branded as Starbucks

6. In 2009, Palm introduced the Pre, a smartphone that competes with the iPhone. Describe how it may have worked through the four steps of market opportunity identification discussed in this chapter. (The fifth step is discussed in Chapter 4).

7. Based on Table 3.4, which outlines the major brands of ground coffee in the mid-2000s, what type of new product/service offering could the Massimo Zanetti Beverage Company possibly introduce into this competitive, and declining, market? Support your answer using the market opportunity guidelines discussed in this chapter.

Endnotes

1. This chapter is based on Glen L. Urban and John R. Hauser, *Design and Marketing of New Products,* 2nd ed. (Upper Saddle River, NJ: Prentice Hall, 1999).

2. Online at: http://ecommerce.hostip.info/pages/303/Dell-Computer-Corp-EARLY-HISTORY.html#ixzz24rB4n58C; also: http://www.dell.com/; (accessed August 28, 2012).

3. Joan Magretta, "The Power of Virtual Integration: An Interview with Dell Computer's Michael Dell," *Harvard Business Review* (March-April 1998): 73–84.

4. P. G. Smith and D. G. Reinertsen, *Developing Products in Half the Time* (New York: Van Nostrand Reinhold, 1991).

5. P. Koen, et al., "Providing Clarity and a Common Language to the 'Fuzzy Front End,'" *Research Technology Management* 44(2) (2001): 46–55.

6. E. Von Hippel, "Economics of Product Development by Users: The Impact of 'Sticky' Local Information," *Management Science* 44(5) (1998): 629–644.

7. Online at: http://www.ideo.com/ (accessed June 28, 2013).

8. Online at: http://continuuminnovation.com/ (accessed June 28, 2013).

9. Larry Huston and Nabil Sakkab, "Connect and Develop: Inside Procter & Gamble's New Model for Innovation," *Harvard Business Review* 84(3) (March 2006).

10. Online at: http://www.innocentive.com/servlets/project/ProjectInfo.po (accessed June 28, 2013).

11. Brendan Koerner, "Geeks in Toyland," *Wired* 14(2) (2006, February). Online at: http://www.wired.com/wired/images.html?issue=14.02&topic=lego&img=6 (accessed June 28, 2013).

12. Andreas Neef, "Zpunkt, Summit of the Future 2005," Future of Corporate Innovation. Online at: http://www.clubofamsterdam.com/contentsummit/summit%20website/summit%20 for%20the%20future%20presentations/science%20&%20technology/Summit%20for%20 the%20Future%20Andreas%20Neef.pdf (accessed June 28, 2013).

13. Eric von Hippel, "Lead Users: A Source of Novel Product Concepts," *Management Science* 32(7) (1986): 791–805; G. Urban and Eric von Hippel, "Lead User Analyses for the Development of New Industrial Products," *Management Science* 34(5) (1988): 569–582.

14. Brendan Koerner, "Geeks in Toyland," *Wired* 14(2) (2006, February).

15. E. M. Rogers and F. F. Shoemaker, *Communication of Innovations: A Cross-Cultural Approach,* 2nd ed. (New York: Free Press, 1971).

16. E. M. Rogers, *Diffusion of Innovation,* 5th ed. (New York: Free Press, 1962).

17. Online at: http://www.post-it.com/wps/portal/3M/en_US/Post_It/Global/?WT.mc_ id=www.post-it.com/ (accessed May 15, 2013).

18. Online at: http://www.flickr.com/photos/russellelbert/2897705069/sizes/l/; This photo is licensed under the Creative Commons Attribution-Share Alike 3.0 Generic license.

19. R. G. Cooper, S. J. Edgett, and E. J. Kleinschmidt, "Optimizing the Stage-Gate Process, *Research Technology Management* 45(5) (2002): 18–33.

20. Olivia Toubia, "Idea Generation, Creativity, and Incentives," *Marketing Science* 25(5) (2006): 411–425.

21. R. G. Cooper, *Winning at New Products: Accelerating the Process from Idea to Launch,* 3rd ed. (New York: Basic Books, 2001).

22. H. W. Chesbrough, *Open Innovation: The New Imperative for Creating and Profiting from Technology* (Cambridge, MA: Harvard Business School Press, 2003).

23. J. Howe, "The Rise of Crowdsourcing Forget Outsourcing," *Wired* 14(6) (2006): 176–183.

24. A. F. Osborn, *Applied Imagination,* rev. ed. (New York: Scribner, 1957); J. E. Arnold, "Useful Creative Techniques," in *Source Book for Creative Thinking,* eds. S. J. Parnes and H. F. Harding (New York: Scribner, 1962).

25. Jacob Goldenberg, David Mazursky, and Sorin Solomon, "Creative Sparks," *Science* 285(5433) (1999): 1495–1496; Jacob Goldenberg, David Mazursky, and Sorin Solomon, "Toward Identifying the Inventive Templates of New Products: A Channeled Ideation Approach," *Journal of Marketing Research* 36(2) (May 1999): 200–210.

26. R. U. Ayres, *Morphological Analysis, Technological Forecasting and Long Range Planning* (New York: McGraw-Hill, 1996), 72–93.

27. G. S. Altshuller, *Creativity as an Exact Science* (New York: Gordon & Breach, 1985); G. S. Altshuller, *And Suddenly the Inventor Appeared: TRIZ, the Theory of Inventive Problem Solving* (Worcester, MA: Technical Innovation Center, Inc., 1996).

28. G. M. Prince, *The Practice of Creativity* (New York: Harper & Row Publishers, 1970).

29. Edward de Bono, "Exploring Patterns of Thought: Serious Creativity," *The Journal for Quality and Participation* 18(5) (1995): 15–18.

30. Toubia, Olivier, "Idea Generation, Creativity, and Incentives," *Marketing Science* 25(5) (2006): 411–425.

31. Ibid., pp. 411–425.

32. Ibid., p. 416.

33. J. M. Pethokoukis, "The Deans of Design from the Computer Mouse to the Newest Swiffer, IDEO Is the Firm behind the Scenes," *US News,* 2006, September 24. Online at: http:// www.usnews.com/usnews/biztech/articles/060924/2best_4.htm (accessed June 2, 2013).

34. T. Kelley and J. Littman, *The Art of Innovation: Lessons in Creativity from IDEO, America's Leading Design Firm* (New York: Crown Business, 2001).

35. Continuum Homepage, *Continuum LLC.* Online at: http://continuuminnovation.com/ (accessed June 2, 2012).

36. Jeffery Howe, "The Rise of Crowdsourcing," *Wired* 4.06 (June 2006).

37. InnoCentive Homepage, *InnoCentive, Inc.* Online at: http://www.innocentive.com/ (accessed June 2, 2012).

38. Online at: http://www.innocentive.com/servlets/project/ProjectInfo.po (accessed June 28, 2013).

39. Arnold, *A Source Book for Creative Thinking*, pp. 251–268.

40. M. Diehl and W. Stroebe, "Productivity Loss in Brainstorming Groups: Toward the Solution of a Riddle," *Journal of Personality and Social Psychology* 53(3) (1987): 497–509.

41. Goldenberg, et al., "Creative Sparks."

42. Goldenberg, et al., "Toward Identifying the Inventive Templates of New Products"; Jacob Goldenberg and David Mazursky, *Creativity in Product Innovation* (Cambridge, U.K.: Cambridge University Press, 2002).

43. G. Altshuller, *Innovation Algorithm: TRIZ, Systematic Innovation and Technical Creativity* (Worcester, MA: Technical Innovation Center, Inc., 1999).

44. Online at: http://ebiz.mit.edu/research/papers/103%20EDahan,%20JHauser%20 Dispersed%20Product.pdf (accessed June 28, 2013), p. 18.

45. Goldenberg and Mazursky, *Creativity in Product Innovation*.

46. Ross VW homepage, *Ross VW.* Online at: http://www.Rossvw.com (accessed June 28, 2013).

47. Goldenberg, et al., "Toward Identifying the Inventive Templates of New Products"; Goldenberg and Mazursky, *Creativity in Product Innovation*.

48. Online at: http://www.polo-harlekin.de/want35_11.html (April 22, 2013).

49. Online at: http://en.wikipedia.org/wiki/On_star (accessed April 22, 2013).

50. William Weilbacher, *Advertising,* 2nd ed. (New York: MacMillan, 1984).

51. R. Von Oech, *A Whack on the Side of the Head: How You Can Be More Creative* (p. 6). New York: Grand Central Publishing, 1990).

52. Ibid., 25th anniversary revised ed., 2008.

53. Edward de Bono, *Po: Beyond Yes and No,* Rev. ed. (Harmondsworth, U.K.: Penguin, 1973).

54. Based on Urban and Hauser, *Design and Marketing of New Products.*

55. Ibid.

56. Urban and Hauser, *Designing and Marketing of New Products*, p. 92.

57. Urban and Hauser, *Design and Marketing of New Products.*

58. Based on William J. Abernathy and Kenneth Wayne, "Limits of the Learning Curve," *Harvard Business Review* 52(5) (1974): 109–119.

59. For a discussion of learning effects, see: T. M. Devinney, "Entry and Learning," *Management Science* 33(6) (1987): 706–724; P. S. Adler and K. B. Clark, "Behind the Learning Curve: A Sketch of the Learning Process," *Management Science* 37(3) (1991): 267–281.

60. Based on Endnote 58.

61. U.S. Securities and Exchange Commission, "Annual Report," in *U.S. Securities and Exchange Commission* (1999, June 30). Online at: http://www.sec.gov/Archives/edgar/ data/80424/0000080424-99-000027.txt (accessed June 28, 2013).

62. Urban and Hauser, *Design and Marketing of New Products*

63. Philip Kotler and Kevin Keller, *Marketing Management,* 12th ed. (Upper Saddle River, NJ: Prentice Hall, 2006).

64. For statistical tests and descriptions of alternative measures of switching including first choice, rank order preference, consideration set membership, and logit switching probabilities, see Glen Urban, Philip L. Johnson, and John R. Hauser, "Testing Competitive Market Structures," *Marketing Science* 3(2) (1984): 83–112.

65. J. R. Hauser and Glen L. Urban, "Value Priority Hypotheses for Consumer Budget Plans," *Journal of Consumer Research* 12(4) (1986, March): 446–462.

66. Online at: http://www.booz.com/global/home/what-we-think/global-innovation-1000 (accessed June 28, 2013).
67. Online at: http:// www.merriam-webster.com/dictionary/crowdsourcing (accessed June 28, 2013).
68. Neef, "Zpunkt, Summit of the Future 2005."
69. de Bono, *Po: Beyond Yes and No.*

Creativity Resources

Overcoming Mental Blocks

de Bono, E. 1995. "Exploring patterns of thought: Serious creativity." *The Journal for Quality and Participation* 18 (5): 15–18; Stefanovich, Andy. 2011. *Look at more: A proven approach to innovation, growth, and change.* Sand Francisco: Jossey-Bass. (A personal favorite.)

Incentives and Idea Generation

Toubia, O. 2006. "Idea generation, creativity, and incentives." *Marketing Science* 25 (5): 411–425.

Open Innovation

Chesbrough, H. W. 2003. *Open innovation: The new imperative for creating and profiting from technology.* Cambridge, MA: Harvard Business School Press.

Crowdsourcing

Howe, J. 2006. "The rise of crowdsourcing forget outsourcing." *Wired* 14 (6): 176–183.

Applied Imagination

Osborn, A. F. 1957. *Applied imagination*, rev. ed. New York: Scribner; Arnold, J. E. 1962. Useful creative techniques. In *Source book for creative thinking*, eds. S. J. Parnes and H. F. Harding. New York: Scribner.

Creativity and Science

Altshuller, G. S. 1985. *Creativity as an exact science.* New York: Gordon & Breach; Altshuller, G. S. 1996. *And suddenly the inventor appeared: TRIZ, the theory of inventive problem solving.* Worcester, MA: Technical Innovation Center, Inc.

Inventive Templates

Goldenberg, J., D. Mazursky, and S. Solomon. 1999. "Creative sparks." *Science* 285 (5433): 1495–1496; Goldenberg, J., D. Mazursky, and D. Solomon. 1999. "Toward identifying the inventive templates of new products: A channeled ideation approach." *Journal of Marketing Research* 36 (2): 200–210.

Morphological Analysis

   Ayres, R. U. 1969. *Morphological analysis, technological forecasting and long range planning.* New York: McGraw-Hill, pp. 72–93.

Practice of Creativity

   Prince, G. M. 1970. *The practice of creativity.* New York: Harper & Row Publishers.

Out of the Fuzzy Front End into the Design Phase

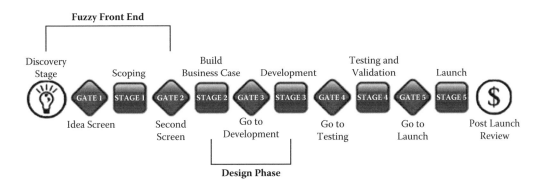

Learning Objectives

In this chapter, we will address the following questions:

1. Which stages are included in the fuzzy front end of the development process and which are included in the design phase[1]?
2. What happens in a Go/No Go gate review?
3. What happens during the Scoping Stage?
4. How can a manager best capture the voice of the customer?
5. What are the components of the Business Case?

Designing the Segway Human Transporter

In 2001, Dean Kamen announced the arrival of his self-balancing, zero emissions, personal transportation vehicle, the Segway® Personal Transporter (PT). Segway's research and development was focused on creating devices that took up a minimal amount of space, were extremely maneuverable, and could operate on pedestrian sidewalks and pathways.[2] Its dynamic stabilization technology with gyroscopes and tilt sensors monitored a user's center of gravity about 100 times a second. When a person leaned slightly forward, the Segway moved forward. When leaning back, the Segway moved back.

Kamen remained secretive with the public on the design of the personal transporter to deter any possible competition. However, an early press release was "leaked" about the invention, touting it as being bigger than the Internet or the PC as a radical new product without explaining what a personal transporter actually was. The innovation, code-named *Ginger*, was to revolutionize city planning, create an upheaval in several existing industries, and be an environmentally friendly transportation solution. If widely adopted, the Segway would lead to urban redesign and renewal, Kamen promised. He predicted in *Time* magazine that the Segway "will be to the car what the car was to the horse and buggy."[3] Yet, Kamen knew that his innovation wasn't perfect. In a secret unveiling to Steve Jobs, CEO of Apple Computers, Jobs responded to Segway employees' enthusiasm about the upcoming launch, "Its shape is not innovative, it's not elegant, it doesn't feel anthropomorphic," he said, ticking off three of his design mantras. Jobs added, "… You don't have a great product yet. You'll only get one shot at this, and if you blow it, it's over."[4]

In 2003, *Wired* magazine reported, "It would be premature to call the most talked about scooter in the history of humankind a huge bust. But, the Segway has always been ahead of its time (Figure 4.1). For a decade, Dean Kamen fiddled and tested and

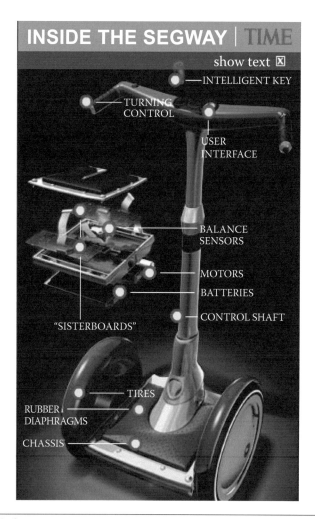

Figure 4.1 Inside the Segway.

tinkered with his invention, finally stage-managing its public unveiling in December 2001. He figured 2002 would be the year that the Segway launched a transportation revolution."[5] Segway's breakout year wasn't even a few months old before bad news started to hit. Supposedly, obvious corporate buyers like Federal Express said "no thanks," and others government agencies remained noncommittal. Kamen's largest customer that year was Walt Disney, which ordered four dozen machines for its theme parks and cruise ships. Maybe Steve Jobs was right about the design.

Introduction

If an organization is successful in identifying sources of new ideas and using the methods outlined in Chapter 3 to generate ideas, many exciting opportunities should be evident. The firm, however, should not just march forward without taking a step back and evaluating the potential success of the ideas put forth during **Discovery**. This step is called a "gate review." During the gate analysis, a go/kill decision will be made whether to take the project to the next stage. Our next stage is **Scoping**, where we capture the **voice of the customer** and then assess the marketplace potential of the product concept. Voice of the customer results are used in finalizing a product definition that will guide the R&D development stage. Concepts that pass through the second screening (Gate 2), move into the **Design Phase**. The first part of the Design Phase is Stage 2, **Build the Business Case**, where formal market and financial assessment are conducted. The outcome of Stage 2 is (1) a well-defined product concept that will be taken into Stage 3, **Development**, and (2) a preliminary business plan to set a roadmap to follow in the next stages of development.

Go/No Go Decision Making

In the last chapter, we discussed how ideas germinated during the Discovery stage are evaluated for opportunity and market attractiveness. Before these ideas move into Scoping, they must first go through a "gate,"[6] a formalized go/no go decision point that ensures that the right project is being done in the right way. Gates occur at every major junction in the new product development (NPD) process typically before a new stage is started. At each of these points in time, the firm should take a hard look at the project: (1) how well it continues to fit in the strategic plans of the firm and (2) what is its probable success. Because new product development can take anywhere from three months to five years or longer to complete, strategic plans can change drastically from the time the project was first initiated. It is better to kill a possible money loser, or at the very least send it back to a previous stage for refinement, than to continue to commit resources to a losing plan.

The go/no go process evaluates three major concerns: quality of execution, business rationale, and quality of future plans.[7] The **quality of execution** concerns how well the project team is conducting the NPD process. However, a well-executed NPD

Table 4.1   Inputs for the Idea Screening Go/No Go Gate Analysis

1. List of potential ideas to take into product design
2. Market analysis of possible segments
3. Competitive analysis for segments
4. Alignment with strategic direction of organization
5. Probable investment needed to capture profitable market share
6. Need to acquire outside expertise/technologies
7. Risk/reward assessment

project that does not meet the **business rationale** should not be continued. Even a well-executed project that does meet the strategic goals of the firm, but also does not have a solid **future plan**, will still be a failure. These plans should consider the market analysis, competitive analysis, and resource commitments.

So, what should happen at a **gate review**? Based on the directive that the gate process must be realistic and easy to use, the go/kill/recycle decision point is not a complicated process. This review is the time in which all the data gathered in the previous stage are evaluated to determine the next steps. Sometimes a decision may be made to send the project back to a previous stage or even to "kill" it. There are three components that must be evaluated at each gate: the "deliverables" going into the gate, the "criteria" by which the deliverables should be judged, and the "outcomes" of the gate. The deliverables are the results of the previous stage (phase) of the NPD process. The goal of the gate review is for the team to analyze these components and make recommendations for next steps. In Gate 1, **Idea Screening**, the inputs would be the list of ideas that have been generated through the Discovery process and the Opportunity Identification assessments accumulated during the market evaluation exercises described in Chapter 3. Table 4.1 shows what some of the deliverables might be going into this idea screening gate.

Although the list in Table 4.1 is specific to the Idea Screening gate, this type of information is required at each gate. The critical information will vary gate-by-gate, but should include both financial and qualitative criteria. Cooper[8] indicates that there should be a list of must-meet criteria (show stoppers) and should-meet criteria, which alone are not show stoppers in themselves, but could lead to a kill decision when many other factors are considered. Table 4.2 shows a list of these types of criteria. Each new project should be evaluated against these company-directed must-meet and should-meet criteria. NPD managers must be able to defend the project at each gate before it is able to progress to the next phase.

Effective gate evaluations should keep in mind the following directives:[9]

1. *The NPD process is a sequential and conditional process:* Neither "go" nor "no go" decisions are irreversible. Incrementalizing the process into discrete steps that are scrutinized before moving forward helps to reduce risk from both a go

Table 4.2 Possible Criteria Requirements in the Go/No Go Evaluation

MUST-MEET CRITERIA
  1. Strategic alignment
  2. Favorable size of market
  3. Within technical capabilities of firm or ability to acquire technology
  4. Significant product advantages compared to competition
  5. Environmental issues met: governmental, economic, technical, sustainable, etc.
  6. Risk versus return hurdles of organization met
  7. Profit potential hurdles met

SHOULD-MEET CRITERIA
  1. Strategic importance of project to firm
  2. Unique product offering compared to competitors
  3. Adequate market size versus growth
  4. Leverage core competencies of firm—marketing, technical, and/or manufacturing
  5. Technically feasible with low risk
  6. Certainty of return and/or profit and/or sales
  7. Favorable competitive situation
  8. Index of attractiveness requirement

Source: From Endnote 10.

and no go perspective. Go gate decisions can be looked upon as "options" on an uncertain future revenue stream that may be exercised in the future or just allowed to lapse.

2. *Look to maintain a balance between errors of acceptance and errors of rejection:* If poor projects make it through the process, already scarce resources are committed to losing projects and possible continual escalation of a less than profitable endeavor can more easily occur (this is called a type 1 error, which is discussed below). Being too restrictive also can result in rejecting good projects (this is a type 2 error, which can be viewed as the error of excessive skepticism, also discussed below). Establishing both qualitative and quantitative guidelines for gate analysis will help avoid these types of errors.

3. *The process is characterized by uncertainty of information and the absence of solid financial data:* Early in the gates, data on forecasts, costs, and capital requirements are limited and sometimes little more than educated guesses. It is not until later when the product is near launch that data becomes more accurate. Managers need to make reasonable decisions without reliable data.

4. *Project evaluation involves multiple objectives and, therefore, multiple decision criteria:* A firm may have many objectives from their new product development process besides profit and growth. Other goals could be industry growth, new partnerships, complementing existing products, or strategic redirections. The NPD manager should strive to align these goals with project progression.

5. *The Go/No Go evaluation process must be easy to use and grounded in the day-to-day operations of the corporation:* A go/no go process that is not easy to use will not be used. Yet, it needs to be realistic so that the results are valid. At the

gate, tough decisions need to be made about advancing or killing a project. Making this process a task that can be accomplished in one or two well-represented meetings should be the goal of a gate evaluation.

A "kill" prognosis does not mean that the product idea is never to be considered again, and, in fact, it may be sent back to a previous phase for greater refinement. This is the iterative nature of the NPD process we discussed in Chapter 1. Likewise, a "go" decision does not guarantee that a new product will make it through to launch, but only that it will go on to the next phase. Sometimes a "hold" may be the outcome of a gate evaluation. This is a common occurrence for many projects during economic downturns. Many companies limit the number and size of the projects that they are undertaking in their new product process, but once the economy recovers, many projects resume.

Idea Screening Gate

We have identified what inputs are required for the **Idea Screening** gate, but what comes out of the gate? Which of these ideas should advance to the next stage? The two key managerial concepts in screening ideas are (1) the selection process and (2) how many ideas to advance to the design phase.

Idea Selection Process

In idea selection, we are trying to avoid two errors: type (1) approving a potentially unprofitable product and type (2) rejecting a potentially profitable product. Avoiding these errors isn't always easy because detailed information about the product design is not normally available in the fuzzy front end of the NPD process and accurate estimation of financial outcomes is not always feasible. The "fuzziness" of the NPD process can easily result in committing one of the errors trying to be avoided. Dean Kamen and the venture capitalists supporting the Segway surely completed a thorough analysis when considering whether to move forward with their personal transporter. How do you think they committed a type-1 error, which is approving a potentially unprofitable product.

To try to minimize these types of errors, a new product manager should have a systematic process of deciding which projects should continue to the design phase. One method of determining which ideas should go into the Scoping stage is through calculating an **index of attractiveness**.[11] The simple procedure divides the expected return by the development cost:

$$I = \frac{T\,C\,P}{D} \tag{4.1}$$

where:
- I =   Index of attractiveness
- T =   Probability of successful technical development

C = Probability of commercial success given that it is technically successful

P = Profit if successful

D = Cost of development

If an idea is technically feasible, the probability of technical success (T) will be high and the cost of development low (D). For the same profitability (P) and chance of commercial success (C), this project would be preferred to an idea that requires a technological breakthrough (low T and high D). This simple formula is one way of trading off the different risks, return, and levels of knowledge of various ideas. After the index (I) is calculated for each project, they are ranked and the projects with the highest index of attractiveness (I) are considered for funding. Final decisions include qualitative judgments that augment the index. What index of attractiveness would you calculate for the Ginger, which was the Segway prototype?

Caution is warranted in using this formula. The model often oversimplifies the problem and biases toward small projects where uncertainty is low. There also is the possibility from "deliberate underestimations"[12] used to marshal support for a project. One explanation for overestimating probability of success might be **entrepreneurial optimism biasness**[13] where failure seems unlikely to those championing the idea. How do you think a company should guard against this type of bias? How do you think Dean Kamen, inventor of the Segway, might have been able to refrain from letting his enthusiasm jade his business decisions? As an inventor, Kamen holds more than 440 U.S. and foreign patents, many of them for innovative medical devices,[14] including an all-terrain wheelchair. This background surely influenced his prediction for probabilities of success. Has your earlier calculation for Ginger's index of attractiveness changed? Did you refine your original estimate of (C), probability of success, with this new information?

While costs of fuzzy product ideas are difficult to estimate, determining probabilities for Equation (4.1) are even more difficult. For example, project originators and those with implementation responsibility tend to give more optimistic estimates than the average evaluator, while those with a "knowledge gap" about technical feasibility tend to give more pessimistic estimates. However, these cautions do not diminish the need for some process to screen ideas. Such a process should allow a critical look at the many aspects of the alternative ideas as they relate to new product development. Furthermore, a process fosters dialogue among the disparate interest groups involved. It seems reasonable to initiate a conversation on costs and probabilities of success with or without a formal procedure. We discuss other methods for evaluating projects later in this chapter.

Number of Ideas

How many ideas should be identified for design work? It is reasonable to assume that there are "good" ideas and "bad" ideas, with their distribution as shown in Figure 4.2. If we consider generating an idea as making a random draw from the

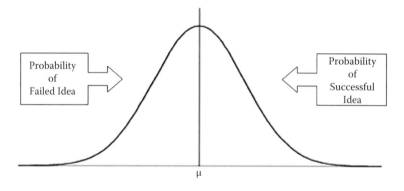

Figure 4.2   Probability of success/failure of ideas.

normal distribution of Figure 4.2 and we generate only one idea each time we develop a product, we will get an average expected reward, μ.

If we could generate two independent ideas and select the "best" one, the expected value would be substantially greater. As more ideas are generated for a development opportunity, the expected reward increases. The overall gain depends on how many ideas are sampled, the variance in the distribution of ideas, the reliability and validity of our methods of finding the "best" idea, and the costs of generating ideas.[15] At this stage of the NPD process, in almost all cases, the rewards of generating several ideas are greater than the costs. It is strongly suggested to generate multiple ideas for each market opportunity or technological approach and not become committed to one idea alone. A common pitfall in new products is selecting the first idea and allocating large amounts of resources to it without considering alternatives that may be better. There, of course, is a threshold on the number of ideas that should be considered. Too many ideas can drain resources from the few "best" ideas. It is up to the new product manager to decide what number of ideas is too little and how many are too much.

Scoping: The First Stage

A few good ideas have made it out of the Idea Screening gate and now await "scoping." In **Scoping**, we focus on the critical, upfront activities that precede the physical development of a new product. Scoping involves preliminary analyses of a preliminary market assessment, a technical assessment, and a business and financial assessment. Based on these tasks, it's easy to see that a cross-functional team consisting of marketing, engineering, and finance representatives is needed to provide input to the feasibility of the product/service/project. Voice of the customers (VOC) studies should occur during the final stages of scoping or during the early stages of Building the Business Case. Because VOC is so important to the entire NPD process, we include it as part of the Scoping stage.

Scoping is meant to be "quick and dirty," meaning that not a lot of time or resources should be taken up in this stage. Scoping is "detective work and desk research: gathering available in-house information (e.g., talking with the sales force, distributors, and technical service people); examining secondary sources (e.g., reports and articles published by trade magazines, associations, government agencies, and research and consulting firms); contacting potential users (e.g., through a phone blitz or focus group); and canvassing outside sources (e.g., industry experts, magazine editors, or consultants)."[16] There is a wealth of information available to the NPD manager with just a few days of solid sleuthing. Here's an example of an early assessment that was conducted by a team of MBA students:

> A healthcare-related company based in the Boston area asked students to conduct a preliminary market assessment. The company provides Web-based practice and clinical management software and services to physicians' offices across the United States. Their core service helps doctors get paid by untangling the mess of healthcare payers. In looking to expand their product offerings, they were interested in patient issues with medical services. They had a rough sense that there are different types of patients, but needed to explore how patient populations differ, how they might be segmented, and what is important to patients when communicating with doctor offices. In particular, they were interested in the types of communication physicians should conduct with their own clients, the patients. The MBA students conducted a series of interviews with patients and also conducted a field study of consumer preferences. The study results showed that there was a viable market for an efficient, 24-hour-a-day Web-based method for patients to interact with doctors' offices for scheduling appointments, retrieving test results, viewing health information, etc. The new service concept was approved by top management and moved to the business case stage of development.

The **preliminary market assessment** consolidates all the findings that were accumulated during the market definition stage, including commercial viability of the product/service, market (industry) attractiveness, competitive environment, and segmentation guidelines. Research should not be limited to the domestic market, but also should include an international component. Table 4.3 shows where supplemental information can be obtained from during Scoping.

In the Scoping stage, we also begin to include technical considerations. The **preliminary technical assessment** includes identifying initial product requirements, identifying feasibility of engineering product design, cost estimation of engineering and resource estimations, and manufacturing feasibility. This assessment also would include assessing the need to out-source certain aspects of the design, such as those involved with open innovation. Patent searches and regulatory considerations should be a component of this part of the assessment as well. Key concerns about the technical viability of the product include:[17]

1. Approximate product/service requirements and specifications
2. Product/service idea technical feasibility—how much research is required, how much existing technology is used?
3. Costs and time resources

Table 4.3   Resources for Secondary Research

1. Internet searches, including YouTube videos
2. Library searches
3. Internal company reports
4. Sale force interviews
5. Key customers interviews
6. Focus groups
7. Trade associations/trade journals/trade shows
8. Service representative (within and/or outside of firm)
9. Channel partners, both international and domestic
10. Government agencies, including Edgar
11. Industry experts
12. Marketing research firms
13. Retail store visits
14. Competitor literature, advertisements

Source: Endnote 18. With permission.

4. In-house technical capability and out-of-house partnering needs
5. Manufacturing facilities and service requirements
6. Intellectual property and regulatory issues
7. Management of key technical risks

In addition to market and technical considerations, financial issues must be assessed. The **preliminary business/financial assessment** includes expected profit projection (sales potential, cost of product), core competency assessment, and strategic alignment assessment. A "back of the envelope" preliminary forecast also should be calculated. The danger of this type of calculation is that managers may be overly optimistic, as in the case of Segway, who expected to be making more than 10,000 units monthly and instead demand was only for less than 100 units monthly.

The preliminary market, technical, and business assessments are meant to be a "sanity check" for the organizations. At this point, it is not prudent to be too dismissive of product ideas if they don't fully pass these preliminary hurdles. Products like the Swiffer and the electric vehicle (EV) may not seem to have long-term potential based on this type of structured assessment. The household cleaning implement market was very mature with few really new innovations. Early assessment would seem to say that there was no potential market to exploit for a new mop-like tool, such as a Swiffer. Likewise, the EV market doesn't appear to be very inviting, even in 2013. Addressing the alternative fuel vehicle market is difficult—future regulations remain unpredictable and extended-life battery technology remains unproved—yet, there are numerous companies that have begun delivering their own EV versions, including Nissan's Leaf, the Chevrolet Volt, and Mini Cooper-EV among others.

Scoping should be followed by another go/no go evaluation to weed out sure failures. At this decision point, only a few new product ideas will be ready to go to the business case development stage of the design process. The number will be dependent

Table 4.4   Scoring Model Worksheet

FACTORS	SCORE 1–5 (1 IS POOR; 5 IS EXCELLENT)
Market Size	
Market Growth	
Competitive Environment	
Regulatory Freedom	
International Potential	
Manufacturing Feasibility	
Financial ROI	
R&D Payback	
Consumer Acceptance	
Engineering Feasibility	
Technological Environment	
Strategic Fit	

upon the type of innovation and the industry. For incremental innovations, there may be five or more; for really new innovations, it will be only one or two.

Scoring Models

Scoring models[19] were introduced in Chapter 2 for strategic portfolio evaluation. They are very useful as well in the Scoping stage to evaluate individual product ideas. Here projects are rated or scored on a number of criteria usually on a ranking of 1 to 5 or 0 to 10 scales. Each project is then given a total score that is used to rank it against other projects. For an example of a scoring card, see Table 4.4. Projects with the highest ranking are considered for the next stage. Sometimes the factors may be weighted. In designing alternative fuel vehicles, regulatory freedom may be more important than technological stability as many states are mandating zero-emission vehicles within the next 10 years. Beyond financial models, scoring models are one of the most frequently used ranking methods by firms. Figure 4.3 shows the criterion that are most often used by firms as reported in a study of 205 industrial companies. Scoring models are easy to use, are useful for facilitating conversation on important factors to consider, and can be easily customized to a company's specific strategic goal.

Voice of the Customer Analysis

After the preliminary assessments, insights into the customer's needs and wants should be obtained. One of the most crucial inputs into the core benefit proposition (CBP) of the product is **voice of the customer** studies.[20] We first gather this type of information during the Scoping stage. Voice of the customer (VOC) strives to record in the customer's own words the benefits of a product or service; it is a description of the customer's needs within a specific product category.[21] The goals of attaining these

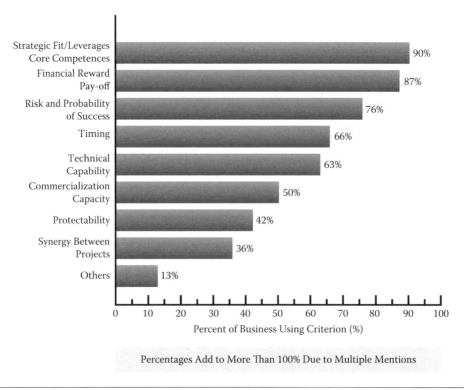

Figure 4.3   Criteria used to rank projects. (From Endnote 22. With permission.)

customer viewpoints are (1) to set a clear definition of the product and its target market, which will be used in the Building the Business Case stage; and (2) to provide engineering with this definition and a list of customer needs to be taken into consideration in the R&D process. If the product is engineering-driven and has already undergone R&D, the VOC is still necessary to inform engineering of the consumer tradeoffs in regards to product attributes. For example, is turning radius more important than stability in a Segway or vice versa.

Quality design begins with first identifying the needs of the customer. These customer needs are statements or phrases, stated in the customer's own words, that enable the design team to focus on delivering the benefits the customer wants. These statements are best obtained first with qualitative studies, such as focus groups, ethnographic research, fly on the wall studies, and lead user interviews. This is followed by quantitative studies, such as concept tests and conjoint studies. Less formal techniques also can be utilized. For example, some Japanese firms simply place their products in public areas and encourage potential customers to examine them, while design team members listen and note what people say. Other firms examine complaint files or talk to user groups or poll representatives and distributors. The most important lesson is to listen to your customers for input to the design process.

Figure 4.4 demonstrates how the voice of the customer is used to identify customers' needs, prioritize those needs, and then incorporate them into the product design.

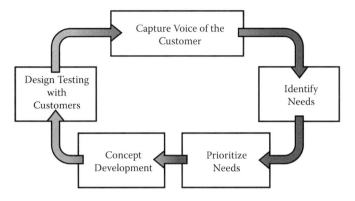

Figure 4.4   Incorporating voice of the customer in the design process. (From Endnote 23.)

This is an ongoing process as products are improved and new innovative products (line extensions) are identified.

VOC studies provide key inputs into the R&D process. However, to be of value, these VOC insights need to be organized and summarized so that they can direct the strategic development of the new product. Figure 4.5 provides a structure with which to organize this information. Qualitative studies inform quantitative studies. Qualitative techniques include experiential interviews, empathic design and user observation, elicitation techniques, benefit chains, and Web-based "eavesdropping." We discuss these below. The goal of the qualitative study is to gain knowledge whether the product idea, even in its fuzzy state, will satisfy the needs and wants of

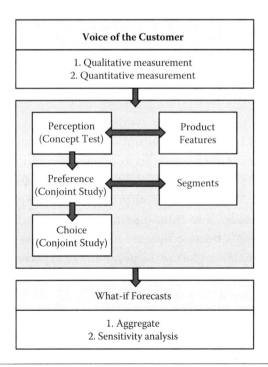

Figure 4.5   VOC studies inform product design and business strategies.

the customers. During these types of studies, importance of specific features, attributes, and benefits should be uncovered. The results from the qualitative study will inform the quantitative study, which takes a clearer, less fuzzy product definition to the consumer for their opinion.

Quantitative studies provide information about product features to the R&D team and provide market information to the business development team. **Perception models**, such as concept tests, might tell Segway designers that the key strategic benefits are a comfortable trip, low cost, foldable for ease of storing, electric refueling, little to no training, etc. **Preference models**, such as a conjoint study, might tell Segway Inc. that low cost is more important than the training period, and a product feature evaluation might tell the designers how to achieve low costs with a new wheel carriage or a different fueling system. Perception and preference studies help the team select which benefits to include in the CBP and how much emphasis to put on each. Product features tell the team what can be done to realize the CBP. **Choice models** let us know how preferences for product features can lead to purchase decisions for our products versus competitive offerings for different segments.

Quantitative studies also can identify the trade-offs that must be made in product design. For example, when customers are interested in buying a personal transporter, they prefer fun to cumbersome, no learning to extensive training, quick refueling to time-consuming refueling, economical to expensive, etc. But, if we are limited in the amount of money we can spend to improve the design, if we are looking only for cost-effective improvements, or we want a strategy consistent with a target market, which of these design features should we change? Or should we even limit ourselves to these improvements? Maybe the greatest impact comes from having a complete redesign? This is where the insights from VOC can help the new product manager guide through the process; exposing the needs and wants of customers will drive the answers.

Experiential Interviews[24]

For incremental designs targeted at an existing or familiar customer base, focus groups[25] provide valuable information. However, focus groups are not without their flaws. The group is exposed to the possibility of social distraction where the customer gets caught up in interacting or not interacting with other members of the group as opposed to the product. This sometimes leads to miscommunication or even lack of communication about the customers' needs because they are unable to express themselves in a group setting. Many companies are exploring the possibility of **experiential interviews** where the customer is interviewed one-on-one and then asked to respond with their personal interaction and perception of the product. In this scenario, the interviewer is able to ask and receive answers about the fundamental issues that the customer is looking to resolve.

When looking for potential interviewees, there is a selection procedure that divides the market so that there is a level of diversity in the types of customers that are interviewed.[26] This is illustrated in Table 4.5 where you can see the different types of

Table 4.5 Customer Selection Matrix for Online Doctor Communication

MARKET SEGMENT	CURRENT CUSTOMERS	COMPETITORS' CUSTOMERS	LEAD USERS	UNTAPPED CUSTOMERS	LOST CUSTOMERS
Cancer or other Long-term Care Patients					
Families with Children					
Routine Care Patients					

customers that are contacted. Also, if there are many different segments of customers that make decisions, in this case doctors and patients, they all will be consulted.

Multiple members of the NPD team should review the transcripts generated during experiential interviews. Professors Abbie Griffin and John Hauser[27] proposed that each team member has the ability to recognize half of the needs that the consumer expresses in a single transcript. When team members work together, they are usually successful at identifying more than 95% of the expressed needs. Companies and interviewees have found that body language and other nonverbal forms of communication are critical components in the expression and understanding of the consumers' needs. Therefore, many companies have taken the liberty of videotaping interviews, so that nonverbal cues that might have been missed during the interview due to verbal communication are perceived and recorded. These video or audio clips are often distributed amongst team members and are commonly referred to as the Face of the Customer (FOC). For example, having the audio of a customer saying, "I use Pages on my MacBook and I need a built-in pointer that does not make me move my fingers from the keypad." This form of feedback is much more useful than a written summary of "Pointing device is important." It is even more useful to see the consumer demonstrate their struggle and frustration with current pointers. Many companies now try to get the design engineers in on the interviewing process so long as it is a cost-effective choice.[28] Another important question often asked is: "How many people should be interviewed?' Griffin and Hauser[29] indicate that 15 to 20 experiential interviews per market segment identify the vast majority of customer needs (Figure 4.6).

Empathic Design and User Observation[30]

Many companies have discovered that no matter how much effort and innovation is poured into the research, design processes, and testing, some understanding and data can only be gathered whilst observing the customer in their natural state of being (work, home, office etc.).[31] This type of study is called an *empathic design*. These types of studies are important when the consumer is not able to verbalize his/her needs, insights, and perceptions or these concepts are just not apparent to the consumer. Empathic design is commonly referred to as "fly on the wall" or ethnographic studies. The approach to empathic design necessitates team members to fully integrate themselves into a potential consumer's environment, so much so that they are able to act as

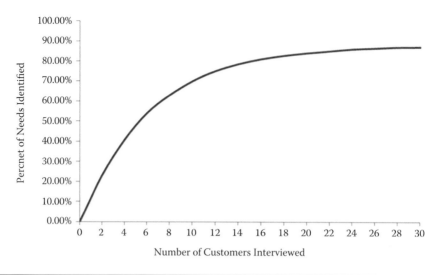

Figure 4.6 Number of experiential interviews required. (From Endnote 32.)

if they were customers. This also allows them to understand the problems that arise and how they can be resolved. If there is any sort of issue in the product in regards to inconvenience, inefficiency, or inadequacy, the designers themselves experience the problem as it happens. Empathetic methods are very good at establishing workplace efficiency and functional design of a product. Empathic research can be done by members of an NPD team as well as by marketing professionals. These studies should be captured in their entirety in video format so that they can be reviewed later by the NPD team.

Intuit, the manufacturers of Quicken, which is currently the most successful personal financial software bundle on the U.S. market, invented the "Follow-Me-Home" program.[33] This program enable Intuit employees to watch their customers from when they received the product to when they have it all set up and fully functioning. Using empathic studies, Intuit has been able to create the most user-friendly systems including features such as push buttons, auto-fill of commonly used data, and check that bear a remarkable resemblance to the paper equivalent. Even more impressive, Intuit took the initiative in making maintenance checks to improve printer and printer drive functions. Empathic design allowed Intuit to see firsthand why people were displeased or not buying their products. Even though it was not technically their responsibility, they went ahead and tried to solve what they perceived as the customers' and products' problems. More specifically, they pinpointed the software and hardware problems that might have eventually driven the product off the market. This kind of attention to customer needs makes Intuit one of the most successful firms in an incredibly competitive market.

Elicitation Techniques[34]

In addition to exploring customers' stated needs, NPD teams often seek to understand customers' underlying meanings and values. When used in VOC studies, cultural anthropology[35] is used to study the latent significance in products. This sort of research extends beyond the emotional requirements and enters the social strata. What are the customers' core social values? Cultural anthropology can be used to influence product design by discovering the social meaning and context of the product and keeping that meaning constant. For example, if customers are interested in electric vehicles because they are concerned about the environment, they object to using dirty fuel (i.e., coal) to produce electricity to charge the batteries.

Zaltman's metaphor elicitation technique (ZMET*)* is a type of cultural anthropology that employs self-expression. This self-expression is visual and is meant to express the latent values and meanings that guide customers in product selection. ZMET often uses visuals to determine consumers' latent needs. It stresses metaphors as being fundamental to learning and communication. The application of ZMET can be described as two major processes: data collection through unstructured interviews and data analysis by coding as conducted by the researchers. ZMET was used in a study of the adoption of 3G mobile banking services.[36] Some of the findings revealed that a relative advantage of using 3G mobile banking services is to increase one's self-prestige; however, issues of privacy are considered enough of a risk that could hinder the ready acceptance of these services in the marketplace. These underlying values may not have revealed been with other VOC techniques.

Benefit Chains

Benefit Chains[37] determine why customers have certain requirements that have not been met by existing products. For example, studies may show that customers are looking for a small, lightweight, notebook computer, but the latent requirements that lead to these needs may not be immediately evident. Are customers such workaholics that they need to accomplish amazing amounts of work, therefore, needing performance? Do they need to take it everywhere they go, and require it be light? Or do they want to spend less time in the office and find it easier to do their work out of their house on a laptop? The latent beliefs that define these needs will differ from customer to customer and being able to identify these needs and variations will enable more accurate product solutions. The workaholic customer demanding performance might also need other product features, such as long battery life. Whereas, someone who is looking to use their laptop for leisure purposes might need programs that are easy to use. Elderly customers might need larger keypads and fonts. A benefit chain analysis can help to determine these needs.

Figure 4.7 shows a benefit chain for voice-over-the-internet telephone services. In the interview conducted, the consumer revealed his need to make international calls without toll charges. The family values of the individual is such that he wants to reach

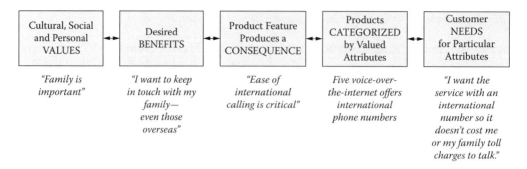

Figure 4.7 Benefit chain structure for voice-over-the-internet telephone service. (From Endnote 38. With permission.)

out to his family often without incurring large telephone bills. This benefit chain, thus, links value (family ties) to needs (low cost international costs) that can be satisfied with product features (internet telephony).

A related method includes a "means-end chain" model of customer choice behavior.[39] The **means-end chain** technique for measuring such benefit chains begins with Kelly's[40] repertory grid technique. The means-end chain describes how a product interacts with the consumer through its "attributes," "benefits," to deliver "values," as shown in Figure 4.8. It approaches the "chain" in the opposite direction compared to the benefit chain, which looks at values first to derive the final features/attributes of a

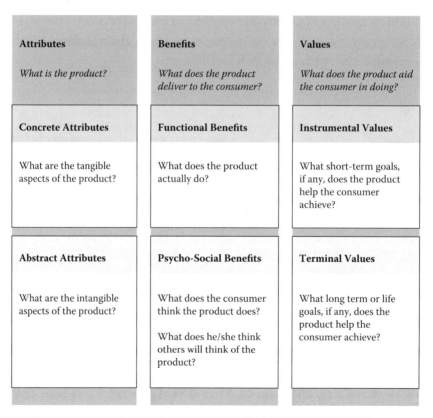

Figure 4.8 Means–end chain. (From Endnote 41.)

new product. Within each of the three groupings, two subgroups exist that draw out how a new product as its basic core (attributes) can provide benefits to the consumer so that the consumer values the product enough to adopt it into his/her life. These six sections allow marketers to analyze not only how a product interacts with the consumer, but how the consumer interacts with the product. Coffee can become more than just a beverage. Dr. Eda Gurel-Atay, who studied this type of chain, presents an example of coffee in her research presented in the *Journal of Advertising Research*.[42] "Consumers who endorse fun and enjoyment in life may want a cup of coffee for its rich, pleasant taste. Meanwhile, people who value a sense of accomplishment may rather use coffee as a mild stimulant. People who value warm, loving relationships with others may want a cup of coffee to share in a social manner. Perspective and personal beliefs greatly influence behavior."[43] The goal for the new product manager is to identify the segments that value the product in unique ways and to deliver attributes and benefits that fulfill these values. If a new product is not valued by a consumer, no matter how amazing the attributes and how beneficial the company feels the product is, sales will not materialize. This apparently was the problem faced by Segway—attributes and benefits did not deliver value to the consumer.

Laddering is another technique for finding the benefits an end user may see in a new product. Consumers are typically presented with three products. After potential consumers have evaluated this set of three products, they are then asked what feature they like the most. They then are continually questioned as to why they have that preference until some core beliefs are discerned.[44] "Laddering refers to an in-depth, one-on-one interviewing technique used to develop an understanding of how consumers translate the attributes of products into meaningful associations with respect to self, following means-end theory" (Gutman, 1982). Laddering involves a tailored interviewing format using primarily a series of directed probes, typified by the "Why is that important to you?" question, with the express goal of determining sets of linkages between the key perceptual elements across the range of attributes (A), consequences (C), and values (V). These association networks, or ladders, referred to as perceptual orientations, represent combinations of elements that serve as the basis for distinguishing between and among products in a given product class."[45] This technique is illustrated in Figure 4.9.

These benefit chains are useful for the manager in qualitatively determining what attributes of a new product/service will provide the most value to a consumer/end user. Without value, consumers are not willing to adopt or pay for the new product.

Web-Based "Eavesdropping"[46]

The Internet is also a great way of culling information about customers' needs and desires by watching how they behave while on a Web site. This voyeuristic method of capturing unmet customer needs by observing customer interactions was tested with an Internet-based sales recommendation system for trucks.[47] A virtual sales representative tries to figure out what the customer is interested in while a virtual engineer watches how they react to

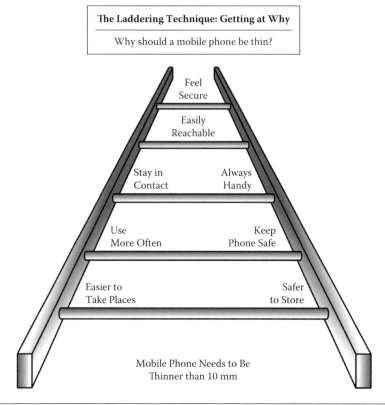

Figure 4.9 Laddering example for a mobile phone. (From Endnote 48. With permission.)

different features, particularly those with which the consumer seems the most unhappy. The virtual engineer then talks to the customer about why they feel that their needs are not met and how to meet them without compromising other positive aspects of the product. As more and more people have become comfortable with avatars and virtual worlds, such as Second Life,[49] this method has begun to gain a larger following.

In less structured formats, companies, such as Nielsen BuzzMetrics,[50] have been organized around helping firms monitor consumer-generated, word-of-mouth in blogs, message boards, and review sites. Their Web site claims: "The premise behind services like these, as well as companies' own internal Internet-monitoring programs, is that online discussions—be it in forums, on blogs, or elsewhere—are a modern replacement for customer satisfaction surveys or focus group reports, which can take months to compile and analyze." Figure 4.10 gives an example of how the Internet can be monitored for new product trends, such as discussions regarding the iPhone. We see that in mid-June 2009 discussions regarding the iPhone peaked, which is not unexpected since Apple launched an upgraded version of the iPhone on June 19. We see that compared to the iPhone (peak of 1.1% of all blog posts monitored), little discussion has been ongoing for 3G mobile banking (peak of 0.00250% of blog posts). The lack of discussion on 3G mobile banking can be taken as an opportunity for a new service or could be viewed as a weakness because of the need to educate consumers on mobile banking benefits.

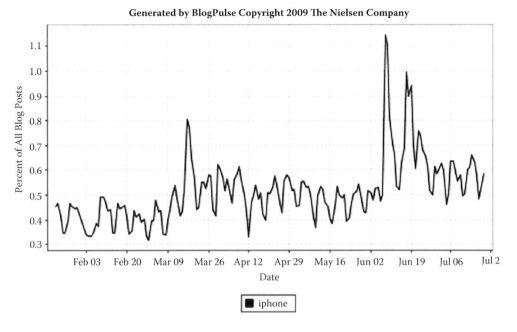

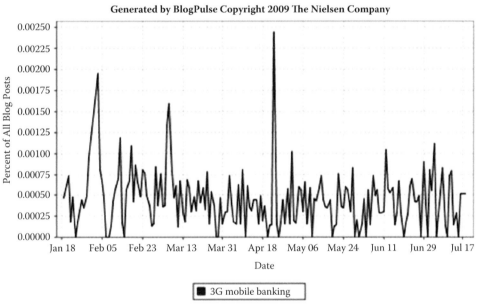

Figure 4.10   "Pulse" on the Internet for iPhones and 3G mobile banking: Discussions for January 18–July 17, 2009. (From Endnote 51. With permission.)

Evaluating the Data

After gathering VOC data, it must be evaluated. At the most detailed level, a team might identify 100 to 300 phrases that describe customer needs. Some will be redundant, but many will relate to distinct, but subtle, differences in the potential product. For example, Table 4.6 lists 18 of the 100+ customer needs for the Segway's balancing

Table 4.6 User Needs for Segway Balancing System[a]

NO.		NEED
1	Balancing System	Easy to use
2	Balancing System	No special skills required
3	Balancing System	No maintenance required
4	Balancing System	No computer interface required from user
5	Balancing System	Less than 30 minutes of training required
6	Balancing System	Uses existing LeanSteer technology
7	Balancing System	Smooth ride over any terrain
8	Balancing System	Steerable over any bumps or rough surfaces
9	Balancing System	Ease of turning
10	Balancing System	No special balancing skills on inclines
11	Balancing System	Warns of misuse
12	Balancing System	Safe in all weather conditions (i.e., rain, snow)

[a] Hypothetical evaluation.

PRIMARY NEEDS	SECONDARY NEEDS	TERTIARY NEEDS
Easy to Use	Easy to turn	0° radius turn
		Minimum leaning
		No training needed
		Feels balanced when turning
		Smooth turns, not jerky
Smooth Ride	Shock Resistant	No difference as terrain changes
		No knee stress from standing
		No special effort over rocks
	Wheels	Durable
		Long life
		Easily replaced
		All-terrain
Durable Mechanics	Easy to maintain	Low/no maintenance
		No greasing
		No computer updating
		No loose parts
		Long life
		Long warranty

system—the gyroscopic design should have "0° turn radius," "minimal lean required," "no special training," "feels balanced," etc. Other needs for Segway users would be a smooth ride that is shock resistant and wheels that are durable. However, the more durable the wheels are, the more rigid they will be making them less shock resistant. Ideally, the engineering team will find a creative solution to satisfy both needs, but if they cannot, they will need to trade off one benefit versus the other. The trade-offs consumers are willing to make (shock absorbent versus durability) must be considered by the engineering team developing the product.

Building the Business Case

After Scoping and passing through a second screening gate (following the same guidelines for gate analysis as previously discussed), we now can enter the **Design Phase** with a few well-selected new product concepts for further evaluation. In the design phase, there are two primary steps being conducted: (1) building the business case and (2) product development. In this chapter, we focus on the *Building the Business Case*. The *Development* stage, which focuses on designing the product and producing prototypes, will be discussed in Chapter 6.

During **opportunity identification** and **scoping**, a considerable amount of market and engineering information was accumulated. The business case begins to combine all of this information into a cohesive strategy for the new product. This is when the product really begins to come to life from a business perspective. The primary difference between a standard business plan and a new product business plan is the single-minded focus of the plan—that of describing the strategy for a single product (or product line) compared to a portfolio of products. The key components of the new product business plan are **situational analysis, product definition, project justification,** and **project plan:**

1. Situation Analysis
   - Market analysis
     – Consumers, users, and marketing participants, including global and global perspective
     – Buying processes
     – Distribution structure
     – Market targets/segments, with positioning for each
   - Competitive analysis
     – Current competitors' strategies
     – Direct and indirect competitors
   - Voice of the customer (user needs and wants) studies
     – Focus group findings, lead user analysis, etc.
     – Concept test
       – Purchase intent
     – Perceptual maps
       – Probable positioning
     – Conjoint study
       – Consumer preferences and market share simulation results
   - Technical feasibility/requirements
     – Existing capabilities
     – Capabilities to acquire
       – Partners required
     – Technical risks
     – IP strategy

  - Supply issues
  - Manufacturing capabilities
  - Capital requirements
- Opportunities and problems
  - Key exploitable market opportunities
  - Key problems
  - Key environmental or exogenous factors
    - Governmental regulations
    - Environmental factors
    - Supplier issues

2. Product Definition
   - Core Benefit Proposition
     - Product/service details
     - Comparisons with competition
   - Product platform/architecture (if applicable)
3. Project Justification
   - Sales forecasts in dollars and units
   - Market shares on sales, profits, and budgets
   - Contribution to profit, with pro forma income statement
   - Future capital expenditures, including cash requirements
   - Risk statement
4. Project Plan
   - Overall guiding statement, including key quantitative and qualitative objectives
   - Overview of launch strategy

Each company will likely have its own format, but the key components will be similar. We briefly discuss each component.

Situational Analysis

The Situational Analysis sets up the environment into which the new product will be brought. The five components of the situational analysis begin to "paint a picture" of the market, and technical and strategic environments that co-exist.

Market Analysis: Information collected in the market definition phase and the scoping phase used to determine consumer profiles (both global and **glocal**[52]), the consumer buying process, distribution channels, and target segments.

Competitive Analysis: Research conducted during the opportunity identification phase that recognizes the primary competitors. Indirect competitors may not have been clearly considered.

Voice of the Customer Studies: Studies, including focus group studies, concept testing, conjoint studies, and qualitative ethnographic studies, that identify customer needs and wants. The purpose of the VOC studies is to receive feedback from customers on how the proposed product may or may not meet their needs and wants. This phase is detailed below.

Technical Feasibility Requirements: Information gathering in the preliminary technical assessment can be used as the starting point for delineating technical feasibility issues. It will be necessary to work with engineering/R&D to gather the information required to determine the technical capabilities available in-house and those that will need to be obtained through open innovation endeavors. Manufacturing and supply issues also should be assessed for this section of the business plan.

Opportunities and Problems: Any special issues not falling within the other categories should be considered in this section of the plan. For example, with the Segway, the company spent considerable amounts of money lobbying with cities in order to allow the Segway to be used on sidewalks. With an electric vehicle, governmental regulations also play a large factor in the potential success of EVs.

Product Definition

An important output of the business plan phase is to create a product definition,[53] which will be delivered to the development stage for use by the R&D engineers. The product definition, with product features, attributes, product drawings, technological requirements, and probable costs, provides a framework from which engineering can begin development.

A fundamental component of the product definition is to identify the *core benefit proposition*, which delineates the value of the product to the end user. The CBP states the product's unique benefits to provide to customers as well as those benefits required to meet and/or surpass competition. It must be clear and concise, striking immediately to the essential characteristics of the product's strategy in terms of the benefits it delivers to users. It forms the cornerstone upon which all elements of the marketing strategy are built and the vision that underlies the engineering design. Here are some examples of core benefit propositions:

1. American Express Credit Cards: Accepted everywhere; prompt replacement and complete protection if lost; prestige.
2. Hewlett-Packard LaserJet: Quietly prints documents with excellent print quality on several media and in several types. Easy to use and maintain, reliable, and flexible.
3. Silkience Self-Adjusting Shampoo: A shampoo that provides the appropriate amount of cleaning treatment (automatically) for different parts of your hair. Cleans the roots without drying the ends of your hair.

4. Personal-Care Hospital: A full-service hospital committed to the personalization of healthcare. Provides high-quality primary care to the patient, increased accessibility of the staff, and a friendly, caring, "first name" atmosphere.
5. Sear's DieHard Battery: Longer-lasting battery with more power output.

Each of these CBPs is succinct and emphasizes the most important attributes of the product or market. Sometimes, the intended market is explained fully in the CBP and sometimes it is made clear in the market definition. The CBP is written from the perspective of the customer, and outlines the needs of the customers as well as characteristics that will make them most likely to purchase. This is not just advertising, but describes how and why the customer will benefit from the product. It also is not a description of the product, but a reflection of the intended customer benefits. This new all-encompassing product definition should be supplied to the research and design as well as engineering team in order to aid in the process of designing the product.

The CBP is simple to state, but not simple to attain. It is the end result of a careful new product design process that identifies the needs and priorities of the customer, develops a feasible design to deliver those benefits, and coordinates a marketing effort to communicate those benefits. The CBP forces the design team of management, marketing, engineering, production, and other functions to reach a consensus on basic benefits and services. This finalized product definition will be provided to R&D/engineering to guide the development process.

Project Justification

Project justification includes financial information about revenue streams, capital/operational/development expenditure requirements, as well as foreseeable risks with the project. Before committing funds to the launch of a new product, which often may be millions of dollars, top management requests a forecast of customer response sales. In the preliminary business analysis, "preliminary" forecasts were considered. Now more formal calculations are required in the business case development.

Forecasting information may be obtained through VOC studies, such as the concept test or conjoint studies. In these studies, the new product design includes a description of the product features that are linked to a prediction of how the new product will be perceived by each customer in the sample. The forecast of sales is an aggregation of the predictions of behavior for individual customers. We discuss forecasting methods and considerations in later chapters.

Project Plan

The business plan culminates in a project plan—a future directive for which to use in the development of the new product. Included in this summary should be an R&D plan, a preliminary marketing plan (4Ps and launch strategy), and a production plan.

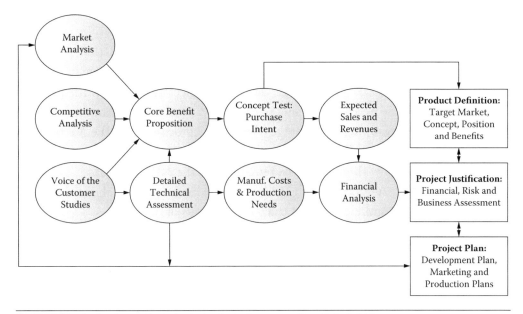

Figure 4.11   Factors to consider in developing the business case. (From Endnote 54.)

The development plan should include a project time line for project completion. The preliminary market plan will include a synopsis of the 4Ps (price, product, promotion, place), positioning information, and intended launch dates for production planning. The production plan will include information about factory and capital requirements. Any other crucial issues not addressed in the other parts of the plan will be addressed here.

Figure 4.11 shows how all the different components come together to inform the product definition, project justification, and project plan requirements.

Chapter Summary

The Discovery Stage of the new product development process should result in numerous new product ideas, but all of these cannot go into development. Scoping is used to narrow down the most promising innovations. The voice of the customer techniques are used to solicit customer input into formalizing a winning product idea into a concrete product definition. Although many companies feel that the VOC is not always useful because they feel that consumers don't always know what they need or want, many studies have shown that truly understanding how product features materialize into customer values is essential for developing products that are successful in the marketplace. The insights gathered through scoping and voice of the customer analysis are culminated in a business plan, which is the road map for the project as it moves from the fuzzy front end of the new product process into the development phase.

Glossary

Benefit chains: Determine why customers have a particular need that is not yet addressed by existing products. It seeks to identify consumers' values, the benefits obtained from these values, the consequences arising from the benefits, and how these needs can be met with product attributes or features.

Build the business case stage: The stage of the NPD process where the NP business plan is developed. This business plan becomes the road map by which the team will follow throughout the remainder of the NPD process.

Business rationale: A concern in the go/no go gate process that evaluates how well the product continues to meet the strategic goals of the firm.

Choice models: A quantitative method, such as conjoint analysis or discrete choice surveys, that determines consumers' preferred choices from a set of products/services.

Core benefit proposition: States the unique benefits that the product provides to customers as well as those benefits required to meet and/or surpass competition. It forms the cornerstone upon which all elements of the marketing strategy are built and the vision that underlies the engineering design.

Design phase: Comprises Stage 2, *Build the Business Case*, and Stage 3, *Development*.

Development stage: The stage in the NPD process where R&D and/or engineering is conducted to develop the product/service. The outcome of this stage is typically a product/service ready for mass production.

Empathic design: Insights can only be gained by observing customers in their natural habitat.

Entrepreneurial optimism biasness: Situations in which failure seems unlikely to those championing a new product idea.

Experiential interviews: The needs and desires of customers are explored in one-on-one interviews in which the customer describes his or her experience with the product class. The interviewer probes deeply into the underlying, more stable, and long-term problems that the customer is trying to solve.

Future plans: A concern in the go/no go gate process that evaluates how well the product meets the future plans of the firm, remains competitive in the marketplace, and stays within budget.

Gate review: The process in which the product/project is evaluated for continuation to the next stage. It is where the go/kill/recycle decision is made for the product.

Idea screening gate: The gate at which product ideas are screened to determine if they should advance to the Scoping stage.

Index of attractiveness: Divides the expected return by the development cost as a measure of how attractive the new product is compared to other potential new products.

Laddering: Qualitative technique for finding the benefits an end user may see in a new product.

Means-end chain: The means-end chain describes how a product interacts with the consumer through its "attributes," "benefits," to deliver "values." It takes the opposite approach to the benefit chain, which starts with values to derive attributes.

Perception models: A quantitative method, such as concept testing, that gathers consumers' perceptions about the features, benefits, or attributes of a new product/service.

Preference models: A quantitative method that gathers consumers' preferences for features, benefits, or attributes of a new product/service.

Preliminary business/financial assessment: Includes expected profit projection (sales potential, cost of product), core competency assessment, and strategic alignment assessment. A "back of the envelope" forecast also should be calculated.

Preliminary market assessment: Consolidates all the findings that were accumulated during market analysis, including commercial viability of the product/service, market (industry) attractiveness, competitive environment, and segmentation guidelines.

Preliminary technical assessment: Includes identifying initial product requirements, identifying feasibility of engineering product design, cost estimation of engineering and resource estimations, and manufacturing feasibility. This assessment also would include assessing the need to out-source certain aspects of the design, such as those involved with open innovation.

Product definition: A component of the new product business plan where a clear product/service definition, including the core benefit proposition for the product/service, is spelled out.

Project justification: A component of the new product business plan that addresses the financial considerations of a new product/service. It would include financial information about revenue streams, capital expenditure requirements, as well as foreseeable risks with the project.

Project plan: A component of the new product business plan that addresses a future directive including a R&D plan, a preliminary marketing plan (4Ps and launch strategy), and a production plan.

Quality of execution: A concern in the go/no go gate process that evaluates how well the project team is conducting the NPD process.

Scoping stage: The NPD stage in which a preliminary market assessment, a preliminary technical assessment, and a preliminary business/financial assessment are conducted. In this stage, voice of the customer also is collected.

Scoring models: A method for scoring an individual project to determine its attractiveness compared to other potential products.

Situational analysis: A component of the new product business plan that paints a picture of the market, technical, and strategic environments co-existing for the product idea.

Voice of the customer (VOC): Strives to record in the customer's own words the benefits of a product or service; it is a description of the customer's needs within a specific product category.[55] Voice of the customer (VOC) studies should be a combination of qualitative and quantitative research.

Web-based "eavesdropping": A voyeuristic method of capturing unmet customer needs by observing customer interactions on Internet-based discussion boards, social network postings, blogs, etc.

Zaltman's metaphor elicitation technique (ZMET): Suggests that the underlying values and meanings that drive customers toward specific product choice decisions may be uncovered through a process of visual self-expression.

Review Questions

1. What stages are involved in the Fuzzy Front End? The Design phase?
2. What is the purpose of a gate? What is involved in a gate review?
3. What is involved in the Scoping phase of the NPD process?
4. Why should managers start new product development projects by studying consumers rather than engineering designs?
5. Compare the different methods of obtaining the voice of the customer presented in this chapter. Discuss the advantages and disadvantages of each.
6. How does the mean-end chain increase the probability that the company delivers products that consumers are willing to purchase?
7. How should companies use voice of the customer (VOC) in their development processes?
8. What are the primary components of Building a Business Case?

Assignment Questions

1. If you could advise Dean Kamen, inventor of the Segway, at the Scoping stage of the development process, what would you have reported back to him? (Yes, hindsight is 20/20, but put yourself into the 1990s when the product was still yet an idea).
2. Develop a set of must-meet and should-meet criteria for a "gate" analysis for:

   - A new iced coffee drink being sold by McDonald's Corporation
   - An EV-mini van with seating for six
   - Mobile dentist office
   - 3G mobile banking

3. A new concept has been proposed for a gravesite memorial stone that combines data and traces left in the Internet by a human being in the course of his or her life. A computer program written especially for this purpose filters personal data, defining the individual shape of a personalized memory stone, which is predetermined by the life of the deceased. (See Figure A.001. Gravesite sculpture using Internet traces of the Individual.)
   - Draw out a benefit chain structure (as demonstrated in Figure 4.6) for this new device.
   - Layout a means-end chain (as demonstrated in Figure 4.7) for this new device.
4. Design a "laddering questionnaire" concerning online learning and administer it to a classmate.
5. In 2011, Mobisante, a Seattle-based medical company, introduced an ultrasound for use with smartphones. The device is intended for ultrasound imaging, analysis and measurement in fetal/OB, abdominal, cardiac, pelvic, pediatric, mucoskeletal, and peripheral vessel imaging. The smartphone-based ultrasound system can leverage both cellular and Wi-Fi to send images for diagnosis, second opinion, or to a Picture Archiving and Communication System (PACS) for storage. This device has recently been used by in-the-field emergency providers.

   - Describe how an empathic design evaluation could be conducted for this product.
   - How should the information gathered through the empathic design be used to design the product?
6. Nokia is looking to compete in the burgeoning smartphone market. Eavesdrop on the Internet to give them an idea of what type of new product they should consider next.

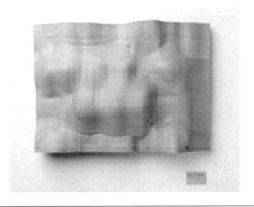

Figure A.001   Gravesite sculpture using Internet traces of the individual.

Endnotes

1. Several sections of this chapter are based on Glen L. Urban and John R. Hauser, *Design and Marketing of New Products*, 2nd ed. (Upper Saddle River, NJ: Prentice Hall, 1993).

2. Online at: http://inventors.about.com/od/sstartinventions/a/segway.htm (accessed June 8, 2013).

3. "Segway's Breakdown," *Wired* 11.03 (March 2003). Online at: http://www.wired.com/wired/archive/11.03/segway.html (accessed June 8, 2013).

4. Online at: http://hbswk.hbs.edu/archive/3533.html (accessed June 8, 2013).

5. "Segway's Breakdown," *Wired*.

6. Robert G. Cooper, *Winning at New Products: Accelerating the Process from Idea to Launch*, 3rd ed. (New York: Basic Books, 2011).

7. Ibid.

8. Ibid.

9. Ibid.

10. Ibid.

11. Urban and Hauser, *Design and Marketing of New Products*.

12. E. Mansfield, J. Rapoport, J. Schnee, S. Wagner, and M. Hamburger, *Research and Innovation in the Modern Corporation* (New York: Norton, 1971), p. 213.

13. David A. Armor and Shelley E. Taylor, "When Predictions Fail: The Dilemma of Unrealistic Optimism," in *Heuristics and Biases: The Psychology of Intuitive Judgment*, ed. Thomas Gilovich (Cambridge, U.K.: Cambridge University Press, 2002).

14. Online at: http://www.ideafinder.com/history/inventors/kamen.htm (accessed June 30, 2013).

15. Urban and Hauser, *Design and Marketing of New Products*.

16. Robert G. Cooper, *Winning at New Products*, 3rd ed. (Cambridge, MA: Perseus Publishing, 2001), p. 108.

17. Cooper, *Winning at New Products*.

18. Edgar is the U.S. Securities and Exchange Commission's online Web site for publicly traded firms financial data and annual reports. Online at: http://www.sec.gov/edgar.shtml (accessed June 30, 2013).

19. For more on scoring models, see D. L. Hall and A. Naudia, "An Interactive Approach for Selecting IR&D Projects," *IEEE Transactions on Engineering Management* 37 (1990):126–133; J. F. Bard, R. Balachandra, and P. E. Kaufmann, "An Interactive Approach to R&D Selection and Termination," *IEEE Transactions on Engineering Management* 35 (1988); R. G. Cooper, *Winning at New Products: Accelerating the Process from Idea to Launch* (Reading, MA: Addison-Wesley, 1993).

20. Ely Dahan and John Hauser, "Product Development—Managing a Dispersed Process," in *Handbook of Marketing*, eds. Barton Weitz and Robin Wensley (London: Sage Publications, 2002).

21. Abbie Griffin and John R. Hauser, "The Voice of the Customer," *Marketing Science* 12(1) (1993): 1–27.

22. Robert G. Cooper, Scott J. Edgett, and Elko J. Kleinschmidt, "Portfolio Management for New Product Development: Results of an Industry Practices Study," *R&D Management*, Industrial Research Institute, Inc. 31(4) (2001): 22, Fig. 14.

23. Based on Dahan and Hauser, "Product Development," pp. 179–222.

24. Ibid.

25. Richard A. Krueger and Mary Anne Casey, *Focus Groups: A Practical Guide for Applied Research*, 4th ed. (London: Sage Publications, Inc., 2009).

26. Gary Burchill and Christina Hepner Brodie, *Voices into Choices: Acting on the Voice of the Customer* (Madison, WI: Oriel Incorporated, 1997).

27. Abbie Griffin and John R. Hauser, "The Voice of the Customer," *Marketing Science* 12(1) (1993): 1–27.
28. Dorothy E. Leonard-Barton, Edith Wilson, and J. Doyle, "Commercializing Technology: Imaginative Understanding of User Needs," paper presented at the Sloan Foundation Conference on the Future of Research and Development, Harvard University, Boston, MA, February 1993.
29. Griffin and Hauser, "The Voice of the Customer."
30. This section is based on Dahan and Hauser, "Product Development."
31. Leonard-Barton, Wilson, and Doyle, "Commercializing Technology."
32. Griffin and Hauser, "The Voice of the Customer."
33. John Case, "Customer Service: The Last Word," *Inc. Magazine* (April 1991): 88–92.
34. This section based on Dahan and Hauser, "Product Development."
35. Gary Levin, "Anthropologists in Adland," *Advertising Age* (February 2, 1992): 3, 49.
36. Morna S.Y. Lee, Peter J. McGoldrick, Kathleen A. Keeling, and Joanne Doherty, "Using ZMET to Explore Barriers to the Adoption of 3G Mobile Banking Services," *International Journal of Retail & Distribution Management* 31(6) (2003): 340–348.
37. Based on Dahan and Hauser, "Product Development."
38. Ibid.
39. Johnathan Gutman, "A Means-End Chain Model Based on Customer Categorization Processes," *Journal of Marketing* 46 (Spring 1982): 60–72.
40. George A. Kelly, *The Psychology of Personal Constructs*, Vol. 1 (New York: W. W. Norton, 1955).
41. Online at: http://advertisingaphasia.blogspot.com/2010/07/what-can-means-end-chain-teach-us-about.html (accessed June 8, 2013).
42. Eda Gurel-Atay, Guang-Xin Xie, Johnny Chen, and Lynn Richard Kahle, "Changes in Social Values in the United States: 1976-2007, Self-Respect Is on the Upswing as 'A Sense of Belonging' Becomes Less Important," *Journal of Advertising Research* 50(1) (2010): 57–67.
43. Online at: http://en.wikipedia.org/wiki/Values_scales (accessed June 8, 2013).
44. Dahan and Hauser, "Product Development."
45. Thomas J. Reynolds and Jonathan Gutman, "Laddering Theory, Method, Analysis and Interpretation," *Journal of Advertising Research* 28(1) (Feb/Mar 1988): 11–31.
46. Based on Dahan and Hauser, "Product Development."
47. Glen L. Urban, *Listening in to Customer Dialogues on the Web* (Cambridge, MA: Center for Innovation in Product Development, MIT, 2000).
48. Based on Dahan and Hauser, "Product Development."
49. Second Life is a virtual world developed by Linden Lab that is accessible via the Internet. Online at: http://secondlife.com/ (accessed June 30, 2013).
50. Online at: http://www.nielsen-online.com/products.jsp?section=pro_buzz&nav=1 (accessed June 9, 2013).
51. Online at: http://www.blogpulse.com/trend?query1=3G+mobile+banking&label1=&query2=&label2=&query3=&label3=&days=180&x=34&y=9 (this Web site is no longer available for public viewing).
52. Glocal environment refers to developing new products with the local environment in mind. This may mean developing more than one version of a product: one for domestic distribution and one for foreign distribution.
53. Urban and Hauser, *Design and Marketing of New Products.*
54. Revised based on: Cooper, *Winning at New Products.*
55. Griffin and Hauser, "The Voice of the Customer."

THE CONCEPT TEST

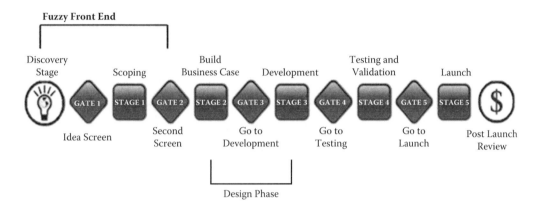

Fuzzy Front End

Discovery Stage — Scoping — Build Business Case — Development — Testing and Validation — Launch

GATE 1 · STAGE 1 · GATE 2 · STAGE 2 · GATE 3 · STAGE 3 · GATE 4 · STAGE 4 · GATE 5 · STAGE 5 · $

Idea Screen — Second Screen — Go to Development — Go to Testing — Go to Launch — Post Launch Review

Design Phase

Learning Objectives

In this chapter, we will address the following questions:

1. When should a company conduct a concept test?
2. What are the ways that a concept description can be shown to survey respondents?
3. How is the purchase intention question used in the concept test? What is the top-box rule?
4. What are product diagnostics? What are attribute diagnostics?
5. How can a positioning statement be created from the results of a concept test?

What Do Consumers TH!NK?

TH!NK, manufacturer of Norway's pioneering electric car, knew that reaching the U.S. marketplace could be important to the success of the company. "The United States is quickly overtaking Europe as an attractive market for EVs [electric vehicles] and is an ideal location to engineer and build EVs," TH!NK CEO Richard Canny announced in a March 2009 press release.[1]

The TH!NK *City* is a small, two-seater or 2 + 2-seater electric vehicle with a top speed of 65 mph and an in-town range of just beyond 100 miles on a full charge. Recharging time takes between three to eight hours.

TH!NK planned to sell the car but lease the battery as a way to overcome one of the biggest uncertainties from owning an electric car, which have typically been: What is the battery life and how much will it cost to replace? The battery is by far the most expensive component of any electric vehicle. By charging a "mobility fee" of $100 to $200 a month that also included services like insurance and wireless Internet access, the company could keep the price of the car in the mid-$20,000 range. To reach this selling price, TH!NK management knew that the vehicle must be manufactured in the United States for that market and in 2010 had plans to open a plant in Indiana.

The car was designed, engineered, and produced to have the lowest possible carbon footprint with recyclable plastic body panels and a fully recyclable interior. TH!NK vehicles were designed to be smart, flexible, and continuously updated to deliver state-of-the-art urban mobility. As confirmation of the unique design, the TH!NK *City* had been nominated for England's prestigious Britt Design Award in 2009.

TH!NK wants to change the way cars are made, sold, owned, and driven. The company will sell cars online, built to order. Test drives will be done through car-sharing franchise, such as partnerships with Zipcars.[2] If you like what you see, you customize and order your TH!NK *City* online. That means no showrooms or pushy salespeople. However, this type of venture is a big gamble, particularly given the SUV-loving Americans.[3] What CEO Canny was still uncertain about was how would the U.S. consumer respond to the innovative car? Would they find it a "fun" concept as the European consumers did? Would the charge time of eight hours be a deterrent to these consumers? What would they think about a "mobility fee?" How do consumers compare TH!NK *City* to the Nissan LEAF and the Chevy Volt? He felt a concept test could help him answer these questions.

Introduction

In this chapter, you will learn how to conduct a concept test, which is a consumer research study that is typically conducted during the **Scoping** stage of the NPD process. During the **Discovery** stage, we began to visualize a new product that could meet the needs and wants of a customer. In the Scoping stage, we refined the product idea and identified several types of qualitative studies that we could conduct in order to get the voice of the customer (VOC). However, many questions remain unanswered with these *qualitative* studies. We use *quantitative* studies to try to fill in these blanks.

For example, during a focus group in the Scoping stage, when customers indicated they didn't want to have to worry about battery life of an electric vehicle, were they implying they were interested in leasing the battery? When they indicated they were dubious of the Big Three U.S. automobile (GM, Ford, Chrysler) manufacturers meeting their needs for green vehicles, did this mean they would embrace cars made by European manufacturers? A concept test can help to make sure that the product design laid out in the business development stage *accurately* captures consumers' needs and wants and delivers a value proposition that beats competitors' offerings. The intent of the concept test is to help managers determine whether a concept, as designed thus far, merits further development.

What Is a Concept Test?

Once a product definition has been clearly defined during the business development stage to yield a technically feasible concept, the product concept can be shown to the customer in the form of a model, a set of drawings, a storyboard, a specifications sheet, a mocked-up print/video advertising, or, possibly, a virtual prototype. This is known as a **concept test**. The intent of the test, which is typically conducted *prior* to undertaking product development tasks, is to determine consumer interest in the product as conceived. The goal of this research is three-fold: (1) to produce a preliminary sales volume forecast; (2) to determine that consumers'/end users' needs and wants are correctly understood and translated into the product definition, and, if not, to recommend engineering design revisions; and (3) to create a positioning statement for the new product.[4] According to Robert G. Cooper, "The concept test is not a prospecting study, but rather a test or validation that the proposed product concept is indeed a winner; that is, intent to purchase is established. Note that at this early stage you still don't have a developed product. The purpose of this concept test is to see if you're heading in the right direction."[5]

Figure 5.1 provides an example of a **concept description** for a mobile printer. Zink, a spin-off from the now defunct Polaroid, saw a need for small, portable inkless printers. They commissioned a concept test to estimate the sales volume potential and customer acceptance of key product features of a mobile printer for the U.S. and Japanese market.[6] A concept test will ask consumers to rate the quality perception of different features, such as image quality, image size, and portability, as in the case of Zink. If portability is more important to consumers than image quality or image size, this would lead to one type of printer compared to if image quality was most important. The results of the concept test are used to make adjustments to engineering

Print Photos: Anytime, Anywhere

A companion to your camera phone or digital camera, the pocket-sized Mobile Printer connects either wirelessly or with a cable directly to any camera phone or digital camera and prints full-color 2"x3" digital photos in less than 30 seconds. Printing on the go has never been easier. Before leaving home, be sure the printer's rechargeable battery is charged and you've loaded the printer with photo paper (it holds 10 sheets). While you are out and about, if you take an image on your camera phone or digital camera that you want to print, send it wirelessly or wired to the Mobile Printer and in 30 seconds you will have a colorful, durable, and high resolution photo.

New Ink-Free Printing Technology

The Mobile Printer uses a new digital printing technology that prints without ink cartridges or ribbons. Instead of using ink cartridges, the color is embedded in the photo paper itself. Before printing, the ink-free paper looks just like regular white photo paper. The Mobile Printer activates the embedded color, producing colorful, durable, high resolution images and text.

Mobile Printer Features

- Compatible with all digital cameras and camera phones
- Prints 2" X 3" borderless, high quality photos in under 30 seconds
- Uses new photo paper that doesn't require ink cartridges or thermal ribbons
- Photo paper available in both regular and sticky-back formats
- Also prints stickers, business cards, coupons, event tickets, maps, internet content and more
- Battery operated via integrated rechargeable battery
- Everything you need is included with the Mobile Printer - Printer, battery, cable, software, and starter pack of photo paper

Figure 5.1   Mobile printer concept test description. (From Endnote 7. With permission.)

requirements before finalizing product design. Once a commitment has been made to a product or service design, it is often difficult to make major changes without incurring substantial expenses. A concept test can help to avoid this pitfall.

When a concept has a number of **attributes** or **features** offered, it is useful to probe which attributes/features contribute to purchase likelihood. Probing for new product perceptions may be obtained through the use of open-ended questions, such as: "You indicated you would probably buy the mobile printer. What do you find most useful about this printer?" The concept test acts as a diagnostic tool to guide positioning in the marketplace by showing how a new product idea fares against competing products.

To capture purchase probability, respondents are asked their **probability of purchase** (or trial) for the product. Consumers may like the idea, but if they are not willing to purchase the product, it will fail in the marketplace. A purchase intention question can be used to project an early sales forecast of the new product. Product ideas with poor sales potential can be killed before major resources to engineering R&D are committed.

The fundamental difference between the concept test and qualitative VOC studies is that these studies are typically used for discovery and prospecting. The intent of early VOC research is to flush out winning new product *ideas*. Possible results of the early VOC study could indicate that the primary buyers of an electric vehicle will be soccer moms who want an environmentally friendly car to use for local errands that will replace the larger family car. A subsequent concept test could identify, for example, that only 2% of this potential market would purchase the TH!NK vehicle as currently described and that charge times of less than three hours and 100 miles between charges are preferred by consumers. The focus in the concept test is on the product attributes and firming up an understanding of consumer needs and wants.

Conducting the Concept Test

Many companies will conduct concept tests themselves instead of using an outside agency because it provides them with another opportunity to meet with customers and to build long-term relationships. There are six general steps in conducting a concept test:

1. Determine goal of concept test
2. Select a survey population
3. Select most appropriate survey format
4. Prepare the concept statement
5. Develop and administer the questionnaire
6. Interpret and report results

We will now look at each of these steps in detail.

Step One: Determine Goal of Concept Test

As previously noted, there are typically three goals of a concept test: concept diagnostics, forecasting, and positioning. The primary goal of a concept test is to obtain insights into how consumers feel about the attributes/features/benefits/value of the proposed new product/service. Because the focus is on product attributes, the results from the study provide insights on adjustments to the product design before resources are committed to the project.

Concept Diagnostics   The goal of concept diagnostics is to determine how successful the product/service will be in the marketplace under its *current* design. Remember, we are still in the fuzzy front end of the NPD process and the product design will continue to evolve especially after evaluating the VOC study. Questions that should be answered from the diagnostic analysis include:

- Will the product sell as currently designed?
- Should the product be advanced to the next stage, Development, which requires substantial commitment of resources?
- How can the product concept be improved?

- Which features/attributes are important to the customers?
- Does importance vary by segment; by product/service use?
- Does the proposed price provide value satisfaction?

Forecasting   The concept test also will provide an opportunity to obtain an early forecast of the product *as currently designed.* A goal of the concept test is to estimate the sales or trial rate of the new product. One method of validating the product idea is to ask an "*intent to purchase*" question. The results from this question of how likely a consumer is to buy this product are then used to develop a preliminary forecast for sales or trial of the new product. Questions that can be answered with the purchase intent question include:

- What is individual probability of purchase or trials?
- What are likely quantities of purchase?
- What are the probabilities of repeat purchase? What is the frequency of repeat purchase?
- What is the price ceiling?

Positioning   A concept test also can assist with market **positioning** decisions. This is done by including questions in the survey about competitors' offerings. You may recall that the positioning statement is one of how a product/service/brand is viewed in the minds of the target market compared to the competition. Importance ratings can be used to capture competitive product or brand preferences to compare these preferences to those of the new product offering. Questions that can be answered about positioning include:

- Is our product better on attribute A or attribute B in comparison to the competition?
- Should we position our product/service around attribute A or attribute B?
- Can clusters of consumers be identified?
- Should different positioning statements be created for the different segments?

Step Two: Select a Survey Population

Concept tests can be completed with a small sample of potential customer or a very large number. The sample size depends upon the goals of the manager. We next discuss guidelines.

Sampling Guidelines   When collecting data from consumers, it is critical that the survey population mirrors the target market in as many ways as possible. An overly receptive survey population can result in overcommitment of resources to an otherwise unfavorable product idea. In selecting the survey population, potential customers from each target segment should be surveyed because a product idea may address multiple market segments. For example, for TH!NK, the electric vehicle (EV) manufacturer, it was thought that early adopters would be previous Toyota Prius owners. How could a firm reach this segment? What other segments might be early adopters of the EV? It is important that the customers surveyed actually represent the intended target market. At this phase in the development process, surveying lead users or innovative consumers can very easily bias the survey results, unless of course, this is the primary intended audience.

Reaching the right survey respondent is extremely important as well in an industrial setting. Care should be taken to understand who the customer is and who the end user is; typically, they are not the same individual. For example, Caterpillar, the world's largest maker of construction and mining equipment, sells products to major companies through dealers. The end user (equipment operator) may be better suited to answer a concept test survey as opposed to the corporate buyer or dealer. The operator, however, would not be able to respond to questions about pricing, so a separate study may be required to administer to buyers if pricing is a major concern. Sometimes feedback from multiple sources will be needed to get a full analysis of the market place.

Although there are no exact or simple formulas for determining sample size, Urlich and Eppinger offer some recommendations as seen in Table 5.1. Small groups may include focus groups, small targeted subpopulations, or maybe even mall intercepts with a small group of customers. Larger groups are typically administered through Web surveys or through consumer panel surveys.[8] Sometimes a small sample population will be interviewed, followed by a larger sample population. This is usually done if there exists some uncertainty about the target market, or perhaps, technological uncertainty that makes describing the concept difficult. Running a survey with a smaller population can reduce this uncertainty.

Step Three: Select Most Appropriate Survey Format

Surveys can be conducted face-to-face, by telephone, mail, electronic mail, or on the Internet. Each method brings its own challenges. A popular method is face-to-face,

Table 5.1 Determining Survey Sample Size

FACTORS FAVORING A SMALLER SAMPLE SIZE	FACTORS FAVORING A LARGER SAMPLE SIZE
• Test occurs early in concept development process.	• Test occurs later in concept development process.
• Test is primarily intended to gather qualitative data.	• Test is primarily intended to assess demand quantitatively.
• Reaching potential customers is relatively costly in time or money.	• Surveying customers is relatively fast and inexpensive.
• Required investment to develop and launch the product is relatively small.	• Required investment to develop and launch the product is relatively high.
• A relatively large fraction of target market is expected to value the product.	• A relatively large fraction of target market is not expected to value the product.

Source: Endnote 9.

which benefits from the ability to further probe consumer comments with an interactive open-ended format, but it proves to be expensive and the results are dependent upon the quality of the survey giver. Common web-based surveys can be less expensive and also can capitalize on virtual reality for displaying the concept test.

There are a number of growing trends in concept testing including testing of multiple products at the same time using extreme-value concept testing.[10] Another trend is in using Internet-based product concept testing methods that incorporate virtual prototypes of new product concepts.[11] Dahan and Srinivasan developed and tested a Web-based method of parallel concept testing using visual depictions and animations, in this case, for bicycle pumps. It is expected that Web-based market research methods will grow in power and applicability as more sophisticated technologies arise for both computers and mobile phones. With further development and testing, these virtual concept testing methods have the potential to reduce the cost and time devoted to not only concept testing, but also prototype development.

Step Four: Prepare the Concept Statement

There are eight ways in which to present the concept description to potential users. Table 5.2 illustrates these methods. We discuss the advantages and disadvantages of each below.

Words Only   When using words only to describe a product concept, your presentation may be purely factual such as:

> "TH!NK *City* is an electric vehicle with zero local emissions. It is powered by lithium batteries, traveling up to 112 miles on one charge, with a top speed of 62 miles per hour. The car is equipped with ABS brakes, airbags, and three-point safety belts with pretensioners."[12]

Or it can be described as:

> "TH!NK *City* is a modern urban electric vehicle with zero local emissions that is sure to ignite the fun factor in everyday driving. It is powered by state-of-the-art, highly powerful lithium batteries. TH!NK *City* allows you to zip around

Table 5.2   Execution Types for Concept Description

MODE	TONE	
	FACTUAL	PERSUASIVE
Words Only		
Visual Only		
Words & Visual		
(Virtual) Reality		

Source: Endnote 13.

town or on the highway with a top speed of 62 miles per hour, traveling up to 112 miles on one charge. Put some excitement back into your driving routines with this sporty, technology-proven, environmentally friendly car."

It is easy to see how the words presented can significantly impact a consumer's decision to purchase a product. Be aware, since the firm uses the results of the concept test to predict the success of the product in the marketplace, *overselling* the product during the concept test can have serious impacts of promoting a potential product failure. There must be a delicate balance in giving the survey respondent enough information to make an informed decision and being overly persuasive. Factual presentations deliver only the core idea. Persuasive presentations deliver the positioning concept.

Visuals Only   Using visuals only is usually not recommended. However, there might be occasions when it is the best alternative. For example, a student group wanted to test the idea of a "*digital waiter,*" similar to the system that is being used at a restaurant in Nuremberg, Germany. Guests order meals from a touch screen located at their table and food is delivered by a "minirailway" from an upstairs kitchen. This new service uses a computer monitor on top of each table at the restaurant instead of having the traditional wait staff. Since there is limited interaction with personnel, part of the concept test was to determine how restaurant guests responded to the computer presentation. There would be no written instructions at the restaurant beyond the computer screen, so the team felt that a visual only presentation was most realistic. Visuals may be in the way of sketches, photos, 3-D renderings, or storyboards, such as that shown in Figure 5.2.[14]

Words and Visuals   Most often, the **full proposition**, which is a combination of words and visuals, is used. This is the most common form of a concept description. Figure 5.1

Figure 5.2   Visual presentation of concept description. (From Endnote 14. With permission.)

demonstrates an example of a full proposition for Zink's mobile printer concept. Just as when using words alone, the more imaginative the writing, the greater the possibility exists of improving the purchase intention response. However, research by Lees and Wright[15] has shown that the type of concept statement formulation (simple or embellished) had resulted in only minor variations on the purchase intention responses for products/service in the early stages of product design. However, this may not be the case when actual prototypes are used in the concept test. Persuasive writing isn't as impactful when the respondent can actually pick up the product and play with it. Caution of overestimating purchase intentions should be taken, especially when using prototypes.

(Virtual) Reality   Because words and pictures aren't always enough, some companies use actual prototypes (reality) for concept tests. The problem with a prototype is that sometimes it works better than the final product and sometimes it is worse. The main problem is that prototypes are expensive. Model mock-ups created out of wood or polymer foams are a less expensive method, but not as realistic, which is why manufacturers of high ticket items, such as automobiles, industrial equipment, and hospital equipment, are turning to virtual reality.

Videos, simulations, interactive media, and virtual reality are all being used to convey concept ideas. This can be an expensive endeavor, but cheaper than building a prototype of a car. The Internet is being used more often for concept testing because of the ability to use virtual reality modeling language (VRML) to produce

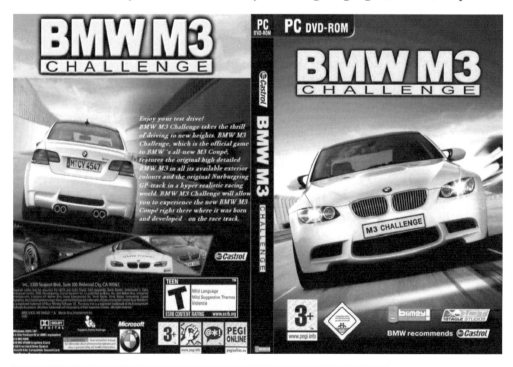

Figure 5.3   Example of virtual reality in auto promotion. (From Endnote 16. With permission.)

highly sophisticated virtual prototypes. Automobile companies and manufacturers of other high-end products often use videos or computer simulations to convey a concept (Figure 5.3). BMW has used a computer game to introduce potential buyers to its M3 coupe. Game players can "drive" the car using a PC steering wheel or the keyboard. Although, the game has primarily been used as a promotional outlet, recent research is being conducted to determine how these types of simulations could be used within a concept test. The downside of the electronic format is that it may bias the sample population toward owners of computers who tend to be more technologically sophisticated compared to nonowners. Additionally, the "sexiness" of virtual reality may result in overly positive responses to a new product.

Information Acceleration[17] Cooper indicates that for the needs-identification-and-translation process to work well, two assumptions must be true:

1. The customer understands his or her needs and is able to verbalize these during the needs-and-wants study [VOC research].
2. The new product manager interprets these needs correctly and does a sufficient job in translating these needs into the final product specs.

So, what happens when the customer can't articulate his or her needs well enough to provide an assessment of the new product/service? In 2009, if you asked the mainstream consumers what features they want in an electric vehicle, many of them would respond that they don't know. They didn't understand how an electric vehicle works or how it differs from a hybrid–gas vehicle, let alone from a plug-in electric vehicle. Because this is new technology they just aren't familiar with, it's difficult for the customers to voice their needs, wants, and desires. This was true for the microwave, the personal computer, and automatic teller machines (ATMs) when they were first introduced to the marketplace.

One method for introducing really new product concepts to consumers is **information acceleration** (IA),[18] which uses the power of the Internet to simulate information to a potential buyer of a new product. The IA technique places consumers in a virtual buying environment. After presenting a consumer with a simulated showroom visit and advertising, measures of purchase (trial) intentions, product perceptions, and preferences can be obtained for really new or radical innovations. With IA, the survey respondents are in control of what they view, for example, they can virtually test the product, discuss the product with a salesperson, join blog forums to learn more about the products, or view advertisements. The respondents choose what they want to see and in what sequence. This is called "future conditioning" of a consumer by providing them information about a product or service they have never before encountered. In a research study, Dahan and Srinivasan[19] report, "As virtual prototypes cost considerably less to build and test than their physical counterparts, design teams using Internet-based product concept research may be able to afford to explore a much larger number of concepts. Virtual prototypes and the testing methods associated with them may help reduce the uncertainty and cost of new product introductions by allowing more ideas to be concept tested in parallel with target consumers." It has been shown that both static and animated virtual prototype tests produced market shares that closely mirrored those obtained with the physical products.

Impact of Concept Presentation   In determining which execution style to use from Table 5.2, it's important to keep in mind how survey takers will respond to the purchase intention question. In general, moving from factual to persuasive produces higher purchase intention scores. It has been argued that emotional appeals delivered visually are more effective in communicating services value than verbal appeals.[20] But, more importantly, comparing concepts evaluated using executions across different cells in Table 5.2 can lead to invalid results. For the mobile printer example provided earlier in this chapter, the research firm used actual prototypes when conducting the in-person concept test in the United States, but used word and visual representation in an online survey in Japan. The results of the two studies cannot be compared because the implementation differed across the studies. When comparing projects against each other in this phase of development, it's important to know the execution style used to present the concept to the respondent.

Step Five: Develop the Questionnaire and Conduct the Survey

After determining which type of concept execution to present to survey respondents, it is necessary to determine what types of data to collect. In a concept test, there are typically four types of data that are collected: (1) purchase intention questions, (2) overall product diagnostic questions, (3) specific attribute questions, and (4) respondent profile questions. The four types are summarized below.

Purchase Intention Questions   The **purchase intention questions** are used to determine probability of purchase or trial of the product/service. Based on the concept description, survey respondents are asked if they would purchase the product/service and, if applicable, how often (frequency). Figure 5.4 shows an example of a "purchase

For the mobile printer just described, how likely would you be to buy this product?

⊙ Definitely would buy

⊙ Probably would buy

⊙ Might/Might not buy

⊙ Probably would not buy

⊙ Definitely would not buy

Figure 5.4   Purchase intention questions.

intention" question. Although a 5-point scale is most common, other scales that are used are 6, 7, and 11 points. A 7-point scale is shown in the Appendix at the end of the chapter.

For nondurables goods, particularly consumer packaged goods, the frequency of purchase is important to know in order to estimate a sales/trial forecast. Some packaged goods, such as chewing gum, may be purchased more often than Clorox® Anywhere® Hard Surface™ Daily Sanitizing Spray. For durable goods, such as an automobile, a question about next intended purchase period is more meaningful. Examples of different types of purchase frequency questions are in Figure 5.5. Any questions that will help to estimate frequency of purchase and quantity, if multiple units might be involved, should be asked as part of the purchase intention section.

Overall Product Diagnostics Managers want to understand why consumers have responded to the purchase intention question in the way they did. If they definitely intend on purchasing a product, what is it they liked? And if they definitely intend on not purchasing the product, what could be improved? A set of **overall product diagnostic questions** that query survey respondents about their perception of the overall value of the product are the second type of question included in the concept test. The goal of these questions is to determine if the overall product concept is understood and valued by the potential customer. Questions about current practices and a consumer's willingness to change could be included in this section as well. Based on the "digital waiter" concept previously described, Table 5.3 provides examples of questions that can be asked.

A 5- to 11-item Likert scale is usually used when asking these types of questions. Open-ended questions also may be beneficial when you are interested in what changes

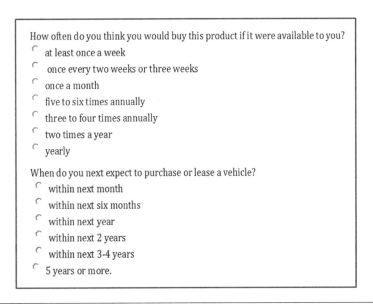

Figure 5.5 Frequency questions for the concept test.

Table 5.3   Overall Diagnostic Questions

Likert Scale[21] (strongly disagree—strongly agree)
- I find the digital waiter highly unique.
- I find the digital waiter to be believable.
- I find the digital waiter practical for ordering food and drinks.
- I find the digital waiter useful for receiving immediate service in a restaurant.
- I find the digital waiter important for solving the possibility of incorrect orders.
- My interest in the digital waiter is extremely high.
- If a digital waiter were available in a restaurant, I would use it instead of relying on wait staff.
- I find the digital waiter's design appealing.
- I believe the digital waiter would be easy to use.

Open-ended items
- I think the digital waiter is better than my ordering from wait staff because: _____.
- What types of activities would you like to do on the computer while you wait for your food?
- What problem do you see with ordering your food on a computer?

a customer might make to the product to improve it or find alternative uses for the product that might not have been considered by in-house designers. For example, the product manager of Febreze thought this new product would be used mainly by smokers; instead, it is used mainly by families with small children and pets. College students seem to be big users, too. Open-ended questions regarding usage of odor eliminators might have identified these benefits. Some managers feel the most useful information comes from open-ended questions.

Specific Attribute Questions   A manager also wants an understanding of how specific attributes or features are perceived. Questions should be used to probe which attributes and features contribute to purchase intention. **Specific attribute questions** also can be used to determine which features or attributes could be improved upon before freezing the product design. Attitudes about the importance of features can be obtained as well. Some examples of attribute diagnostic questions for the digital waiter concept test can be seen in Table 5.4. Appendix A at the end of the chapter,

Table 5.4   Attribute/Feature Diagnostic Questions

Rate the digital waiter on the following aspects (Likert scale: poor—excellent):
- Accuracy when making special orders.
- Speed when placing of food orders.
- Speed when requesting things I need for my meal, such as water or ketchup or more coffee.
- Security when processing payment.
- Speed when processing payment.
- Viewing pictures of dishes available.
- Usefulness in displaying important information about dish ingredients, especially for allergies.

Open-ended items
- Identify the features you like best about the digital waiter.
- Identify the features you like least about the digital waiter.

which has an example of a concept test for the TH!NK *City*, also shows a set of attribute diagnostic questions.

Profiling Variables   Reaching the right stakeholders is important for gathering the true reaction to the proposed product concept. **Profiling variables** refer to data about the respondents' demographics, purchasing behaviors, or possible resistance to adopt a new product. There may be multiple target markets or different types of stakeholders that will have different opinions about the new product, so it's critical to be able to identify the different segments. A set of questions is required to identify the different responses from different groups. For example, we would expect that college students would have no problem embracing a digital waiter, but it might not be as valuable to an older generation of senior citizens who don't interact with a computer on a daily basis. Demographic and psychographic (life styles, attitudes, values, opinions) questions are the most frequently asked profile variables.

To Include Pricing or Not   A question that frequently comes up with concept tests is whether the purchase price should be included as part of the concept description. Pricing information can dramatically influence the results of a concept test. If the price is within what is *typical* for a similar product, price is usually not included in the description. In this case, it is suggested that a separate question in the diagnostic section explicitly ask the respondent how much they expect to pay for the product described. If consumers' expectations are substantially different than the intended pricing, the development team may need to re-evaluate either the price or the product offering.

Step Six: Interpret and Report the Results

As previously noted, the outcome from the concept test should include diagnostic information, a preliminary forecast, and a positioning statement. In this section, we evaluate how to interpret the results from the study.

Diagnostic Information   As previously noted, the intent of the diagnostic questions is to determine how successful the product/service will be in the marketplace under its *current* design. This information is captured through the overall product diagnostic questions and the specific attribute questions. These questions combined with the probability of purchase question can paint a very telling picture of the potential success of the product. If the manager has included open-ended questions in the survey, this could provide great detail on the likes and dislikes of the consumer. The information gathered should be used to refine the product description that will be delivered to R&D/engineering. In this analysis, the manager should determine if consumers'/end users' needs and wants are correctly understood and should translate these desires into the product definition, and, if they have not been met, to recommend engineering design revisions. If a product receives a poor evaluation from potential buyers, it is likely best to send the idea back into the Discovery stage or even kill the idea altogether. At this point in the process, large resources have not been expended on the project and it is better to kill a potentially failed endeavor before serious resources are committed.

Using the Purchase Intention Questions   Research has shown that there is a strong correlation between how a respondent answers the purchase intention question and the actual trial of the proposed new product.[22] The responses to the top two choices, "definitely would buy" and "probably would buy," are combined to give an indicator of the trial success of a product/service. This is called the **top-two box score**. Concepts with high top box scores typically have higher trial rates. What is considered "high" is dependent upon the firm or industry as we further discuss below. When the purchase intention (PI) question is used in product test (a test of actual product, instead of the predevelopment proposed product), high top box scores are an indication of whether a new product merits launch, whereas the PI question in the concept test provides an indication of whether the product should go into development. This is an important distinction.

The relationship between top box response and actual trial is most prominent for durable goods compared to packaged goods.[23] Studies[24,25] have shown that high top box scores are correlated more with purchase (1) for existing products than for new ones, (2) for durable goods than for nondurable goods,[26] (3) for products than for services, (4) for shorter than for longer time horizons, and (5) for specific brands compared to product category levels.

There are general rules of thumb for what is considered a good top-two box score. For example, based on their experience with hundreds of brands, Taylor and colleagues[27] claim that a concept test for consumer packaged goods should receive 80 to 90% favorable answers (two top boxes percentages combined) before moving on to the next stage. However, the cutoff is highly firm- and industry-specific. The Zink mobile printer study showed a top-two box of 51%, which the company rated as "above average" compared to its benchmark. Each company will set their own benchmark. Research by Schwartz[28] also has shown that the cutoff for *only* the first top box (definitely would purchase) varies by product category. He provides these cutoffs: fragrances, 9%; detergents, 12%; on average all categories, 19%; food, 20%; cleaning products, 28%.

It should be kept in mind that purchase intention is not effective at measuring cannibalization, brand comparisons, or attribute preferences. **Conjoint analysis** is a better method for examining these types of issues. Conjoint analysis is a market research technique in which consumers choose between three or more hypothetical product offerings, each described by a set of attributes. With each respondent responding to 10 or more of these choice sets, the researcher can run an analysis to determine the individuals' preferences for the products features or attributes. Conjoint analysis lets the market research know which attributes are most important to potential buyers.

Sales Forecasts Based on Purchase Intent

Purchase intentions are routinely used to forecast sales of existing products and services. It appears that purchase intentions can provide better forecasts than a simple extrapolation of past sales trends.[29] From past experiments and experience in the product category, a manager can often translate responses to purchase intention questions into estimates of purchase probability. For example, Figure 5.6[30] illustrates the average relationship between **stated intentions** and actual purchase probabilities over the first six months of market experience. Stated intentions are what someone *says* they will do. Actual purchase probabilities indicate what they *actually* do. Note from Figure 5.6 that some people indicate they will *not* buy the product and then actually do. However, the percentages in Figure 5.6 should not be used blindly because there is a large variation across product categories, and six months does not typically capture the full trial potential. In the study that led to Figure 5.6,[31] the trial percentages varied from lows for cordless irons and stay-fresh milk to highs for home computers and pump toothpastes.[32] In a study of new services, it was found that the probability of actual purchase varied from a value of 0.45 for respondents answering "definitely would buy" to 0.05 probability of actual purchase for respondents answering "definitely would not buy" when the service was offered with promotion.[33] It is clear the "norms" for translating intent into purchase must be developed carefully within the specific market and launch practices.

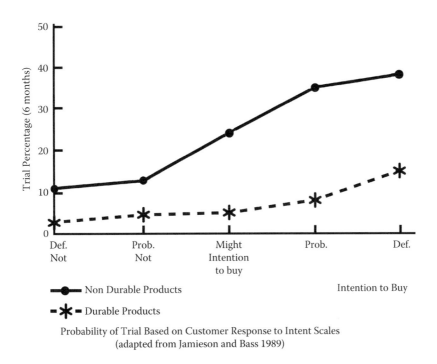

Probability of Trial Based on Customer Response to Intent Scales
(adapted from Jamieson and Bass 1989)

Figure 5.6   Purchase intent relationships. (From Endnote 30. With permission.)

Forecasting New Product Sales from Likelihood of Purchase Ratings

Responses to purchase intention questions are used to estimate forecasts of trial. An example will demonstrate how to best evaluate this data. Suppose for the mobile printer concept presented in Figure 5.1, hypothetically the following scores were obtained:

Definitely would buy: 19%
Probably would buy: 26%
Might/might not buy: 25%
Probably would not buy: 17%
Definitely would not buy   13%

These results provide us with a top-two box of 45% (19% + 26%), which might be acceptable for the company. Research has shown that stated purchase/trial intent does not always result in actual purchases.[34] If the company has several years of results from past concept tests. If we assume that Zink has found that 90% of respondents who marked "definitely would buy," 40% of those responding with "probably would buy," and 10% of those checking "might/might not buy" will actually purchase the product once it is available.[35] (This assumes that the customers are aware of the product and it is available to them.) Using a weighted average approach, we can adjust for this discrepancy that consumers don't always buy even though they say they will.

$$\text{Trial probability} = \beta_1 \, (\% \text{ definite}) + \beta_2 \, (\% \text{ probable}) + \beta_3 \, (\% \text{ mights}) \qquad (5.1)$$

such that:

$$0.30 = (0.90)(0.19) + (0.40)(0.26) + (0.10)(0.25)$$

Applied to the mobile printer case, this yields an overall estimate that 30% of the sample population would buy Zink's mobile printer if they were aware of it and the equipment was available. This score is considerably lower than the 45% based on an unadjusted top-two box rule.

While one cannot say for certain how intent-to-purchase scales will translate to actual purchase for a new product category, the above example provides some initial estimates. In each industry, studies of past products or managerial judgment can be used to derive the β coefficients in Equation (5.1). In some industries, such as packaged goods, market research suppliers have calculated norms for translating these measures across a wide range of products and sample screening procedures. Several market research companies specialize in collecting intentions data and offer proprietary models that use weighted levels of intent based on past studies as one input to forecast sales of new packaged goods (e.g., Nielsen BASES, GfK, Ipsos, and the Kantor Group, to name a few). Marketing research firms charge significant fees to companies who are interested in using historical databases of stated *intent to purchase* results compared to actual sales data. When possible, a new product team also should collect measures of satisfaction and need fulfillment to provide additional support to the decision on whether the product is likely to be a winner or a loser.

We can use the top box rule as well to calculate predicted sales. Sales volume with repeat volume can be estimated by looking at the projected size of the target market along with the top box scores. For example:

$$\text{Sales volume per household in time period} = MS*Tr*(1+(R*Occ*Units)) \quad (5.2)$$

where

MS = Estimated size of target market

Tr = % of household expected to try the product (weighted top box percentage)

R = % of triers who will repeat purchase (information obtained from surveys such as a concept test)

Occ = Number of occasions of repeat purchase within set period (information typically obtained from concept test)

Units = Expected # of units bought each purchase occasion (information obtained from concept test)

The purchase frequency and number of units per purchase occasion should be included as questions in the concept test so that this forecast estimate can be made.

We could use this same method for durable goods, which are typically only bought once per buying period. In this case R = 0. For example, if there were 25 million digital cameras expected to be sold in 2010,[36] MS = 25. If we use the 30% purchase intention calculated earlier, Tr = .30, and 1 printer is bought for a digital camera, this leads to expected sales of 7.5 million mobile printers annually. This assumes full awareness and availability for the product. With revenues of $80/unit and costs of $55, the expected profit would be $187.5 million. We will discuss further different forecasting methods in later chapters.

Creating a Positioning Statement

After evaluating the purchase intention questions, the product diagnostic questions need to be evaluated. Responses to specific diagnostic questions can be used to provide engineering with feedback on features or attributes that need to be refined prior to finalizing the product design. The specific attribute questions also can be used to conduct "**quadrant analysis**."[37] Mapping consumer responses to the attributes against each other can be used to create **positioning statements** for the new product. For example, after running the statistical analysis, we may find that convenience and ease of use are the most important attributes for consumers regarding our product, the digital waiter. We then can create the perceptual map as shown in Figure 5.7. If respondents rate the product within quadrants 1 and 4, design changes are required on one of the attributes. If the product falls within quadrant 3, it may require killing the product or at least redesigning it. Products falling into quadrant 2 are viewed

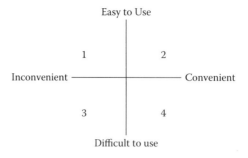

Figure 5.7 Quadrant analysis map 1.

favorably by respondents and are ready to enter the next stage (at least based on these two attributes). Similar quadrant analysis maps can be created for other attributes.

The specific attribute questions also can be used to create a positioning statement for the new product. Figure 5.8 shows how the digital waiter may rate compared to other restaurant ordering options. We see that the digital waiter is viewed as not being as easy to use as compared to ordering from wait staff, but it is more convenient than either self-service counters or using wait staff. "Convenience" is the positioning that can differentiate the digital waiter from its competitors. We discuss the creation and use of perceptual maps in later chapters.

Concerns with Concept Tests

Prior to sinking valuable resources into the research and development process, the concept test is the final test that validates that the proposed product concept is a winning solution with consumers/end users. A concept test, however, is not immune from making wrong judgments. For example, when Clorox® Anywhere® Hard Surface™ Daily Sanitizing Spray went through a concept test, it received high purchase intention ratings. Yet, in test markets, it sold far fewer units than was forecast. In reviewing what went wrong, it was found that the concept description presented to consumers was well received; however, when it came to actually purchasing the product, the benefits seemed implausible. Consumers wondered how Clorox could produce a product that could be used anywhere—on a baby's high chair tray, around pets—and not be

Figure 5.8 Perceptual map for digital waiter.

toxic? Unbelievability in the product claims led to fewer sales at launch than projected. When really new products are being developed, it is particularly important that qualitative studies also accompany the quantitative concept test to ensure that consumer latent concerns about a product are revealed.

Concept tests for radically new products can be complicated as well. Consumers may not be familiar with radically new technologies or services—its benefits and shortcomings. TH!NK faces this situation. Although most consumers recognize the environmental benefits of an electric vehicle, they are typically unaware of how battery life may impact maintenance of the vehicle. Life-style changes are difficult to explain using a product description. The upside of using concept tests with really new and radical innovations is that there is a school of thought in market research that argues that consumers can't always articulate what they need and want. For example, through the late 1970s and into the 1980s, microcomputers, now called personal computers (PCs), were developed for household use with software for personal productivity, programming, and games. However, the average household had little understanding why microcomputers would be useful to them. Showing a consumer a concept description can help them to visualize how the technology could be used, and provides them with an opportunity for insightful feedback for improving new technology.

Concept tests make sense when customers can adequately judge the value of the product. This occurs when the product is tangible or can easily be described. There are a number of situations when concept tests will not give adequate results:

- When the prime benefit is a personal sense, such as smell or taste.
- When an innovation is so radical that the consumer does not understand its value.
- When concepts embody new art and entertainment such as movies, books, and TV shows.
- When the product asks consumers to move from a comfortable status quo, such as moving from personal banking tellers to ATMs.
- When consumers are not aware of a problem, such as the initial introduction of personal computers. No one knew what to do with the "over-designed" typewriter.

Concept tests also can give misleading results. "For example, consumers may systematically overstate their intentions to purchase socially desirable products (e.g., healthy products, recyclable products) and may systematically understate their intentions to purchase socially undesirable products (e.g., cigarettes, high-fat foods, etc.)."[38] Concept tests are not foolproof, but are useful for obtaining consumers' initial reactions to a new product.

Chapter Summary

In this chapter, we have introduced the concept test, which is a voice of the customer survey typically conducted during the scoping or build the business case stage of the new product development (NPD) process. There are several goals of a concept test:

(1) to determine that consumers'/end users' needs and wants are correctly understood and translated into the product definition, and, if not, to recommend engineering design revisions; (2) to produce a preliminary sales volume forecast; and (3) to create a positioning statement for the new product. If the company wants to develop a positioning statement, it can collect information about competitors' products to create a perceptual map. Perceptual maps are considered in detail in the next chapter.

Glossary

Attributes: Refers to the core that allows the basic functioning of the product/service, i.e., a jackknife compared to a full feature Swiss Army knife.

Concept description: A product idea shown to the customer in the form of a model, a set of drawings, a storyboard, a specifications sheet, a mocked-up print/video advertising, or possibly a virtual prototype.

Concept Test: A marketing research survey that queries potential users on the probability of buying a new, not-yet developed product as explained in the concept description. It may include other questions about likes and dislikes on the product idea.

Conjoint analysis: Market research technique in which consumers choose between three or more hypothetical product offerings, each described by a set of attributes. A useful technique when the importance of an attribute is of interest or price elasticity is important.

Features: Refers to items that provide added capabilities to a product or service, i.e., a full-featured Swiss Army knife compared to a jackknife.

Full proposition: Combination of words and visuals that provide the concept description in a concept test.

Information acceleration: Using the power of the Internet to simulate information to a potential buyer of a new product.

Intent to purchase: See purchase intention question below.

Overall Product Diagnostic Questions: Survey questions asked in the concept test that get at respondents' perception of the overall value or perception of the product. It is a focus on the "whole" product/service compared to a specific feature or benefit.

Positioning statement[39]: A value proposition that identifies the target audience, the product and its category, a specific benefit, and its differentiability from the nearest competitive alternative.

Probability of purchase: The likelihood that a potential customer will buy the product as described in the concept test. Can be determined using the two-top box rule.

Profiling variables: Refers to data about the respondents' demographics, purchasing behaviors, or possible resistance to adopt a new product.

Purchase intention question: A question in the concept test used to determine probability of purchase or trial of the product/service.

Quadrant analysis: Evaluates the quadrants of a perceptual map to determine the best positioning option for a new product/service/brand.

Specific attribute questions: Questions in the concept test used to determine which features or attributes could be improved upon before freezing the product design.

Stated intentions: Consumers' statements whether they intend on buying a new product/service. This rarely equates to actual behavior. Stated intentions are typically inflated for environmentally friendly goods and deflated for "sin" products, such as cigarettes.

Voice of the customer (VOC): Strives to record in the customer's own words the benefits of a product or service; it is a description of the customer's needs within a specific product category.[40] Voice of the customer studies should be a combination of qualitative and quantitative research.

Review Questions

1. What are the goals of a concept test?
2. What are the different types of questions that should be asked in a concept test?
3. What is the difference between diagnostic attributes and specific attributes?
4. How can the results from a concept test be used to create a forecast of future sales?
5. How can the results of a concept test be used to create a positioning statement?
6. What are some of the downsides of conducting a concept test?

Assignment Questions

1. Create a concept test product description for your favorite automobile. Consider the target market. How would you change the concept test if you were conducting it in the United States versus Europe versus Asia? Be sure to include target sample population, size of sample, factual versus promotional. Give the concept test to five friends and see their purchase intention response.

2. Write a summary on how you would administer a concept test for:
   a. A new mobile phone
   b. Online banking service
   c. A top of the line refrigerator
   d. A new movie
   And then indicate if you would include price or not in the survey questions. Why?

3. Choose an industrial product and write a product description of it using only words, and then using words and visuals. Then choose a consumer product and complete the same task. What are the differences in the two descriptions?

4. Take the latest tablet (such as the iPad) and come up with a list of diagnostic attribute questions that you might want to ask a new user. Next, create a list of specific attribute questions.

5. Calculate an early forecast for a new line of sneakers from Nike. In a concept test they conducted, the top box (definitely would purchase) was 22%, the second box (probably would purchase) was 37%. Nike has historically 53% repeat purchase and most customers buy twice a year. Defend any assumptions that you make.

Appendix: Concept Description Example for Th!nk Electric Vehicle

TH!NK *City* is a modern urban electric vehicle with zero local emissions and an energy efficiency three times that of a traditional combustion engine car. A choice of sodium or lithium batteries allows the vehicle to accommodate your driving style, traveling up to 112 miles in one charge, with a top speed of 62 miles per hour.

TH!NK *City*'s body is made of recyclable acrylonitrile butadiene styrene plastic, designed for city driving, ideal for anyone who would like to avoid visible scratches and little dents. The dashboard can be completely recycled. The fabric, body, supports, air ducts, adhesives, and fixings are designed using the same recyclable materials. The plastic bodywork and other plastic panels are unpainted, reducing both energy consumption and toxins, while also making the panels easier to recycle. The batteries are returned to the supplier at the end of their useable life. The car is equipped with ABS brakes, airbags, and three-point safety belts with pretensioners. The frame is designed to absorb energy and distribute it away from the passenger's compartment.

The car's interior is designed to give the driver a pleasant environment, where the emphasis is on providing a clear view of the traffic. The interior is spacious, with instruments that are easy to read. The glass rear hatch allows the driver to see right down to the bumper of the car behind when reversing making parallel parking easier.

TH!NK City Specifications:

Number of seats: Two (two rear seats including three-point seat belts as an optional extra)
Number of doors: Three, including rear hatch
Front-wheel drive system
Plug-in electric power train
Colors: Black Jungle, Blue Sky, Red Energy, City Citrus

Standard Equipment:

Two front airbags   Central locking
Disable key for passenger airbag   4 kW electric heater
ABS brakes   Electric windows and mirrors
Regenerative brakes   Two loudspeakers and aerial
Power steering

Extra Equipment:

Electrically heated windscreen   Bluetooth
Full length sunroof   Alloy wheels
Radio/MP3   Winter tires
Radio/CD with MP3   Steel wheels
Radio/CD

Purchase Intention Questions

Based only on the product concept statement for the TH!NK *City*, how likely are you to purchase this vehicle? (Choose one)

○ Very Likely   ○ Somewhat Unlikely
○ Likely   ○ Unlikely
○ Somewhat Likely   ○ Very Unlikely
○ Undecided

When do you next expect to purchase (or lease) a vehicle? (Choose one)

○ Never   ○ Within next two years
○ Within next month   ○ Within next three to four years
○ Within next six months   ○ Five years or more
○ Within next year

Endnotes

1. Online at: http://www.think.no/think/Press-Pictures/Press-releases/Think-Announces-U.S.-Factory-Plans (accessed July 1, 2010). The company is no longer operating as of July 9, 2013.

2. Zipcar is a car-sharing program. Online at: http://www.zipcar.com/ (accessed June 10, 2013).

3. Online at: http://money.cnn.com/magazines/business2/business2_archive/2007/08/01/100138830/index.htm (accessed June 10, 2013).

4. For additional information regarding concept tests, refer to: Charles Merle Crawford and C. Anthony Di Benedetto, *New Products Management* (Nokia, India: Tata McGraw-Hill Education, 2008); Robert G. Cooper, *Winning at New Products: Accelerating the Process from Idea to Launch,* 3rd ed. (New York: Basic Books, 2011); Robert Dolan, *Concept Testing,* HBS 9-590-063; Jay L. Weiner and Barry Zacharias, *Pricing New-to-Market Technologies: An Evaluation of Applied Pricing Research Techniques, Ipsos Insight,* Ipsos Novaction & Vantis.

5. Robert G. Cooper, *Winning at New Products: Accelerating the Process from Idea to Launch*, 3rd ed. (New York: Basic Books, 2011).

6. This is from a report by Godfrey Research, copyright 2007. "… this report may not be reproduced in any form without written permission of ZINK Imaging, Inc., Bedford, MA."

7. Ibid.

8. Nielsen manages one of the most well-known consumer panels (http://en-us.nielsen.com/), although many others exist.

9. Karl T. Ulrich and Steven D. Eppinger, *Product Design and Development* (New York: McGraw Hill, 2004), p. 148.

10. Ely Dahan and Haim Mendelson, "Extreme-Value Model of Concept Testing," *Management Science* 47(1) (Jan. 2001): 102–116; V. Srinivasan, William S. Lovejoy, and David Beach, "Integrated Product Design for Marketability and Manufacturing," *Journal of Marketing Research* (1997): 154–163.

11. Ely Dahan and V. Srinivasan, "The Predictive Power of Internet-Based Product Concept Testing Using Visual Depiction and Animation," *Journal of Product Innovation Management* 17(2) (March 2000): 99–109, (11).

12. Online at: http://www.autobloggreen.com/photos/thinks-u-s-plans-march-2009/1426719/ (accessed June 30, 2013).

13. Based on Robert Dolan, Concept Testing, *Harvard Business Review* 9-590-603.

14. Photo source from http://sbaggers.de/main-ger/?sid-hom (accessed June 30, 2013).

15. Gavin Lees and Malcolm Wright, The Effect of Concept Formulation on Concept Test Scores, *Journal of Product Innovation Management* 21(6) (November 2004): 389–400, (12).

16. Online at: "http://jual-gamespc.blogspot.com/p/blog-page/16.html. BMW M3" http://jual-gamespc.blogspot.com/p/blog-page/16.html. BMW M3 is a registered trademark of BMW AG.

17. Glen L. Urban, John R. Hauser, William J. Qualls, Bruce D. Weinberg, Jonathan D. Bohlmann, and Roberta A. Chicos, "Information Acceleration: Validation and Lessons from the Field," *Journal of Marketing Research* 34(1) (1997): 143–153.

18. Ibid.

19. Dahan and Srinivasan, "The Predictive Power of Internet-Based Product Concept Testing."

20. L. Ha, "Advertising Appeals Used by Services Marketers: A Comparison between Hong Kong and The United States," *Journal of Services Marketing* 12(2) (1998): 98–112.

21. Rensis Likert, "A Technique for the Measurement of Attitudes," *Archives of Psychology* 140 (1932): 1–55.

22. Dolan, "Concept Testing."

23. Manohar U. Kalwani and Alvin J. Silk, "On the Reliability and Predictive Validity of Purchase Intention Measures," *Marketing Science* 1(3) (Summer 1982): 243–286.

24. Vicki Morwitz, Joel Stickle, and Alok Gupta, "When Do Purchase Intentions Predict Sales?" Working paper (2006). Online at: http://ssrn.com/abstract=946194 (accessed July 1, 2011).

25. C. Robert Newberry, Bruce R. Klemz, and Christo Boshoff, "Managerial Implications of Predicting Purchase Behavior from Purchase Intentions: A Retail Patronage Case Study," *Journal of Services Marketing* 17(6) (2003): 609–620.

26. Durable goods are "tangible goods that normally survive many uses" and nondurable goods (packaged goods) are "tangible goods that normally are consumed in one or a few uses." Philip Kotler and Kevin Lane Keller, A Framework for Marketing Management, 5th ed. (Upper Saddle River, NJ: Prentice Hall, 2011).

27. J. Taylor, J. Houlahan, and A. Gabriel, "The Purchase Intention Question in New Product Development: A Field Test," *Journal of Marketing* (January 1975): 90–92.

28. D. Schwartz, *Concept Testing* (New York: AMACOM, 1987).

29. J. Scott Armstrong, Vicki G. Morwitz, and V. Kumar, "Sales Forecasts for Existing Consumer Products and Services: Do Purchase Intentions Contribute to Accuracy?" *International Journal of Forecasting* 16(3) (2000): 383–397.

30. Glen L. Urban and John R. Hauser, *Design and Marketing of New Products*,. 2nd ed. (Upper Saddle River, NJ: Prentice Hall, 1993).

31. Linda F. Jamieson and Frank M. Bass, "Adjusting Stated Intention Measures to Predict Trial Purchase of New Products: A Comparison of Models and Methods," *Journal of Marketing Research* (1989): 336–345.

32. For other studies, see Alvini J. Silk and Manohar U. Kalwani, "Measuring Influence in Organizational Purchase Decisions," *Journal of Marketing Research* (1982): 165–181; Donald G. Morrison, "Purchase Intentions and Purchase Behavior," *Journal of Marketing* (1979): 65–74.

33. William J. Infosino, "Forecasting new product sales from likelihood of purchase ratings." *Marketing Science*, 5(4) (Autumn 1986): 372–384. Special issue on consumer choice models.

34. V. G. Morwitz, "Methods for Forecasting from Intentions Data," in *Principles of Forecasting: A Handbook for Researchers and Practitioners*, ed. J. Scott Armstrong (New York: Springer, 2001), pp. 33–56; J. S. Armstrong, V.G. Morwitz, and V. Kumar, "Sales Forecasts for Existing Consumer Products and Services: Do Purchase Intentions Contribute to Accuracy?" *International Journal of Forecasting* 16(3) (2000).

35. These numbers are hypothetical in order to demonstrate the technique. They are not meant to reflect an average for any company.

36. Online at: http://cameras.about.com/b/2010/03/04/report-25-million-digital-cameras-to-be-sold-in-u-s-in-2010.htm (accessed July 3, 2013).

37. Dolan, "Concept Testing."

38. Morwitz, Steckel, and Gupta, "When Do Purchase Intentions Predict Sales?"

39. Online at: http://marketinghighground.wordpress.com/2011/03/04/what-is-a-positioning-statement/ (accessed July 3, 2013).

40. Abbie Griffin and John R. Hauser, "The Voice of the Customer," *Marketing Science* 12(1) (1993): 127.

PERCEPTUAL MAPS

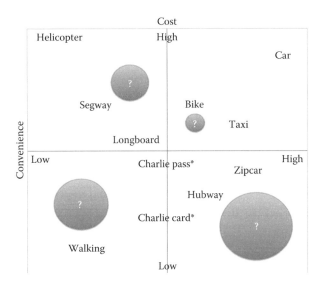

Learning Objectives

In this chapter*, we will address the following questions:

1. What is the difference between a determinant gap map and an attribute ratings gap map?
2. How are the different types of maps used?
3. How can a firm use a perceptual map in positioning new products?
4. How does a value map differ from a perceptual map?
5. When might a firm use a snake plot to map attributes?
6. What is the difference between a feature, function, and benefit?

Positioning Palm's Smartphone, The Pre

At the industry-anticipated 2009 Consumer Electronics Show (CES) in January 2009, Palm introduced the Pre, Palm's new smartphone. At the time of the introduction of the Pre, Apple's iPhone and Blackberry's Storm & Bold had the largest

---

* Based on Urban and Hauser, Design and Marketing of New Products (Upper Saddle River, NJ: Prentice Hall, 1998).

market share of smartphones due to their state-of-the-art touchscreen technology and ability to allow on-the-go connection to the Internet. Smartphones need to be as versatile as desktop computers and to easily connect to the Internet. It is reported that there are some 4 billion mobile phones in use around the world, and, in 2009, Palm reported that only about 10% of them are smartphones.[1]

Palm realized that users wanted to sync their mobile phone applications with their computer and with Internet applications, such as Facebook, LinkedIn, and Twitter—social networking sites. Palm saw this as an opportunity to provide a better smartphone. But, to do so, Palm would have to compete with Apple's iPhone and Blackberry's line of popular offerings. Supported by nearly 50,000 apps, more than 21 million iPhones had been sold by January 2009, making it very versatile and attractive to consumers. Due to Apple's brand name and market presence, iPhone was the primary product that Palm needed to compete against.

Palm had been successful with its personal digital assistant (PDAs) line. Their first hit was the Pilot, which pretty much created the PDA category. It enabled people to organize all their personal information on a computer and then synchronize this data with the portable PDA. The Palm Treo combined a PDA with a mobile phone to compete in the smartphone category. Jon Rubinstein, executive chairman of Palm, saw Palm's capabilities to easily synchronize to other PC applications as a major competitive difference. "We looked at Palm's DNA and said, 'What made it great?' Synching—from day one [of development], Palm has been about synching." He says, "People keep their data all over the place. It's no longer spread all over their computer. It's spread throughout the [Internet] cloud." Millions of bytes of data are stored on the Internet: e-mail, pictures, videos, and social networking information. The problem is that all these data streams are increasingly hard to manage. The promise of the Pre's Web OS is that it can take all these feeds and wirelessly combine them into one comprehensive contact list, without duplicates. On the Pre, this is known as Synergy."[2]

The Pre functioned as a camera phone, a portable media player, a GPS navigator, and an Internet client (with text messaging, e-mail, Web browsing, and local Wi-Fi connectivity) demonstrating good versatility compared to other smartphone offerings. It allowed multitasking and seamless integration of virtually every online, social, and information-management application. In a head-to-head review of the iPhone 3GS and the Palm Pre, *Consumer Reports* (the magazine) had found that the Pre and the BlackBerry Storm bested the iPhone 3GS in messaging, while the Pre was considered a "superior multitasker" with its card deck handling of multiple applications. Additionally, Pre's hand gesturing functionality, which seemed to replicate iPhone-like technology, wowed the industry crowd.[3] As of mid-2009, it had appeared that Palm had been successful in positioning its new product in a competitive space. Figure 6.1 demonstrates a hypothetical perceptual map for smartphones in 2009, including the Pre.

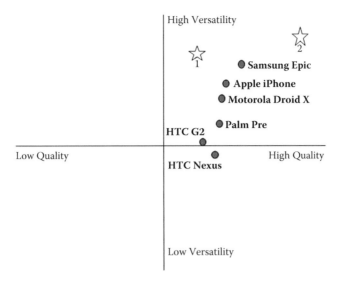

Figure 6.1 Perceptual map for smartphones. (From Endnote 4.)

Introduction

As we have seen in previous chapters, the unifying concept in the design process is the core benefit proposition (CBP) that helps to guide the firm's efforts to build quality products that create value for customers. However, new products are not introduced in a vacuum as we have seen with the Palm Pre. Palm had to understand intimately not only its own products, but also those of its competitors. In this chapter, we present perceptual maps as a tool for understanding the relationship of the CBP in reference to competitors' offerings and how a firm's positioning of a product can impact customers' choice processes. We also outline technical methods for creating perceptual maps in order to facilitate positioning of a new product in a competitive marketplace.

Customers Buy Based on Perceptions

On any day when you turn on the TV, read the newspaper, or pick up a magazine you find many claims. You might turn on the television and notice a commercial image of a beautiful young woman in a flowing dress crossing a bridge over the Seine River. In the background is the haunting melody of an aria. Your mood is that of an elegant, romantic trip to Paris, a mood the sponsor wants you to attribute to Veuve Clicquot champagne, a mood you may not have otherwise attributed to low-end sparkling wines. Flipping channels to another program, you notice that Palmolive dish soap gives you that "pure and clean, without unnecessary chemicals," and that Glad bags are "strong and thick, with lots of room." Turning off the TV, you decide to catch up on your *Wired* magazine and find that Toshiba has "the world's most powerful" gaming laptop, that Fujitsu offers portable scanners that are "faster and more advanced than ever," or that Microsoft offers software that operate at the "speed of light." In *Business Week,* you notice that British Airways offers "luxury for less." Of course, you might be successful enough in your career to deserve your own private corporate jet, in which case you might choose a Gulfstream G150 for having "the longest range at the fastest speed." More down to earth, the Bank of America is the "bank of opportunity," whereas Franklin Templeton Investments suggests that you "gain from our perspective."

Positioning is critical for a new product. Not only must a new product deliver the benefits the customer needs, but it must do so better than the competition. When it was initially introduced, Sure deodorant by Procter & Gamble was positioned as drier than other deodorants with a commercial message that encouraged consumers to try Sure under one arm and their current favorite deodorant under the other arm. Procter & Gamble was "sure" that consumers would notice the difference. By placing free samples of the product in the hands of a large percentage of households, the introduction succeeded because the product delivered superior dryness as promised to those trying the new product. It would have been difficult for Sure to have succeeded with a "me too" parity position in the crowded deodorant market. It was much better to surpass existing products in both the CBP and the fulfillment of the CBP. In order to do this, Procter & Gamble understood what benefits and features register with consumers.[5] Understanding how to position a product so that its value proposition is clearly conveyed is essential for the success of a new product. In the previous chapter, we used a concept test to gather information about customers' perceptions of a new product. In this chapter, we learn how to use this data to establish a positioning statement for the product.

Benefits and Value

The new product manager must be able to convey to customers what value and benefits a new product offers compared to the competition's offerings. In your introductory marketing class, you learned that consumers buy products and services based on perceived benefits. But benefits are often expensive to deliver, and, because of the resulting price, expensive for the customer to purchase. The related concept of **value** is one of "benefits versus price paid." Firms strive to provide value to their customers so that they are willing to pay the price to obtain the benefits from a product offering. In this chapter, we discuss two tools to assist in benefit/value positioning: perceptual maps and value maps.

Perceptual maps visually summarize dimensions (also called factors), such as "ease-of-use," "versatility," "ergonomically pleasing," that customers use to perceive and judge the firm's and competitors' products and services. The figure after the chapter title provides an example of a perceptual map for the smartphone market of 2010.[6] To create meaningful maps, we must know the number of dimensions, the names of those dimensions, which customer needs make up the dimensions, where competitors are positioned, and where competitors don't perform well on these dimensions, thereby opening a possible entry point for a new product. By focusing on key benefits and forcing an explicit comparison to competition, perceptual maps provide a method for ensuring that the CBP provides a differential advantage.

Value maps augment the information in perceptual maps. Value maps portray the benefit versus price tradeoff by displaying products on a "benefit per dollar" basis. For example, Apple's iPhone might have a position as being cool, easy to use, and ergonomically attractive, but Palm's Pre sells well because, in some cases, customers perceive it as providing better value for the price.[7] Similarly, private label aspirin might be perceived as a better value than a branded aspirin, such as Bayer. Consumers feel they are getting the same product at a lower price, thus, getting better value. Value includes a dollar assessment so if product "A" delivers four units of "effectiveness" for a price of $2, then it delivers two units of "effectiveness per dollar." Another product might deliver three units of "effectiveness" for $1 and be perceived as a value leader because it delivers an additional unit of "effectiveness per dollar" compared to product A. Value maps provide a discipline to help recognize the interrelationship between benefits and price within a given market segment. When used in conjunction with perceptual maps, value maps provide important strategic information. We first turn to perceptual maps and examine their role in providing the key managerial concepts and then discuss their evolution into value maps.

Perceptual Maps

Perceptual maps are used to display the position of a product on a set of primary **factors**. Factors are the features, functions, and benefits that describe a product or service. Table 6.1 provides a typology of these factors. Figure 6.2 is a perceptual map that displays brands of pain relievers on the factors (benefits) of "effectiveness" and "gentleness." "Effectiveness" reflects more detailed needs of strong, fast, long-lasting relief and the ability to make headache pain go away fast. "Gentleness" represents perceptions that the product would not upset one's stomach, causing heartburn, or result in a nervous jittery feeling. Tylenol is the most gentle on the stomach, relative to other brands. There appears to be a positioning opportunity for a CBP of gentle and effective as shown in the area labeled "gap." If a product and marketing campaign were developed to make and fulfill credible claims on both gentleness and effectiveness, a unique position could be achieved. Indeed, Extra Strength Tylenol is a product that was introduced in an attempt to fill this gap. There is also a gap at high Gentleness, but poor Effectiveness. This is certainly not a place that a manager would want to position a product. It is important to note that the size of the gap is *not* an indication of the size of the market.

Perceptual mapping is not restricted to frequently purchased consumer goods. Services and industrial products also can be represented by perceptual maps. Figure 6.3 shows consumers' perceptions of transportation services along three primary needs: "speed-and-convenience," "expense," and "psychological comfort."

Table 6.1 A Typology of Factors Describing Products or Services

PRODUCT FACTORS		
FEATURES: A PRODUCT/SERVICE CAPABILITIES THAT SATISFIES A SPECIFIC USER/BUYER NEED.	*FUNCTIONS:* HOW A PRODUCT/ SERVICE WORKS	*BENEFITS:* VALUE RECEIVED FROM A PRODUCT/SERVICE FEATURE OR FUNCTION
• Ingredients • Services • Esthetic characteristics • Performance • Trademarks • Components • Materials • Price • Warranty	• Esthetics • Ergonomics • Process • Mechanical • Software application • etc.	• Uses • Savings (time, effort, energy) • Sensory enjoyment • Nonmaterial well-being • Economic gain • Socially responsible • Health focus
Examples: Slim line, HD interface, 20 oz. weight, unlimited calling, open 24 hours	Examples: Sprays on, twists, pinch gesture for smartphones, first-in-first-out, capacitive stylus	Examples: Convenience, reputation enhancing, cost savings, cholesterol reducing

Source: Endnote 8.

1. Speed-and-convenience reflects the ability of a mode of travel to provide on-time service that gets consumers quickly to their destinations without a long wait.
2. Expense includes cost of using a transportation method, including annual maintenance.
3. Psychological comfort includes attributes such as relaxing, no worry about assault or injury, and no annoyance from others.

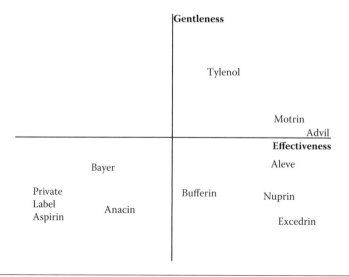

Figure 6.2 Perceptual map of pain relievers.

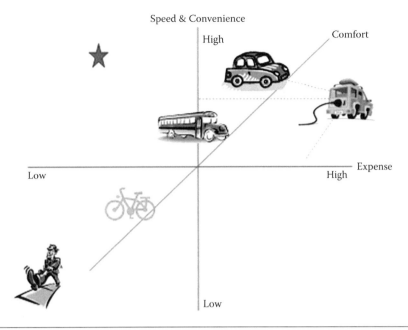

Figure 6.3 Perceptual map of transportation options. (From Endnote 9.)

If you were a transit manager or community planner trying to increase utilization of public transportation, Figure 6.3 suggests that you would need to drastically improve consumers' perceptions of public transportation with respect to both psychological comfort and speed-and-convenience. You might try to modify the existing bus system or introduce a new type of service that is quicker, more convenient, and easier to use.

If you were a manufacturer of electric vehicles (EVs) or a community planner trying to increase utilization of fuel efficient vehicles, Figure 6.3 suggests that you would need to drastically improve consumers' perceptions on two dimensions of the EVs: expense and speed-and-convenience. On the third dimension, comfort, the EV rates better than other eco-friendly transportation. You might try to modify the existing EV technology or introduce a new type of recharging service that is quicker, more convenient, and easier to use.

Product positioning is also relevant for industrial products as illustrated for solar-powered air conditioners in Table 6.2, which lists the needs associated along two dimensions: "risks" and "benefits." In industrial buying, there is often a further managerial consideration that more than one person may be involved in the buying process. The perceptual structure may differ for different groups of vested users: plant managers, production engineers, top managers, and heating–ventilation–air-conditioning (HVAC) consultants. In this case, a good product positioning must balance the needs of the four interest groups.

Table 6.2   Primary Needs for Air-Conditioning Systems as Perceived by Corporate Engineers

RISKS	BENEFITS
Reliability	Reduced Pollution
Field tested	Energy savings
Development cost	Protection
Noise level	Modernness

Types of Perceptual Maps

There are three types of perceptual maps: determinant gap map, attribute rating gap map, and overall similarities gap map. We discuss these different types of maps next and provide guidelines as to when they are most useful. The determinant gap map is based on qualitative analysis, whereas the other two types of maps are based on quantitative analysis. Let's first discuss the determinant gap map.

Determinant Gap Map   Determinant gap maps are one type of perceptual map; instead of basing the product positioning based on **customer input**, **managerial input** is used to create the map. Determinant gap maps can be used to compare different products or brands against one another. Figure 6.3 demonstrates a determinant gap map based on studies conducted in the transportation industry with the goal of comparing EVs to alternative transportation choices. Each mode of transportation was scored on the three dimensions of "speed and convenience," "expense," and "comfort," and then plotted. It is at a manager's discretion where on the map the product or brands are placed. Managers are intimately familiar with product designs and competitors' and substitute offerings, and with this managerial knowledge can make the judgment calls necessary to build a determinant gap map.

In Figure 6.3, there are two evident gaps. Gap 1 exists for a transportation option with low expense, high-speed convenience, and high comfort. Car-sharing options, such as the Zipcar service in the United States, is looking to position itself in this gap. Car-sharing is referred as *autodelen* in Dutch, *autopartage* in French, *bildeling* in Danish, *auto condivisa* in Italian, *bilpool* in Swedish, and *car clubs* in the United Kingdom. There is evidently a need for this type of product/service worldwide. Gap 2 sits at high expense, low comfort, and low speed convenience. Naturally, this is not a space in which companies want to operate.

One way of determining which attributes to map is to use the affinity diagram technique, where the new product team sorts the customer needs themselves. While there are many variations in the methods by which a team of people can sort customer needs, a typical session might go something like the following. Suppose that we begin with 20 customer needs and 4 team members. Each customer need is written on a 3-in. × 5-in. index card or a Post-it® note and each team member sorts the customer needs into most important to least important from his or her own perspective. One team member begins by selecting a seed card from his or her pile, reading it aloud, and placing it on the table or the wall. The remaining team members select cards from their piles that they feel express a similar customer need, read them aloud, and place them next to the seed card. If the team agrees that the new need is similar to the needs in the pile, the card stays there; if not, a discussion ensues resulting in either a new pile or an agreement that the card is indeed similar. Piles continue to be generated through team discussion and, if necessary, new "seeds" are selected until all of the customer needs are sorted. The team then discusses the piles and imposes a hierarchy reflecting their judgment on how the detailed needs cluster to strategic (primary) needs. Titles are added to the groupings for clarification. Finally, the team reviews the structure, rearranging and retitling as appropriate. The final attributes chosen should be both differentiable and important so as to act as "determinant" factors in helping to position the product/brand in the competitive marketplace.

The advantages of a managerial sort are that it is quick, inexpensive, and feasible for all organizations, and it helps team members reach a consensus on the importance of customer needs. It's also possible to easily develop multidimensional maps when more than two factors are important. If a mode of transportation is speedy, it is typically also seen as convenient. EVs have the same speed as a traditional combustion engine, but EVs may not be as convenient because they currently need recharging every 100 miles or so. The disadvantage of managerial sort is that the map represents the management's viewpoint, but not that of the customer. The manager will need to have a very good understanding of competing products in the marketplace in order to make comparisons. The manager may be biased toward his/her own products and can make a product opportunity look more beneficial than it really is. When the consumer's perspective is important, attribute gap maps or overall similarity gap maps are useful. We discuss these next.

Attribute Rating Perceptual Map **Attribute rating (AR) perceptual gap maps** are based on input from the buyers and/or users of a product or service. Whereas **determinant gap maps** are based on the viewpoints of new product managers, AR perceptual maps are based on the marketplace perceptions of consumers. Both maps can complement each other as the former represents the perception of the manager and the latter represents the perception of the buyer or user. The first step in AR perceptual gap mapping is to identify a set of attributes, features, benefits, or functions that describe the product category being considered. Customers' perceptions of the available choices on each of these attributes are collected as described in the concept test survey in the previous chapter.

Figure 6.4 shows average customer ratings on 20 needs related to smartphones. These perceptions were obtained from students at a private university using a five-point Likert scale.[10] Figure 6.4 plots the average ratings. This "**snake plot**," so-called because a line connecting the ratings snakes down the page, allows you to interpret visually the perceived positions of competing products. This map indicates how

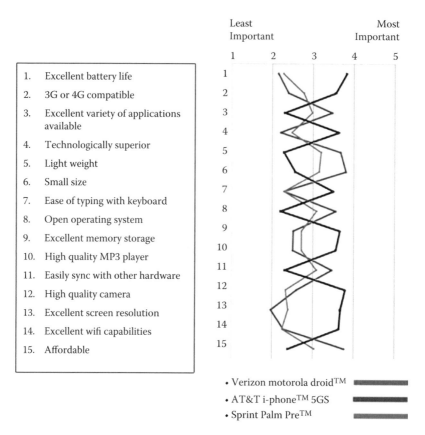

1. Excellent battery life
2. 3G or 4G compatible
3. Excellent variety of applications available
4. Technologically superior
5. Light weight
6. Small size
7. Ease of typing with keyboard
8. Open operating system
9. Excellent memory storage
10. High quality MP3 player
11. Easily sync with other hardware
12. High quality camera
13. Excellent screen resolution
14. Excellent wifi capabilities
15. Affordable

- Verizon motorola droid™
- AT&T i-phone™ 5GS
- Sprint Palm Pre™

Figure 6.4 Snake plot of perceptions on smartphone technology-based features.

college students in the Northeast region of the United States perceive different types of smartphone technology-based features.

Careful examination of Figure 6.4 indicates the detailed strengths and weaknesses of each mobile feature. However, it is difficult to identify with a snake plot a clear strategic positioning strategy. While the detailed list of customer needs will certainly influence the adoption of any given technology-based feature, Figure 6.4 is difficult to interpret. However, we can see that the AT&T i-phone is rated of high quality, but not very affordable, and the Droid is seen as easy to use and lightweight.[11] Snake plots are too messy for easy strategic positioning analysis. Aggregation of the data is required to ease strategic planning, which is why marketing managers create maps based on the **underlying factors**, which are a set of attributes (usually 4–5) that summarize the list of features. In this example, one factor could be "quality" and another "versatility." (Can you identify other factors?) The factors are then used to create an attribute rating perceptual map, such as the one shown in Figure 6.1.

A statistical technique, "factor analysis," is used to reduce the list of features in a small number of meaningful attributes. In developing the attribute rating perceptual gap map, we use a data reduction technique. In data reduction, factor analysis consolidates a large number of variables by looking at the interrelatedness of the factor and explains these variables in terms of their common underlying factors. The goal is to reduce the variables into a few dimensions that can be used to more easily explain the data. For example, using this method, the 15 features as listed in Figure 6.4 were grouped into six primary needs: quality, versatility, ease-of-use, affordability, open source, and technical strength. Factor analyses can be run on one of several statistical packages (SPSS, SAS, Systat, Statgraphics, etc.).

While customers may evaluate products on a larger number of needs, we expect that these evaluations can be summarized strategically by a smaller number of dimensions. Even for complex products, such as automobiles, we expect that customers remember the basic benefits of different attributes and summarize those attributes by that set of benefits. For example, for a sports car, there might be 10 to 15 strategic needs that are summarized by three basic benefits, such as "reliability," "style," and "performance." When selecting a strategic positioning vis-a-vis competition, we want to be sure that the core benefit proposition provides the strategic positioning and reflects the basic benefits that customers use to summarize their perception of products. Factor analysis seeks to identify these basic benefits.

Determining the Number of Factor Dimensions   You may be wondering how the number of factor dimensions is determined? A factor analysis computer program identifies factors by examining the matrix of correlations of the detailed attributes. It extracts factors one-by-one in order of variance explained, as shown in Figure 6.5. Thus, each

TOTAL VARIANCE EXPLAINED

FACTOR	INITIAL EIGENVALUES			EXTRACTION SUMS OF SQUARED LOADINGS		
	TOTAL	% OF VARIANCE	CUMULATIVE %	TOTAL	% OF VARIANCE	CUMULATIVE %
1	5.21	26.05	26.05	5.21	26.05	26.05
2	3.26	16.30	42.35	3.26	16.30	42.35
3	2.00	9.99	52.34	2.00	9.99	52.34
4	1.66	8.30	60.64	1.66	8.30	60.64
5	1.42	7.08	67.71	1.42	7.08	67.71
6	1.39	6.97	74.68	1.39	6.97	74.68
7	.957	4.79	79.47			
8	.819	4.09	83.56			
9	.730	3.65	87.21			
10	.574	2.87	90.08			
11	.531	2.66	92.74			
12	.344	1.72	94.46			
13	.310	1.55	96.01			
14	.238	1.19	97.19			
15	.016	1.07	100.000			

Extraction Method: Principal Component Analysis.

Figure 6.5   SPSS output for factor analysis. (From Endnote 12.)

successive factor explains less variance than the previous factor. In our example, factor 1 explains 26.05% of the variance in our data set, but factor 2 only explains 16.3% of the variance. The 15 factors taken together explain all of the variance. In our student survey on smartphones, we found six factors, explaining almost 75% of the total variance. It is up to the marketing analyst's judgment to decide how many factors are needed to explain enough variance so that the basic information is still retained, but in a simpler form. There are three classical rules for determining the number of factors: the "cumulative variance explained," the "scree" rule, and the "eigenvalue greater-than-one" (EGO) rule.

The **cumulative variance-explained rule** is the starting place to determine the number of underlying factors in the perception data. The general rule is that the factors retained should explain 70% or more of the **cumulative variance** of the attributes, although some managers use 80%, or even 90%, as the cutoff. In the SPSS output shown in Figure 6.5, we see that one factor explains 26.05% of the variance, whereas five factors explain 67.71% cumulative variance. Six factors explain 74.68% cumulative variance, thus, the cumulative variance-explained rule would indicate six factors underlie the 15 attributes.

Figure 6.6 illustrates the **scree plot**, which is a visual presentation of the variance explained. The **scree rule** uses the **eigenvalues** provided in Figure 6.5 to plot the amount of variance an additional factor explains. Where the next factor explains marginally less variance, the plot levels off. This leveling off is shown in Figure 6.6 after the fifth factor or sixth factor, which accords with the cumulative variance-explained rule where we retained 6 factors. The scree rule says to drop all further components after the one starting at the "elbow." The scree rule takes its name from geology where the scree is the pile of rock that accumulates at the bottom of a rock slide. As shown

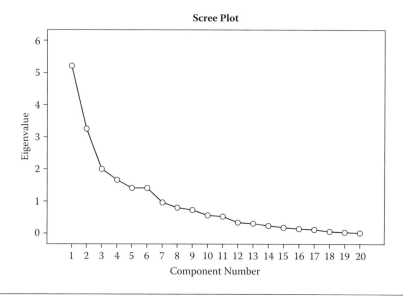

Figure 6.6   Scree plot. (From Endnote 13.)

in Figure 6.6, the elbow could be at factor (called component by SPSS) five, six, or maybe even seven. Researcher bias may be introduced due to the subjectivity involved in selecting the elbow.

The eigenvalue greater-than-one (EGO) rule is similar in concept to the scree rule. The eigenvalue for a given factor reflects the total variance in all the variables that is accounted for by that factor. For example, in Figure 6.5, the eigenvalue of 5.21 is the variance of the 15 variables that are accounted by factor 1. Because the attributes are standardized to have variance 1.0, a factor is no better than a single attribute if it cannot explain more variance than 1.0. Thus, if the eigenvalue of a factor drops below 1.0, the EGO rule suggests that it not be retained. In Figure 6.5, there are six factors that are greater than 1.0.

In practice, any one of the three rules described above can be used to select the appropriate number of factors, but the decision must be tempered by managerial judgment. It is good practice to consider retaining the number of factors indicated by a combination of the rules and to consider retaining slightly more or slightly fewer factors. The final selection should balance the variance—explained with the ease-of-interpreting the factors and the usefulness of each factor toward developing the perceptual map. In our case, the three rules were consistent in identifying six underlying factors; this is not always the case.

Identifying the Factor Dimensions   The determinant gap map in the opening vignette suggests that customers summarize smartphone technologies by the basic benefit dimensions of "versatility" and "quality." If we already knew these two factor dimensions, we would just look at the correlation between factors in order to decide which attribute scales align with which factor dimensions. We don't know this information before hand and, as Figure 6.4 demonstrates, it can be difficult to observe all factor dimensions directly. Instead, we use the factor loadings resulting from our factor analysis to identify the factors.

The computer program outputs a set of correlations, called **factor loadings,** that represent the correlations between the attribute variables (our list of features in our survey) and the underlying factors that the algorithm identified. Figure 6.7 shows the **rotated factor loading matrix** for our smartphone data from SPSS. The columns define the factors (or components as labeled by SPSS); the rows refer to how a variable loads on each factor. The loadings can be interpreted as the correlation coefficients and measure which variables explain a factor and to what degree. The square of the loadings multiplied by 100 represents the percentage of variance in the variable (row) explained by the factor (column). A negative factor loading represents a negative correlation with the factor. Because the factor loadings are squared to determine variance explained, negative loadings have the same interpretation as positive loadings. To determine if a variable is "loading" on a factor, we use a cutoff of 0.4. Any variable with a loading of absolute value of 0.4 or higher is said to be correlated with that factor. For details on evaluating factor loadings, see Churchill and Iacobucci.[14] In Figure 6.7, we can see

FACTOR LOADING MATRIX[a]

	FACTOR					
	1	2	3	4	5	6
1. Excellent battery life	.267	.375	.647	−.269	−.255	.167
2. 3G or 4G compatible	.154	.106	.200	−.281	−.325	.514
3. Excellent variety of apps available	.565	.097	.173	.359	.147	.130
4. Technologically superior	.273	.244	.134	−.595	.333	−.355
5. Light weight	.124	.263	.721	.181	.203	−.255
6. Small size	.011	.534	.240	.086	−.317	.363
7. Ease of typing with keyboard	.337	.075	.469	.203	−.072	.240
8. Open operating system	.152	.377	.168	−.310	−.045	.653
9. Excellent memory storage	.495	.221	.187	.272	.111	−.210
10. High quality MP3 player	−.201	.593	.185	.161	.322	−.273
11. Easily sync with other hardware	.604	.312	.139	−.302	.248	−.106
12. High quality camera	.053	.432	.366	.157	−.008	.050
13. Excellent screen resolution	−.231	−.149	.297	−.577	−.369	.180
14. Excellent wifi capabilities	.076	−.073	−.076	−.614	.242	−.062
15. Wearable	−.277	.288	.005	−.236	.783	.099

[a] These values are hypothetical and not based on any data set.

Figure 6.7 Factor *loading* matrix.

how the 15 attributes load on each of the six factors. Attributes 3, 9, and 11 all load on factor one. Attributes 6, 10, and 12 load on factor two. Following this technique, each attribute is assigned to one of the six factors. In this manner, the 15 attributes are reduced to 6 underlying factors, hence, the name *data reduction*.

The next step is to name each factor for ease of evaluation. By looking at Factor 1 in Figure 6.7, we name it *versatility* because of the attributes that load on it (i.e., "excellent variety of apps available," "excellent memory storage," and "easily sync with other hardware") and factor 2 can be called *quality* because of the attributes that load on it (i.e., "small size," "high-quality MP3 player," and "high quality camera"). Notice that *versatility* is the name that the new product team uses to describe this group of attributes; the attribute versatility is not directly evaluated by the customer. The name given to the factor is entirely up to the new product manager. Nonetheless, the name given via factor analysis should be a good strategic summary of how the customers view the features presented to them.

Producing the Attribute Rating Perceptual Map   Once the manager has decided on the number of factors and has named them using the factor loadings, the next step is to produce the perceptual map. The statistical software packaged computer algorithm uses the correlations among the attribute variables and the factor loadings to compute a set of "regression" weights, called *factor score coefficients*. These coefficients are used to compute the factor scores. The factor scores are estimated by multiplying the factor

FACTOR SCORE COEFFICIENT MATRIX

	FACTOR (COMPONENT)					
	1	2	3	4	5	6
1. Excellent battery life	.178	.072	.395	−.294	.222	−.128
2. 3G or 4G compatible	−.045	.019	.125	−.003	−.210	.255
3. Excellent variety of apps available	.376	.001	.040	.170	.143	.108
4. Technologically superior	.014	.031	.029	.422	.051	−.049
5. Light weight	.057	.027	.350	.032	−.170	.069
6. Small size	.008	.298	.030	.066	−.043	−.179
7. Ease of typing with keyboard	.118	.049	.278	−.031	.011	.066
8. Open operating system	.016	.013	.047	−.089	.133	.383
9. Excellent memory storage	.298	.146	.020	.182	.072	−.062
10. High quality MP3 player	−.063	.312	.006	.019	.124	−.092
11. Easily sync with other hardware	.325	.126	.052	−.058	−.120	−.002
12. High quality camera	−.107	.346	.115	.071	−.155	−.145
13. Excellent screen resolution	.015	.107	−.081	.283	.081	.039
14. Excellent wifi capabilities	−.015	−.007	−.002	.361	.119	−.185
15. Wearable	−.020	−.061	.008	−.052	.377	−.179

Extraction Method: Principal Component Analysis. Rotation Method: Varimax with Kaiser Normalization.

Figure 6.8 Factor *score* matrix.

score coefficients (Figure 6.8) times the attribute weightings and summing over attributes[15] (Figure 6.9). The factor scores (which are not the same as the factor loading matrix) for each product are then used to produce a perceptual map similar to the one seen in Figure 6.1. This map is called an attribute rating perceptual map because it uses consumers' responses to rate the attributes.

Estimating the Position of a New Product   After the factor scores have been plotted on the AR perceptual gap map, it can be used to determine the positioning for a new product. In examining Figure 6.1, the Pre is perceived as being of midquality and midversatility among the smartphones evaluated. If a new product team develops a product in the first starred (☆) position in Figure 6.1, a new product offering might be able to capture a significant share of the smartphone market at high versatility and midquality, presumably at a low price, in order to maximize value. Likewise, a new product opportunity could be at a high versatility/high quality positioning shown as the second starred position in Figure 6.1. Samsung has been working on moving its smartphone offering in this direction in order to compete with Apple's iPhone. By searching for the gaps within a perceptual map, a firm can determine where opportunities exist in the marketplace. Not every gap indicates market potential as in Figure 6.1; a firm would not want to introduce a product in the low versatility/low reliability area of the map if it wanted to be competitive. To develop creative strategies and the CBP, the new product team uses the perceptual map to internalize and visualize the market along its

DESCRIPTIVE MEANS/STD DEVIATIONS

	MEAN	STD. DEVIATION
1. Excellent battery life	3.41	0.875
2. 3G or 4G compatible	3.75	0.803
3. Excellent variety of apps available	3.34	1.066
4. Technologically superior	3.44	0.878
5. Light weight	3.44	1.045
6. Small size	4.50	0.803
7. Ease of typing with keyboard	3.44	0.982
8. Open operating system	3.81	1.203
9. Excellent memory storage	4.03	0.861
10. High-quality MP3 player	2.94	1.190
11. Easily sync with other hardware	3.50	1.136
12. High-quality camera	4.00	0.718
13. Excellent screen resolution	3.63	1.040
14. Excellent Wi-Fi capabilities	3.69	0.896
15. Wearable	3.75	1.016

Figure 6.9   Descriptive statistics (means).

primary dimensions. It then searches for a gap to meet its strategic goal; a systematic new product process helps us to reach that goal.

Factor Analysis Summary

Factor analysis has been used successfully in thousands of product categories. It is an effective technique to identify structure within a complex set of detailed attributes. To summarize, the steps to conduct a factor analysis include:

1. Obtain the underlying perceptions of consumers/end users by factor analyzing customers' ratings of product attributes across products.
2. Select the appropriate number of dimensions by combining analytical rules (e.g., eigenvalue cutoff, scree, and EGO) with good managerial judgment.
3. Interpret and name the factor dimensions by examining the factor loadings.
4. Estimate the positions of the existing products, the test products, and the concepts by using the factor scores or the factor score coefficients to create a perceptual map.
5. Examine the gaps in the map to determine the potential opportunities and examine the weaknesses and the strengths of the concepts to identify potential improvements.

Factor analysis is easy to use and has a proven track record. The behavioral assumption of how customers react to rating scales is intuitively appealing and, empirically, factor analysis usually leads to results with strong face validity. The major disadvantage of factor analysis is that it depends on a well-specified set of customer needs (attributes) and it can only identify structure within a set of needs, not beyond them. Furthermore, it is dependent on managerial interpretation (different individuals may see different opportunities) so it is beneficial to have several managers participate in analyzing the maps.

Identifying a New Dimension (Factor)

As discussed in the previous section, most products are evaluated within the perceived structure of the segment. Sometimes, however, a more revolutionary approach based on identifying a new dimension (factor) may be a better choice. If we identify a new dimension that is important to consumers and position uniquely upon it, the firm could potentially enjoy high market share and/or profits.

Identifying a new dimension is difficult. New dimensions are based on latent needs, i.e., needs that customers would view as important if they could be fulfilled. In many cases, a new technology fulfills these needs and makes them more salient to customers. In essence, the new technology "creates" the need. For example, in soft drinks, the use of saccharin and aspartame created the low calorie dimension. In 1963, the Coca-Cola Company helped to successfully expand the diet soda market with Tab, which was originally sweetened with saccharin. Prior to this time, diet sodas were mainly thought to be only for diabetics. Coca-Cola followed in the early 1980s with Diet Coke, which was originally sweetened with aspartame; it's now sweetened with NutraSweet. A latent need for low calorie soda opened up a multibillion dollar market. As another example, the English retail brand The Body Shop created a dimension of ecological responsibility in health and beauty aid retailing. At The Body Shop no products are sold that are tested on animals, all products are biodegradable and environmentally safe, and packages can be refilled at the store. This strategy positions The Body Shop well on the "environmentally conscience" dimension and differentiates them on the new sustainability dimension.

Innovative products may be an opportunity for a new positioning within the current structure, the identification of a new dimension, or an additional segment. When a firm sees no opportunities for improvement in the existing positioning structure, it can strive to add a dimension by identifying previously unmet existing needs. For example, when Anita Roddick started The Body Shop in the late 1970s "social responsibility" and "natural ingredients" were not associated with the cosmetics industry. Dame Roddick created this new dimension, which is now a standard in the industry. One caution here is that managers often think they are revolutionizing a category when consumers feel nothing but a mild perceptual perturbation. Customer reaction must be measured to determine what is really occurring. Minor product changes, though they may be intended to create a new dimension, are instead often reflected in the existing perceptual structure. For example, introducing a retractable hardtop on a car may not create a new dimension, but rather lead to a change of position on the sportiness dimension.

The factor analysis technique may be one way of identifying latent needs that could result in a new dimension for a product. An attribute may not be associated with any other attributes. In our example in this chapter, "wearable" loaded on factor 5 by itself (see Figure 6.7). It is possible that there is an opportunity to identify a unique positioning for jewelry with embedded smartphones. Thus, it is useful to examine carefully any attribute that does not load on any factor; it should not be neglected without further investigation. These unique attributes sometimes represent opportunities for identifying new customer needs. Using an overall similarities gap map, which we discuss next, is another method of identifying latent needs.

Overall Similarity Gap Maps and Other Mapping Techniques

In this section, we will evaluate a couple of other mapping techniques that may be useful to the new product manager. They include overall similarity gap maps[16] and value maps.[17]

Overall Similarity Gap Map

Both the determinant gap and the attribute rating gap maps are based on product or brand attributes. In some cases, new product managers want to enhance creativity by trying to look beyond the attribute structure and identifying opportunities that are not based directly on attribute ratings. One such technique is based on consumers' ratings of how similar/dissimilar products, concept, or brands are to/from one another. This technique can be used to develop **overall similarity (OS) gap maps**.

In creating an OS gap map, survey respondents are first asked to evaluate existing products and product concepts with respect to similarity. A technique known as *similarity scaling*, of which methodologies include discriminant analysis, principal component analysis, cluster analysis, and multidimensional scaling, is used to determine the similarity between variables. The data collection to develop Figure 6.10 asks customers to provide a metric measure of similarity. Customers can be asked to rank each product in relationship to each other product or to rank all pairs of products in terms of similarity. These measures are tabulated for each customer or averaged across customer groups to produce a proximity matrix whose entries represent the similarities or dissimilarities among the products. Similarity scaling obtains the underlying dimensions from respondents' judgments about the similarity of products. This is an important advantage as the underlying dimensions come from respondents' perceptions. It does not depend on researchers' judgments nor does it require a list of attributes to be shown to the respondents.

Consider three U.S.-based business schools—Harvard, Stanford, and Northeastern University in Boston[18]—against a field of other universities. Worldwide, most students

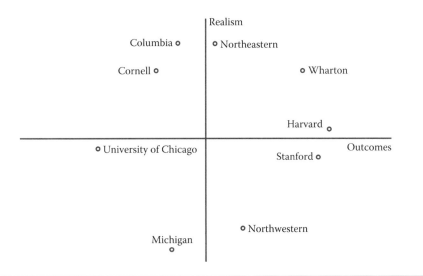

Figure 6.10 Similarity map of business schools. (From Endnote 19. With permission.)

would judge Harvard and Stanford to be the two most similar schools of this set. Northeastern University is likely to be perceived as quite dissimilar from Stanford and even more dissimilar from Harvard. If we mapped these three schools on an overall similarities map, Stanford would be placed close to Harvard and far from Northeastern. This large distance would reflect the dissimilarity among the schools. If instead of asking you just about Harvard, Stanford, and Northeastern, we asked you to also consider Northwestern, Wharton, Michigan, Cornell, and Columbia, you would need to provide more judgments resulting in a more complex map. Figure 6.10 shows a similarity map derived from the judgments of 100 entering students at Northwestern University in Illinois, Chicago. Notice that Harvard and Stanford are indeed close together and that Chicago is on the far left.

In Figure 6.10, the axes are labeled *realism* and *outcomes*. Just as with factor analysis, managerial discretion is used in naming the axes. Using knowledge of the category, the manager names the dimensions to best explain the products' positions. While this may seem arbitrary, it is just this creativity that similarity maps seek to elicit. Sometimes the new product team becomes tied to the attributes that are generated; the similarity map suggests new directions that the customers may not have articulated or that the new product team may not have understood. In summary, new products are positioned in a similarity map by: (1) measuring similarities for the new product and (2) judgmentally shifting the product directly on a perceptual map.

Value Maps and Customer Priorities

Many new products exploit improved technology or features that deliver customer benefits better than competitive products. Other new products exploit process improvements that deliver the same benefits at a lower cost. To evaluate fully such new product opportunities, we consider price as well as the core benefits in creating a position for a new product. Value maps provide a means to incorporate price in a new product's perceptual position. We begin with the concept of customer value.

How many times have you heard phrases such as: "you get what you pay for," "I want my money's worth," "good value for the money spent," "I want the most for my money," "when you buy a … you shop value," and so on. Such phrases reflect a customer's concern for **value**; that is the tradeoff between the benefits of a product and the price paid for the product. The concept of value, or net value, comes from considering a budgeting problem in which a consumer is trying to allocate money across a number of products. For example, a typical consumer may wish to budget money for a car, appliances, home improvements, video equipment, food, clothing, etc. The customer wants to do this in a manner that provides the most "utility." Research by Urban and Hauser[20] shows that a good budgeting rule is to choose first the product with the most utility per dollar, then the product with the next most utility per dollar, etc., until the budget is exhausted. It is useful to consider value in a product's position. One way to do this is with a value map.

Value Maps   Review Figure 6.2, a perceptual map of pain relievers. In that map, Tylenol provides the most gentleness and Advil and Nuprin the most effectiveness. Bufferin is a nice compromise, but Bayer Aspirin and Anacin appear to be dominated by Tylenol and Bufferin—they each provide more gentleness and more effectiveness. Private-label aspirin provides the least of both perceived dimensions and appears to be dominated by almost every brand on the map. If all eight of these brands were priced identically, then few consumers would be rational if they bought Bayer Aspirin, Anacin, or, especially, private label brands. But consumers do buy these brands. They are not irrational; they consider price in addition to effectiveness and gentleness.

Perceptual Dimensions and Price   Suppose that we obtain the prices (per unit dose), rescale the perceptual map such that the lowest brand on each dimension defines the origin of the map, and divide each brand's position by its price. If we do this for the market prices, which were prevailing at the time that the value map in Figure 6.11 was drawn, we see that no brand is dominated by another brand on both dimensions. A rational consumer could find a reason to choose each brand, depending upon the tradeoffs that they wish to make between gentleness per dollar and effectiveness per dollar. To understand better the relationship between benefits and price in a value map, consider a change in the price of Bufferin. If we lower its price, then consumers obtain more value, more effectiveness, and gentleness per dollar. This is shown in Figure 6.11 by a movement to the position shown by a square. If we raise Bufferin's price, consumers get less value, moving the brand into the position shown by a triangle in Figure 6.11. Thus, managers can modify the price (and perhaps benefits) of Bufferin

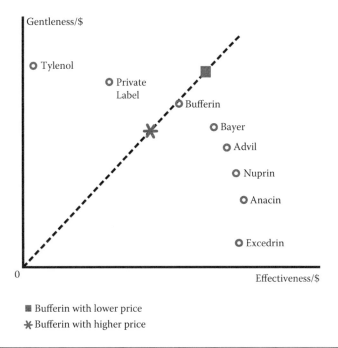

Figure 6.11 Value map for pain relievers. (From Endnote 21. With permission.)

to obtain a value positioning (shown by the square in Figure 6.11) that is likely to attract consumers.

Use of Value Maps   Value maps provide the new product team with a means to set an initial price for their product. To succeed, the new product should provide superior value to the customer. This means that the benefit:price ratio should be high and the product should be positioned such that it is not dominated by another product on the value map.

The exact price depends upon a number of conditions. If the price is set too low, the customers will get a very high value, but the firm may not earn sufficient margins to market the product profitably. (Early in the life of a new product, good margins may be important to recoup R&D expenses or fund product refinement.) However, if there are manufacturing efficiencies for high volume or if the firm expects to exploit an experience curve to lower manufacturing cost, a low price may be justified to penetrate the market. Another concern is competitive response. Low prices often evoke competitive retaliation leading to unprofitable price wars. This would be true if competitors have high fixed costs such that they must defend their volume in order to function profitably. All said, the price decision is extremely complex; the value map is an important input to that decision.

Value maps should be used synergistically with perceptual maps because value maps make strong assumptions. When a value map is drawn it assumes implicitly that the perceptual dimensions are linear such that four units of effectiveness at $2 is the same as two units of effectiveness at $1. This may not always be true, so the team should use inputs from both perceptual and value maps to set the product's core benefit proposition. There are other technical concerns, such as the choice of a "zero point" and the choice of which price to use. In Figure 6.11, we chose the origin of the map to be just below the lower left point of the perceptual map. This choice seems to work empirically,[22] but value map positions can be sensitive to the choice of the zero point.[23] We suggest that the new product team be aware of the sensitivity of the map. The other concern is that there is not always a single price. Package sizes can vary between brands or even within a product line. Some package goods or pharmaceuticals are concentrated, others are not; price can vary by region, etc. In practice, the best choice seems to be the average price per use. For example, in Figure 6.11, the average price per dose based on 100 tablet packages (the most commonly sold package size) was used. In cases where the per dollar dimensions are troublesome, a default approach is to add a price dimension to the perceptual map. This makes the perceptual map more complex, but does capture the price tradeoffs critical in many markets.

Each of the different types of perceptual maps has its unique advantages and disadvantages, which are outlined in Table 6.3.

Chapter Summary

This chapter has discussed methods of perceptual and value mapping and how these maps are used to identify gaps in markets, check positioning of concepts, and determine if a new product fulfills the core benefit proposition (CBP). Perceptual maps help managers understand a product category and recognize opportunities by providing a succinct representation of how customers view and evaluate products in a category. Value maps complete the CBP by enabling managers to set a target price for the new product such that customers perceive it to be of high value. These inputs enhance the creativity of the new product team and encourage marketing, engineering, and R&D to work together to focus their creativity on achieving a successful position in the market. The use of concepts in perceptual mapping is important because they test the CBP and the product's ability to fill a positioning opportunity.

The emphasis in perceptual mapping is on benefits and needs rather than the physical characteristics of the product. Customers make judgments based on perceptions of reality. Early in the new product design process, it is extremely important to identify opportunities, select a target position, and direct the product development process; i.e., establish the CBP. This does not mean that physical characteristics are not important. On the contrary, physical characteristics and psychological cues, such as advertising and salesforce messages, communicate and fulfill the CBP. By identifying the structure of perceptions and by linking it to customer needs and benefits, the new product team better identifies the physical and psychological aspects of the new product.

Table 6.3 Advantages and Disadvantages of Different Perceptual Maps

MAP	DEFINITION	ADVANTAGES	DISADVANTAGES
• Determinant Gap Map	• Perceptual map based on managerial input	• Useful to compare different products or brands against one another • Easy to create • Can be improved with managerial "sorting" of product attributes based on importance	• Does not include VOC • Can be biased based on managerial perspective • Requires an understanding of important product attributes
• Attribute Ratings Gap Map	• Perceptual map based on survey data of consumers' perspectives	• Captures VOC • Based on statistical method—factor analysis • Qualitative studies can be used as input • Can be synergistic with determinant gap map • Easy for consumers to complete • Intuitively appealing for capturing VOC	• Can be time consuming to create • Requires reaching correct sample population • Can be difficult to interpret if unfamiliar with statistical method • Gap may identify poor market opportunity • Requires knowledge about important product attributes
• Overall Similarities	• Perceptual map based on consumers' comparisons of products or brands	• Useful when product attributes are not known • Provides a means to look beyond just attributes • Uses statistical method—similarity scaling	• Requires statistical evaluation • Can be difficult to interpret because there are no set attributes • Biased by the types of products/brands chosen for comparison • May be difficult for consumers to complete
• Value Map	• Perceptual map that includes price perspective	• Considers price in the attribute evaluation • Captures VOC • Synergistic with other perceptual maps	• Requires knowledge of product prices, which may not be available early in the process • Arbitrary "zero-point"

Glossary

Attribute rating perceptual gap map: Type of perceptual map that plots a product placement against two attributes based on input from the potential buyers and users of a product or service. It requires results from a consumer survey to develop the map.

Benefits: Value received from a product/service feature or function.

Cumulative% variance: An output from the statistical method, factor analysis, that represents the cumulative percentage of variance accounted for by the noted factor and all preceding factors.

Cumulative variance explained rule: A rule used in factor analysis that determines how many factors should be retained in the analysis. A typical cutoff is 70% or more of the cumulative percentage variance explained of the attributes.

Determinant gap map: Type of perceptual map based on managerial input. Can be used to compare different products or brands against one another.

EGO rule: See eigenvalue greater than one rule.

Eigenvalue: A statistical measure of total variance in all variables that are accounted for by a single factor in factor analysis.

Eigenvalue greater than one (EGO) rule: A rule used in factor analysis that determines how many factors should be retained in the analysis. If the eigenvalue of a factor drops below 1.0, the EGO rule suggests that it not be retained.

Factor: The feature, function, and/or benefit that describes a product or service.

Factor analysis: Statistical technique used to reduce the number of meaningful attributes that are evaluated when developing a perceptual map. It has many uses: pattern delineation, data reduction, structure analysis, descriptive classification, scaling, and hypothesis testing.

Factor loadings: Represent the correlations between the attribute variables and the underlying factors that the algorithm identified.

Factor score coefficients: Output from the statistical method, factor analysis, that links the standardized input attribute scales to the factor scores.

Factor scores: When averaged across individuals, they specify the coordinates for a product on a perceptual map.

Features: Refers to items that provide added capabilities to a product or service, i.e., a full-featured Swiss Army knife compared to a jackknife.

Functions: How a product/service works.

Overall similarities gap map: Technique based on consumer's ratings of how similar/dissimilar products, concept, or brands are to/from one another.

Perceptual maps: Maps that visually summarize dimensions (such as ease of use, versatility, ergonomically pleasing) that customers use to perceive and judge products, both the firm's and competitor's products or services. Used to represent the positions of products on a set of primary customer needs or attributes. Three types of perceptual maps: determinant gap map, attribute rating gap map, and overall similarities gap map.

Rotated factor loading matrix: Rotated factor loadings used to try to maximize variance accounted for in factor analysis.

Rotation sums of squared loadings: The distribution of the variance after "varimax" rotation. Varimax rotation tries to maximize the variance of each of the factors, so the total amount of variance accounted for is redistributed over the six extracted factors.

Scree plot: Is a visual presentation of the variance explained in a factor analysis. The scree plot takes its name from geology where the "scree" is the pile of rock that accumulates at the bottom of a rock slide.

Scree rule: Plots the eigenvalues whereas the next factor explains marginally less variance, the plot levels off. The scree rule says to drop all further components after the one starting at the "elbow" of the plot.

Snake plot: A type of perceptual map that captures consumers' ratings of a product/service attributes. Lines are drawn between means, which "snakes" down the page. It allows a visual interpretation of the perceived positions of competing products.

Underlying factors: A set of attributes (usually 4–5) that summarize a full list of product/service attributes, features, functions, or benefits.

Value: Benefit as seen by the consumer versus price paid.

Value maps: Provide a method to help recognize the interrelationship between benefits and price within a given market segment.

Review Questions

1. Discuss the difference between a perceptual map and a value map. Why might a new product team find both useful?

2. In some situations, particularly large industrial products, different members of the buying center might perceive the product differently. Why might perceptions differ? How does this affect new product design?

3. What are the differences in a determinant gap map, an attribute ratings gap map, and an overall similarities gap map?

4. What is the goal of a factor analysis?

5. What are the three rules for "factor" retention in factor analysis?

6. Does the cost of making a product affect its position on a perceptual map or a value map? If so, how? If not, how might you represent cost in the deciding how to position a product on a map?

Assignment Questions

1. Figure 6.3 is a perceptual map for transportation choices. Gap 1 indicates there is no offering that is high in speed and convenience and low in expense. If such a product was developed, could you guarantee that it would be profitable? If so, why? If not, why not?

2. Draw a determinant gap map for (a) the toothpaste market, (b) the personal computer market, (c) a financial service, and (d) a heavy-duty industrial compressor. What attributes are most important to consider and why?

3. How might a hospital/health maintenance organization be positioned in your community?

4. There are several approaches to selecting the best positioning for a new product. One approach might construct a concept description for each possible new product position and have consumers evaluate the various descriptions.

The description with the best rating could then be chosen. A second approach would be to build a preference map that describes the entire perceptual space and then use the model to predict best positioning. What are the advantages and disadvantages of each approach? Which approach would you use for services? For products?

5. You are about to graduate and venture into the ominous job market. Being an aggressive, hard-working, and enterprising person, you decide that your excellent grades and outstanding personality are not enough. A proper perceptual positioning is required at your interview. From the placement office and from talking to recent graduates, you discover the following key dimensions by which employers evaluate potential hires.

   A = Preparedness (0–10, 10 best)
   B = Group Skills (0–10, 10 best)
   C = Technical Skills (0–10, 10 best)
   D = Ability to Handle Pressure (0–10, 10 best)

You know your competition because you have seen them in class and have observed them in social settings. The following ratings represent your judgment of how you will be perceived relative to your peers.

	A	B	C	D
John Smith	7	3	2	3
Tatiana Alli	8	5	7	4
Alexander James	4	9	3	9
You	6	2	8	6

a. If only you and John Smith compete for the job, how should you handle the interview?

b. If only you and Tatiana Alli compete for the job, how should you handle the interview?

c. How would your strategy for the interview depend upon the relative importance that the interviewer places on preparedness, group skills, technical skills, and the ability to handle pressure?

d. How might you position yourself against the competition?

Endnotes

1. J. Quittner, "The Pre: Palm's Plot to Take on the iPhone," *Time Magazine* (June 15, 2009). Online at: http://www.time.com/time/magazine/article/0,9171,1902833,00.html (accessed June 6, 2013).
2. Ibid.
3. Online at: http://www.danielroth.net/archive/2009/03/gestures-of-war.html (accessed June 13, 2013).

4. This perceptual map is purely hypothetical as created by this book's authors to demonstrate the differences in different smartphones.
5. Procter & Gamble sold the Sure® brand to Innovative Brands LLC in 2006, who then sold the brand to Helen of Troy Ltd. in March 2010.
6. This perceptual map is purely hypothetical as created by the authors to demonstrate the differences in different smartphones.
7. Daniel Roth, "Gestures of War," *Wired* 21 (2009, April). Online at: www.danielroth.net/archive/2009/03/gestures-of-war.html (accessed June 12, 2013).
8. Based on C. M. Crawford and C. A. Di Benedetto, *New Products Management*, 9th ed. (New York: McGraw Hill, 2007), p. 132; Crawford and Di Benedetto refer to "factors" as "attributes."
9. Urban and Hauser, *Design and Marketing of New Products*, p. 188, Fig. 9.2.
10. William M. Trochim (ed.), "Likert Scaling," *in Research Methods Knowledge Base*, 2nd ed. (Atomic Dogs Publishers, 2001).
11. These results are hypothetical and are not meant to be a true representation of any product.
12. This analysis was obtained using SPSS. An explanation of the terms are as follows: **Factor–** The initial number of factors is the same as the number of variables used in survey; sometimes called attributes. **Initial Eigenvalues–**Eigenvalues are the variances of the factors. Because we conducted our factor analysis on the correlation matrix, the variables are standardized, which means that each variable has a variance of 1, and the total variance is equal to the number of variables used in the analysis, in this case, 15. **Total–**This column contains the eigenvalues. The first factor will always account for the most variance (and hence have the highest eigenvalue), and the next factor will account for as much of the leftover variance as it can, and so on. **% of Variance–**This column contains the percent of total variance accounted for by each factor. **Cumulative %–**This column contains the cumulative percentage of variance accounted for by the current and all preceding factors. For example, the third row shows a value of 52.34. This means that the first three factors together account for 52.34% of the total variance. **Extraction Sums of Squared Loadings–**The number of rows in this panel of the table correspond to the number of factors retained. In this example, we requested that six factors be retained, so there are six rows, one for each retained factor. **Rotation Sums of Squared Loadings–**The distribution of the variance after "varimax" rotation. Varimax rotation tries to maximize the variance of each of the factors, so the total amount of variance accounted for is redistributed over the six extracted factors.
13. This output is from SPSS factor analysis results.
14. G. A. Churchill, Jr. and D. A. Iacobucci, *Marketing Research: Methodological Foundations,* 8th ed. (Fort Worth, TX: Dryden, 2002).
15. One method is to take the summation of the mean of the variable (Figure 6.9) multiplied by the factor score coefficient (Figure 6.8). Thus, in our example, for factor 1, this is: (3.41 × 0.178) + (3.75 × -0.045) + (3.34 × 0.376) + (3.44 × 0.014) + (3.44 × 0.057) + (4.5 × 0.008) + (3.44 × 0.118) + (3.81 × 0.016) + (4.03 × 0.298) + (2.94 × -0.063) + (3.5 × 0.325) + (4.0 × -0.107) + (3.63 × 0.015) + (3.69 × -0.015) + (3.75 ×- 0.020) = 4.09 These factor scores are then plotted on the AR perceptual gap map.
16. P. Green, "Marketing Applications of MDS: Assessment and Outlook," *Journal of Marketing*, 39 (1975, January): 24–31.
17. Urban and Hauser, *Design and Marketing of New Products.*
18. Northeastern University in Boston is a school based on experiential learning. Online at: http://www.northeastern.edu/experiential-learning/cooperative-education/ (accessed June 1, 2013).
19. Based on P. Simmie, "Alternative Perceptual Models: Reproducibility, Validity, and Data Integrity," paper presented at the Proceedings of the American Marketing Association Educator's Conference, Chicago, August 1978.

20. John R. Hauser and G. L. Urban, "The Value Priority Hypotheses for Consumer Budget Plans," *Journal of Consumer Research* 12 (March 1986): 446–462.

21. Urban and Hauser, *Design and Marketing of New Products.*

22. John R. Hauser and Steven P. Gaskin. "Application of the 'Defender,' Consumer Model," *Marketing Science* 3(4) (1984): 327–351.

23. For drawing a value map directly from sales and price data, such as that obtained from automated supermarket checkout systems, see S. M. Shugan, "Estimating Brand Positioning Maps Using Supermarket Scanning Data," *Journal of Marketing Research* 24 (February 1987): 1–18.

7

Estimating Sales Potential

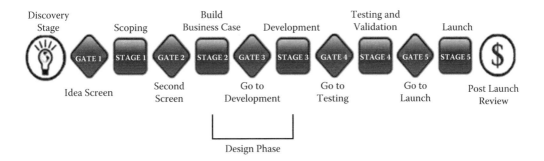

Learning Objectives

In this chapter, we will address the following questions:

1. What forecasting techniques are available for new product managers?
2. What is the difference between diffusion and forecasting analyses? How do diffusion projections differ from forecasting projections?
3. How should the ATAR (awareness, trial, availability, and repeat) model be used?
4. When should the Bass Diffusion Model be used?
5. When should regression analysis be used when forecasting?

Forecasting the Success of the iPad[1]

At its launch in early 2010, Apple pitched the iPad as a revolutionary method for accessing media in its many forms—books, games, movies, magazines, and Web content—all on one device. Many industry analysts felt that the iPad had the potential to redefine how personal computing was conducted. iPad represented an attractive amalgamation of two devices: Internet appliances and tablet PCs, both product areas that had faltered in the past. "I think this is the new Mac, I really do," said Marc Benioff, chairman and CEO of salesforce.com.[2] However, others thought of it as nothing more than a fancy iPhone. "It's ridiculously expensive, way overpriced," said Josh Klenert, a 36-year-old Apple customer. "You may call it a dumb computer or a smart telephone; it's in between. [Yet,] it's a unique, sexier device."[3]

The Tablet PC, a touchscreen computer, had been in the marketplace since the late 1990s. Yet, by 2010, tablets accounted for less than 1% of the personal computer market, according to research firm Gartner, Inc. The computer market never warmed up to tablets because their operating system reportedly was not well designed for tablet computing, the interface did a lousy job with text entry, and the touch screens were often dependent on easily lost styluses. It was reported that "lack of application support, clumsy hardware designs, and price premiums" were barriers for most users.[4] However, more than half the people surveyed by Zogby International, a research firm, said they would use a tablet device, such as iPad, for working outside the office. "Clearly, the iPad has a role to play in the business market," reported Charlie Wolf, a Needham & Co. analyst.

The first wave of reviews praised the iPad's ability to deliver digital books and video quickly, reportedly measuring up against Amazon's Inc.'s Kindle e-book reader and Barnes and Noble's Nook. However, there were several downsides to the iPad. The first version lacked a built-in camera or support for flash software. According to a preliminary forecast from iSuppli Corp., "The device's initial limitations are likely to be overlooked if Apple provides enough content to keep users engaged within the product limitations."

Apple Inc. was notoriously protective of financial information and forecast predictions. Yet, based on market facts, analysts readily made their own predictions. In April of 2010, *Bloomberg BusinessWeek* reported that analysts' estimates on the iPad's sales varied widely.[6]

IPAD FORECASTS

ANALYST	FIRM	IPAD 2010 UNIT SALES ESTIMATE
David Bailey	Goldman Sachs Group	6 million
Katy Huberty	Morgan Stanley	6 million
Andy Hargreaves	Pacific Crest Securities	5.3 million
Ben Reitzes	Barclays Capital	5 million
Gene Munster	Piper Jaffray	2.7 million
Shaw Wu	Kaufman Bros.	2–2.5 million
Toni Sacconaghi[7]	Sanford C. Bernstein & Co.	5 million
Francis Sideco	iSuppli Corp.[8]	7.1 million

Competition was expected to be strong as international competitors began to introduce their own tablets. Motorola, Samsung, Panasonic, and Toshiba, and even Polaroid had plans to enter the category. Nonetheless, being first out of the gate with a low-cost tablet alternative gives Apple a distinct advantage, said Rhoda

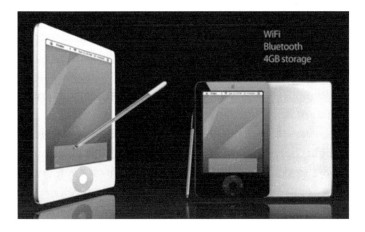

Alexander, director of monitor research for iSuppli. Only time would tell which of the analysts was correct with his/her projections.

Introduction[9]

Our previous discussions have focused on diagnostic information that identifies the "best" product design. But is the "best" product good enough to warrant the cost and time involved in final development and launch? To answer this question, we evaluate the sales potential of the new product in order to develop a sales forecast. The accuracy of the forecasts must be tailored to the decisions being made.

Early in the design process, the product is just an idea, a positioning, or at most a concept description, thus, the investment has been relatively small where many competing potential ideas vie for available resources. In Scoping, the team will have made a rough estimate—an indication of whether the product will be a "bomb," a moderate success, or a spectacular winner. This was the **sales potential**,[10] which is an estimate of company sales reasonably attainable within a given market under a given set of conditions. Now, as we move out of Build the Business Case stage, the evaluation of sales potential becomes more critical and the estimates need to become more exact. Thus, more sophisticated techniques are required to fine-tune the marketing strategy and to make the difficult, but crucial go/no go decision. **Sales forecasts** are required, which is a reasonable estimate of company sales attainable in the given market under consideration with the proposed marketing mix: product, price, place, promotion, positioning. However, because much can still change as the idea is advanced and the product designed during the Development stage, these midprocess forecasts still are "best guesstimates," although refined with voice of the customer and managerial intuition.

While projected sales forecast is an important input to profit potential, this may not be the only consideration; it depends upon the goals of the organization. In private organizations, the main goal is typically profit. However, profit is a complex construct. An organization might develop a new product for its profitability or as a flagship to attract customers to other products in the line, or as a move to preempt competition, or as a means to develop expertise in a market that holds promise for the future. In public services, goals may be different. For example, in healthcare delivery, it is enrollment; in hospitals, it is admissions by illness; in universities, it is enrollment and subsequent hiring placement; and in energy policy, it is efficiency of energy use. In these cases, sales potential must be interpreted more broadly.

Even if the long-run sales are projected to be 100 million units, these sales will not materialize overnight. Sales may start at a low level and grow as more customers try the product, as more stores stock the product, as more uses are found for the product, or, in some cases, as existing products wear out. Thus, the product development team must recognize the difference between forecasts of long-run sales—the sales once the transient phenomena stabilize— and the short-run sales that include the transient phenomena. This is particularly important when decisions to continue marketing a product are made based on early sales data. Due to these multiple variables under consideration, we must look at forecasting from multiple levels.[11]

- *Forecasting level* refers the focal point in the corporate hierarchy where the forecast applies. Common levels include the stock keeping unit (SKU) level, stock keeping unit per location (SKUL) level, product line level, strategic business unit (SBU) level, company level, and/or industry level.
- *Forecasting time horizon* refers to the time frame for how far out one should forecast. New product forecasts could correspond to a single point in the future or a series of forecasts extending out for a length of time. Examples include a one- to two-year time horizon, which is typical for fashion products; two to five years for most consumer product goods; and 10+ years for pharmaceutical products, which are driven by drug patents.
- *Forecasting time interval* refers to the granularity of the forecast with respect to the time interval as well as how often the forecast might be updated. For example, a series of forecasts can be provided on a weekly, monthly, quarterly, or annual basis and updated biannually.
- *Forecasting form* refers to the unit of analysis for the forecast. Typically, new product forecasts early on are provided in a monetary form (e.g., U.S.$) for project approval purchases and later in the process provided in terms of unit volume for production purposes.

Forecasting Techniques

There are numerous forecasting techniques a firm can undertake. The method used will be determinate upon where in the new product development (NPD) process is the forecast being conducted, what is the time line, the type of innovation (consumer packaged good, durable good, industrial good or service, radical, incremental), as well as budget constraints. Kahn,[12] new product forecasting expert, suggests breaking new

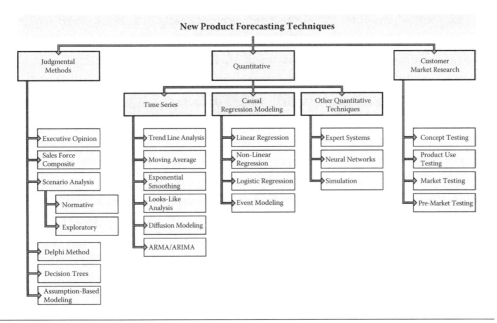

Figure 7.1 New product forecasting techniques (Endnote 12 with permission).

product forecasting techniques into three categories: (1) judgment techniques, (2) quantitative techniques, and (3) customer measurement. Figure 7.1 outlines these different techniques.

Judgment Techniques

Judgment techniques represent models that utilize managerial expertise, judgment, and intuition into formal forecasts. These include:

- *Executive Opinion*: A corporate top-down forecasting technique where the forecast is arrived at through the opinions, predictions, and experiential judgment made by informed executives and experts.
- *Sales Force Composite*: A corporate bottoms-up forecasting method where frontline employees (i.e., sales personnel and customer service representatives) provide forecasts based on customer feedback. These forecasts are then aggregated across all employees to calculate a higher level forecast.
- *Scenario Analysis*: Various scenarios are envisioned to predict the future. Exploratory scenario starts in the present and moves out to the future based on current trends. Normative scenario analysis leaps out to the future and works back to determine what path to take to achieve an expected goal.
- *Delphi Method*: A technique based on subjective expert opinion gathered through several structured anonymous rounds of data collection. In the first round, managers or other experts provide a forecast estimate. After each round, a facilitator provides an anonymous summary of the experts' forecasts from the previous round as well as the reasons they provided

for their judgments. Panel members are encouraged to revise their earlier answers in light of the replies of other members of their panel. Rounds continue until a consensus is reached. Responses remain anonymous so that one individual does not influence the overall final forecast; for example, Steve Jobs would have had a high influence on the forecast outcome of the iPad.

- *Decision Trees*: A probabilistic approach to forecasting where various contingencies and their associated probability of occurring are compared against each other. Conditional probabilities are then calculated, and the most probable events are identified. Decision trees are commonly used in operations research and financial management to help identify a strategy with the greatest probability of reaching a goal.
- *Assumption–Based Modeling*: A technique that attempts to model the behavior of the relevant market environment by breaking the market down into observable factors. Then, by assuming values for these factors, forecasts are generated. These models are also referred to as chain models or market breakdown models. A commonly used one is the ATAR, which we will explore in detail in this chapter.
- *Forecast by Analogy*: A technique that assumes that two different kinds of phenomena share the same model of behavior. For example, one way to predict the sales of the iPad (which was really new to Apple) was to look at the sales of the existing products, the iTouch or iPhone, which "looks like" the iPad in terms of the expected demand pattern for sales.

Quantitative Techniques

Quantitative techniques are broken down into three subcategories: time series, "causal"/regression modeling, and other quantitative methods. **Time series model** uses data from past sales to make project estimates of future sales. **Causal/regression models** consider factors that might impact sales and uses a mathematical model to quantify this relationship.

Time Series Model   Time series methods use historical data as the basis of estimating future. These include:

- *Moving average/*weighted moving average: In financial applications, a simple moving average (SMA) is the (weighted) mean of the previous n data points. A moving average is commonly used with time series data to smooth out short-term fluctuations and highlight longer-term trends or cycles.[13]
- Exponential smoothing: A set of techniques that develop forecasts by addressing the forecast components of level, trend, seasonality, and cycle. Whereas,

in the simple moving average, the past observations are weighted equally, exponential smoothing assigns exponentially decreasing weights over time.[14]

- Autoregressive moving average (ARMA)/Autoregressive integrated moving average (ARIMA):[15] A set of advanced statistical approaches to forecasting, which incorporate key elements of both time series and regression model building, sometimes called Box–Jenkins models. Three basic activities (or stages) are considered: (1) identifying the model, (2) determining the model's parameters, and (3) testing/applying the model. Critical in using any of these techniques is understanding the concepts of autocorrelations and differencing.
- Extrapolation: A method for projecting out a forecast based on past sales. There are many ways of making the projection: linear, nonlinear, continuous, periodic, etc. A sound choice of which extrapolation method to apply relies on a *prior knowledge* of the sales history.
- ***Diffusion models***: Models that estimate the growth rate of a *durable product* by considering the impact of mass media and word-of-mouth influence on the consumer adoption process. In this model, cumulative sales follow an S-shaped curve as the potential market size declines as fewer people have not adopted the innovation. The Bass diffusion model[16] is most widely used with others including the Gompertz Curve[17] and the Logistic Curve.[18] We will discuss the Bass model in detail below.

Causal/Regression Modeling   **"Causal"/Regression Modeling** techniques use statistical analysis to relate an influencing factor to predicted sales. The term *causal* is used very loosely because these models only represent correlations between variables and not true cause-and-effect relationships. For example, a lowering of price may not increase sales, but it is known there is a correlation between price and sales. Kahn[19] reports three popular techniques:

- ***Linear regression***: Assesses the relation between one or more managerial variables and a dependent variable (sales), strictly assuming that these relationships are linear in nature.
- ***Nonlinear regression***: Assesses the relation between one or more managerial variables and a dependent variable (sales), strictly assuming that these relationships are nonlinear in nature.
- ***Logistic regression***: Assesses the relation between one or more managerial variables and a binary outcome, such as purchasing versus not purchasing. You either have bought an iPad or you have not.

Other Quantitative Techniques   The *Other Quantitative Techniques* category contains techniques that are not as commonly used. These include:

- *Expert Systems*: Typically computer-based heuristics or rules for forecasting. These rules are determined by interviewing experts and then constructing numerous "if-then" statements for various scenarios.
- *Neural networks*:[20] An advanced computer program that attempts to decipher patterns in a particular sales time series. They are based on an information processing paradigm that is inspired by the way the brain processes information.
- *Agent-based models*:[21] A computer-based simulation approach that models micro-level consumer choice in order to predict macro-level sales. It allows various exogenous market influences to be added to the model to allow "what-if" scenarios. For example, what if competitors drop their price, how will this impact the firm's sales?

Customer/Market Research Techniques

Customer/Market Research techniques involve market research in order to obtain consumers' intended purchased decisions. The caveat is that intentions do not always represent actual behavior.

- *Concept Testing*: A survey asking consumers to evaluate their probability of buying a new product/service based on a concept description.
- *Product–Use Testing*:[22] A market research process by which customers (current and/or potential) evaluate a product's functional characteristics and performance.
- *Market Components Testing*: A market research process in which consumers evaluate the effectiveness of a marketing campaign.
- *Premarket Testing*: A market research procedure that uses syndicated data (as collected by marketing research firms) and primary consumer research to estimate the sales potential of new product initiatives through forecasting of purchase intent and preference share.
- *Market Testing*: A market research process by which targeted customers evaluate the marketing plan and the new product in a marketing setting.

New Product Forecasting Strategy[23]

It is important to realize that not all of the forecasting methods are appropriate for every situation. Qualitative techniques are quite adaptable to many environments, but can be time-consuming. Quantitative techniques require data, and rely on the critical assumption that current data can predict future states. If meeting this requirement is not feasible, resulting forecasts are not meaningful. Customer/market research tools are time-consuming and expensive to use. A "contingency" approach, therefore, is recommended for applying new product forecasting techniques. The type of forecast to use is contingent upon the situation/product. Kahn[24] presents four new product forecasting situations (Figure 7.2), which we discuss below:

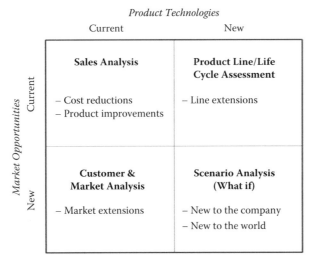

Figure 7.2 Contingent forecasting strategies.

- *Sales analysis* strategy in a *current market* with *current product technology* is where the uncertainties of market and product technology are lowest and includes situations with cost reductions and product improvements. Because these are incremental innovations, historical sales data would be available on past versions of the product. Analysis would focus on looking for deviations and deflections in sales patterns based on previous cost reductions and improvements in the product. Quantitative techniques, such as times series and causal/regression, could be quite useful and manifest objective forecasts.

- *Product life-cycle* strategy in a *current market* with *new technology* for line extensions representing product technology uncertainty. There is an understanding of the marketplace, so analyses would attempt to overlay patterns of previously launched products onto the new line extensions. Judgment, time series, and forecasting-by-analogy techniques with assistance from industry experts and R&D manager who understand the technology are best used.

- *Customer/market analysis* strategy in a *new market* with *current technology*, which includes new use and new market products (market extensions). The purpose of this forecasting strategy would be to reduce risk by understanding the new market. Various customer market research studies might be engaged along with the use of assumption-based models in an attempt to specify market drivers, which would be validated by judgment and customer/market research.

- *Scenario analysis (what-if)* strategy in *new market* with *new product technology*, representing high market and product technology uncertainties akin to new-to-the-company (new category entries) and new-to-the-world products. Scenario analysis would be utilized to envision future directions to be taken. Given a lack of data, potential difficulty in identifying the specific target market, and questions regarding technology acceptance, subjective assessment techniques, such as judgment and agent-modeling or similar simulations would also play a major role here.

These descriptions are not meant to infer that other forecasting techniques can't be used beyond the ones suggested. Customer/market research could be of great benefit in understanding the market for cost reductions, product improvements, and line extensions. Simulation techniques can be utilized to estimate sales forecasts to all types of new products. Additionally, sometimes firms will use more than one methodology. If the estimates are similar, the managers have more faith in the numbers; if they diverge, then it is useful to reexamine the models and compare assumptions and measures until the discrepancies are understood.

Forecasting Using Purchase Intention

Regardless of where you are in the new product process or what forecasting strategy has been decided, getting customer feedback on purchase probability is crucial for making best guesstimates. Several indicators of sales potential can be obtained from the customer. The simplest is a direct purchase intention question, which asks the customer either his/her intent to purchase or his/her probability of purchase. The **intent translation** model is most useful early in the product design process when the new product is still a concept statement or a mathematical position on a perceptual map. This forecast model is an approximation and depends to some extent upon managerial judgment. However, it does give reasonably consistent "ballpark" estimates of purchase and is used to make early evaluations that isolate the concepts or positionings with the greatest potential.

You may recall, in our chapter on concept testing, we developed a "purchase intention" question, as shown in Figure 7.3, where customers were asked to make a subjective estimate of their likelihood of buying a new mobile printer. We used this information to obtain a preliminary forecast of purchase intent for the new product. In this early stage of the NPD process, these measures must be treated with caution. Consumers do not often do what they say they will do. For example, in a study of new services, it was found that the probability of actual purchase varied from a value of 0.45 for those responding to "definitely would buy" to 0.05 for those responding "definitely would not buy," meaning that only 45% of those answering definitely would buy actually did and 5% of those saying they definitely would not buy actually did buy the service when it was offered with promotion.[25]

Table 7.1 shows the empirical results of several studies on purchase probability based on responses to the intent to purchase question. It is clear the "norms" for translating intent into purchase must be developed carefully within a specific market and launch practices. Many market research firms, such as The Nielsen Company (www.nielsen.com) and TNS (www.tnsglobal.com), specialize in forecasting sales based on purchase intention translation models. After conducting hundreds of studies for

Print Photos: Anytime, Anywhere
A companion to your camera phone or digital camera, the pocket-sized Mobile Printer connects either wirelessly or with a cable directly to any camera phone or digital camera and prints full-color 2"×3" digital photos in less than 30 seconds. Printing on the go has never been easier. Before leaving home, be sure the printer's rechargeable battery is charged and you've loaded the printer with photo paper (it holds 10 sheets). While you are out and about, if you take an image an your camera phone or digital camera that you want to print, send it wirelessly or wired to the Mobile Printer and in 30 seconds you will have a colorful, durable, and high resolution photo.

New Ink-Free Printing Technology
The Mobile Printer uses a new digital printing technology that prints without ink cartridges or ribbons. Instead of using ink cartridges, the color is embedded in the photo paper itself. Before printing, the ink-free paper looks just like regular white photo paper. The Mobile Printer activates the embedded color, producing colorful, durable, high resolution images and text.

Mobile Printer Feature
- Compatible with all digital cameras and camera phones
- Prints 2"×3" borderless, high quality photos in under 30 seconds
- Uses new photo paper that doesn't require ink cartridges or thermal ribbons
- Photo paper available in both regular and sticky-back formats
- Also prints stickers, business cards, coupons, event tickets, maps, internet content and more
- Battery operated via integrated rechargeable battery
- Everything you need is included with the Mobile Printer - Printer, battery, cable, software, and starter pack of photo paper

For the mobile printer just described, how likely would you be to buy this product?

○ Definitely would buy
○ Might/might not buy
○ Probably would not buy
○ Definitely would not buy

Figure 7.3 Purchase intention questionnaire from concept test on mobile printers.

their clients, they have built a database of purchase intention probabilities that they can apply to new products.

The intention translation model says that before a customer can buy a product, the customer must become *aware* of it and it must be *available*. For example, suppose that there is an 80% chance that through advertising or word-of-mouth a customer will become aware of Truvia, the stevia-based sugar substitute sweetener, which was developed jointly by Coca-Cola and Cargill, and launched in the United States in 2008. We can assume that there is a 90% chance that the customer's retail grocery store will carry it. The purchase intent questionnaire predicts that there is a 25% chance the customer will buy Truvia. Based on historical data, it can be seen in Table 7.1 that a diet drink mix, which will have a similar customer base, had a 61.5% probability of purchase based on a response of definitely will buy. Using this information, we must adjust the 25% probability of buying by multiplying by 0.615 to become 15.4%. Then the expected probability of actual buying will be (0.8) (0.9) (0.154) = 0.11 for that customer. That is, the estimated number of purchases (for durable goods) or purchase rate (for frequently purchased goods) is given by:

Table 7.1   Purchase Intentions

	DEFINITELY/PROBABLY WILL NOT BUY	MIGHT/MIGHT NOT BUY	DEFINITELY/PROBABLY WILL BUY
NONDURABLE PRODUCTS[26]			
Pump toothpaste	16.7	52.2	52.2
Diet drink mix	17.1	25.0	61.5
Fruit sticks	15.6	17.2	43.5
Stay fresh milk	4.7	0.0	2.8
Salad dressing	12.3	25.9	56.3
DURABLE PRODUCTS			
Home computer[a]	3.8	0.0	42.9
Cordless phone[a]	11.4	9.1	12.5
Touch lamp[a]	2.7	5.6	0.0
Cordless iron[a]	0.0	4.0	0.0
Shower radio[a]	1.3	9.1	0.0
Deep freezer[b]	2.5		11.4
Dishwasher[b]	1.24		8.6
Refrigerator[b]	4.1		10.5
Clothes washer[b]	5.6		18.1
Cars[b]	10.8		33.7

[a] Probability of trial.[27]
[b] Percentage of buyers who said they intended not to buy/percentage of buyers who said they intended to buy.[28]

$$P = \sum_{S=1}^{N} a_w a_v b_s' \qquad (7.1)^{29}$$

where N is the number of customers, s is the number of segments, a_w is awareness, a_v is availability, and b_s' is the adjusted probability of buying for a given segment. If there are 100,000 customers just like this customer, then the expected purchase potential is 11,000 unit sales per period. If customers are not identical, but vary in terms of sales potential, then we must evaluate each segment separately and then sum over all segments.

Awareness is driven by the strength of the marketing campaign that a firm conducts. The goal of an awareness campaign is to have a product/brand enter a consumer's consideration set. A **consideration set** (also called *evoked set* or *relevant set*) is the subset of product/brands that consumers consider when making a purchase decision. Consumers have limited information processing abilities and limit product/brand comparison to a subset of products. This subset is termed the consideration set. One common way to measure a consideration set is to ask a consumer which products they have used, have on hand, or would seriously consider (or definitely not consider).

It is often difficult to estimate the awareness of a new product directly. An alternative method is to estimate **judgmentally unaided recall** awareness and use it to estimate awareness. Unaided recall is often easier for advertising managers to estimate than a consumer's consideration set. Fortunately, unaided recall and consideration are highly related.[30] For example, take out a piece of paper and try to write down all the deodorants you can think of; this is unaided recall. Now suppose you were asked if you have heard of Right Guard, Sure, Ban, Secret, Old Spice, Mennen, Arrid, Soft & Dri, Brut, Spring Tide, Mitchum, Dial, Dry Idea, and NIVEA Calm & Care. If you have, that is aided recall. But how many of those deodorants do you know sufficiently well that you can describe them with respect to the large number of detailed attribute scales that are applicable to deodorants? For new product awareness, we want to know how many customers will become aware of the product at a level that is sufficient so that they have the information to consider the new product seriously; thus, aided recall may be necessary. Did you catch the made-up brand? This is the downside of aided recall; respondents may agree to knowing a nonexistent product.[31] In order to ensure that consumers answer aided recall questions with sincerity, sometimes nonexistent brands are added. If a consumer indicates knowing that nonexistent brand, the results from that consumer are discarded.

Availability, the second component leading to consumer purchase decisions, is based judgmentally on the percentage of retail outlets, online distributors, etc., that will carry the product in the target area. Availability of a product for consumers can usually be estimated with past experience or managerial judgment. For some industrial products, it is the percentage of buyers that are within a feasible delivery area. In healthcare, it might be the fraction of eligible customers in a service area. In transportation, it may be more complex, e.g., what percentage of consumers are within 1/4 mile of the bus line or how many cannot afford an automobile. New product design can have a significant impact on availability of the new product. For example, in 2002 when Coca-Cola introduced "fridge packs," which had a smaller footprint than regular 12-packs, its 12-pack sales increased in supermarkets by 3.3%, which encouraged other retailers to stock Coca-Cola products.[32] Pepsi and other carbonated beverage companies soon followed with similar types of packaging.

With the growing ubiquitous Internet, it's easier to get products into the hands of consumers. **Affiliate marketing** is one way firms have increased the availability of services to consumers using the Internet. Affiliate marketing is a marketing practice in which a business rewards a partner company (the affiliate) for each visitor or customer brought to the sponsoring company's Web site through the affiliate's own marketing efforts. The affiliation industry has four core players: the merchant (also known as *retailer* or *brand*), the network (usually the Internet provider), the publisher (also known as *the affiliate*), and the customer. An example is Amazon's affiliate partners that place a banner or text links on their own site for individual books, or link directly to the Amazon home page. When visitors click through from

A. Increase the distributor's unit volume
   1. Have an outstanding product.
   2. Use pull techniques - advertising, trade and consumer shows, public techniques
      - missionary selling.
   3. Give the distributors a type of monopoly, exclusivity or selectivity.
   4. Run "where available" ads.
   5. Offer merchandising assistance - dollars, training, displays, points of purchase,
      co-op advertising, instore demonstration, store "events," and repair and service
      clinics.
B. Increase the distributor's unit margin
   1. Raise the basic percentage margin.
   2. Offer special discounts - e.g., for promotion or service.
   3. Offer allowances and special payments.
   4. Offer to prepay allowance to save interest.
C. Reduce the distributor's costs of doing business
   1. Provide managerial training.
   2. Provide dollars for training.
   3. Improve the returned-goods policy.
   4. Improve the service policy.
   5. Drop-ship delivery to distributor's customers.
   6. Preprice the merchandise.
   7. Tray pack the merchandise or otherwise aid in repackaging it.
D. Change the distributor's attitude toward the line
   1. By encouragement - management negotiation, sales calls, direct mail, advertising.
   2. By discouragement - threats to cut back some of the above benefits or legal action.
   3. Rap sessions - talk groups, focus groups, councils.
   4. Better product instruction sessions - better visuals, better instructions.

Figure 7.4 Alternative tools and devices for motivating distributors. (From Endnote 33. With permission.)

the associate's Web site to Amazon and purchase a book, the associate receives a commission.

Business-to-business companies often use distributors of some type, usually under some franchise agreement or other partnership, thereby, increasing reach beyond direct-to-customers sales product/service availability. There are numerous ways that a firm can motivate distributors to work with them. Figure 7.4 shows the most prominent: increase the distributor's unit volume, increase the distributor's unit margin, reduce the distributor's costs of doing business, and change the distributor's attitude toward the line.

Repeat Purchasing

Equation (7.1) is best used for durable products and services that are purchased infrequently. For frequently purchased products, the projection of sales is more difficult because of the issue of continued repeat buying. Examples of such products are consumer packaged goods (CPGs), service plans (e.g., gym memberships), or industrial supplies (e.g., buy–rebuy decisions). For products that depend upon repeated purchases, we consider the trial and repeat processes. Because consumers may buy a product multiple times within a few months, management is interested in the long-run market share. The sales of new products that rely on repeated purchases are modeled by two components. The first component is **trial**, and the second component is **repeat purchasing,** which applies only to those customers who have tried the new product. Thus, when we forecast sales for frequently purchased goods or services, we must include terms for trial and repeat.

Trial probabilities are obtained from customers after they have been exposed to the new product, while **repeat shares** are based on the probability of purchasing again after home use of the new product. Note that this model is most relevant late in the design process when an actual product is available to give to customers for in-home use. Earlier in the design process, we replace customer level estimates of trial and repeat with managerial estimates. For example, early in the design process, we might estimate a target repeat rate of 50% to serve as a goal, recognizing that we will modify those estimates when the product design is more complete. In determining the purchase probability, P, we now must incorporate trial and repeat estimations. Trial probabilities can be obtained from the purchase intention question in the concept test, such as we examined in previous chapters. Here, we make sample population estimates of *trial* and *repeat* rather than summing individual estimates:

$$P = a_w t' a_v r \tag{7.2}[34]$$

where a_w is *awareness*, t' is the adjusted *trial*, a_v is *availability*, and r is *repeat purchase*. This model of intent translation is often referred to as the **ATAR model** where:

$$Purchase\ rate = awareness \times trial \times availability \times repeat \tag{7.3}$$

Repeat is determined as:

$$r = (1 + (repeat\ probability^*repeat\ occasions^*units\ purchased\ per\ occasion)) \tag{7.4}$$

If the new product is not in production stage, the long-run share of purchases among triers, *r*, is estimated by analogy to products where repeat purchase data are available. If the new product has similar properties to an existing product/service, these estimates are used for initial forecasting.[35] If the final product is available, these proportions are measured by giving consumers, first, an opportunity to purchase the new product and then, an opportunity, after in-house use, to repeat the purchase of the product. Experience suggests the model is reasonably accurate when used to forecast long-run share.[36] Table 7.2 shows where the new product manager may obtain estimates of awareness, trial, availability, and repeat. We have not yet discussed **product use tests, component tests,** or **market test**, but we will do so in future chapters.

Table 7.2 ATAR Variables

ATAR ITEM	BASIC MARKET RESEARCH	CONCEPT TEST	PRODUCT USE TEST	COMPONENT TEST	MARKET TEST
Market units	Best	Helpful	Helpful		Helpful
Awareness		Helpful	Helpful		Helpful
Trial	Helpful	Best		Best	Helpful
Availability	Helpful				Best
Repeat (adoption)			Best		Helpful
Consumption	Helpful	Helpful	Helpful		Best
Prices per unit	Helpful	Helpful	Helpful	Helpful	Best
Cost per unit				Helpful	Best

Source: Endnote 37. With permission.
Key: Best = best source for that item; Helpful = Some knowledge gained.

ATAR with Cannibalization

Sometimes, the introduction of a new line extension will cannibalize sales of another product. For example, when Diet Coke was introduced, sales of regular Coke decreased. In these types of situations, where cannibalization or product switching may be a factor, r is approximated by the simple switching process shown in Figure 7.5.[38] Customers either repeat-purchase the product immediately or switch back to their existing product(s) and repeat-purchase at a later date. The process will stabilize to a point where the number of customers switching to a new product equals the number

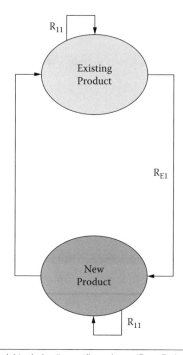

Figure 7.5 Two-state switching model to derive "repeat" purchase. (From Endnote 39. With permission.)

of customers switching away from it. Once the process stabilizes, the repeat rate, r, is calculated by $r = R_{E1}/(1 + R_{E1} - R_{11})$ where: R_{E1} is the proportion of customers who switch to the new product and R_{11} is the proportion of customers who repeat-purchase the new product.[40] For example, if 40% of the customers are observed to switch from existing products to the new product (R_{E1}), and if 50% of the customers are observed to repeat-purchase the new product after trying it, R_{11}, then the long run share among triers is given by $r = (0.40)/(1 + 0.40 - 0.50) = 0.444$. This model correction takes account of those customers who are not 100% loyal to the new product.

Research studies have suggested that one also should measure satisfaction with the product and whether the product fulfills the customers' needs in determining repeat purchase. In one study, it was found that if customers said they were satisfied, they were 14 times more likely to be repeat purchasers than if they said they were dissatisfied.[41] If they felt the product at least fulfilled their needs, they were three times as likely to be repeat purchasers. To illustrate the technique, we assume that a postusage survey found 5% would definitely repurchase Truvia, but 23.7% would probably repurchase Truvia, and 36.1% definitely would not repurchase Truvia. However, historical data tells us that 74% of the "definites," 46% of the "mights," and 13% of the "definitely nots" actually repurchase.[42] Applying these numbers yields an overall estimate that 19.3% of the sample population would repurchase Truvia. That is: $0.193 = (0.74)(0.05) + (0.46)(0.237) + (0.13)(0.361)$.

Probability Scales

Some market researchers prefer probability statements over intent statements. Juster[43] developed a probability statement scale similar to that in Table 7.3 that provides more categories, which are more exactly defined than the intent scales of "definitely will buy," "might buy," and "definitely will not buy."

The probability scale is used in the same way the intent scale is used. Each category is an estimate of some probability of purchase. Several studies have compared the

Table 7.3   Juster's 11-Point Probability Scale

SCORE	VERBAL EQUIVALENT
0	No chance or almost no chance (1 in 100)
1	Very slight possibility (1 chance in 10)
2	Slight possibility (2 chances in 10)
3	Some possibility (3 chances in 10)
4	Fair possibility (4 chances in 10)
5	Fairly good possibility (5 chances in 10)
6	Good possibility (6 chances in 10)
7	Probable (7 chances in 10)
8	Very probably (8 chances in 10)
9	Almost sure (9 chances in 10)
10	Certain, practically certain (99 chances in 100)

accuracy of three-month purchase predictions for a range of durables, services, and fast-moving consumer goods made using a conventional buying intentions scale and the Juster probability scale.[44] These studies confirmed that, overall, the Juster scale was a better predictor of future purchase rates than the conventional buying intentions scale. However, in general, both scales were less successful in predicting purchases of *durables* than purchases of *services* or *fast-moving consumer goods*. With the Juster scale, the higher the predicted purchase rate for an item, the more likely it is that the actual purchase will be accurately predicted. Buying intentions data did not discriminate buyers from nonbuyers well, but the probability scale was better at determining purchasers versus nonpurchasers.[45] In some cases, the new product team may wish to use both a 5-point intent scale and an 11-point probability scale to access convergence. Fortunately, in most cases, it has been found that the two scales provide predictions that are correlated.

Forecast Prediction

Now that we have determined the probability of purchase, we must translate that into a long-run sales forecast. This is done by taking Equation (7.2), *Purchase rate = awareness × trial × availability × repeat*, and multiplying it by the potential market size. This becomes our early prediction of long-run sales for the new product. If we multiply the long-run sales prediction by the contribution margin (revenue/unit − cost/unit), we obtain a prediction of profit:

$$\text{Long run sales} = a_w t' a_v r * \text{market size} \tag{7.5}$$

$$\text{Profit} = \text{Long run Sales} * \text{contribution margin} \tag{7.6}$$

In the Truvia example, we previously indicated that there was 80% awareness, 90% availability, and determined that satisfied customers lead to 19.3% repurchase. If we take a 35% trial rate, this leads to a purchase probability of 6.08%. If the market size for Truvia is every household in the United States, which was 112 million in the 2010 census, this would be 6.81 million packages sold. If the contribution margin is $0.50, this would be a profit of $3.41 million dollars for the first year of sales.

Diffusion of Innovation

As previously noted, intent transformation and probability scales aren't always good at predicting sales of durable products.[46] In the adoption of durable goods (such as home appliances, electronics, lawn and garden appliances), word of mouth has a significant impact on consumer choice. To address this issue, we can use life-cycle models to forecast future growth. These models are based on the assumption that the probability rate of initial purchase at a given time is a linear function of the total number of previous buyers. For example, Table 7.4 shows introduction and takeoff timing for 12 different

Table 7.4 Takeoff Times of Really New Products

CATEGORIES	SAMPLE SIZE	MEDIAN YEARS UNTIL TAKEOFF[a]
All	40	8.5
Pre-World War II	16	15.0
Post-World War II	24	5.5
1965 and later	21	5.0
1980 and later	11	4.0
Kitchen and laundry	14	12.0
Time saving	24	8.5
Leisure enhancing	19	8.0
Electronic	21	6.0
Nonelectronic	19	15.0
High price at commercialization (> $950)	19	9.0
Low price at commercialization (< $950)	21	5.0

Source: Endnote 47. With permission.

[a] Year 1 is year of commercialization.

categories of markets. The actual times to reach peak sales vary. The diffusion model, a type of life-cycle model, was created for predicting the future sales of a durable or industrial product based on the past sales history of a class of products, such as the type we see in Table 7.4.

The **Bass Diffusion Model,**[48] named after Frank Bass who first introduced the method, is a powerful forecasting method that focuses on (a) "innovators" (customers with a high propensity or high need to try a new product in the product category) who buy first and then, (b) through an influence process, encourage others, the "imitators," to adopt. The Bass model captures these two effects with p, the **coefficient of innovation,** and q, the **coefficient of imitation**. Very simply, this forecasting model suggests that consumers will adopt a product based on a combination of how innovative they are in trying new products (the p coefficient) and how they are influenced by their peers (the q coefficient). Mathematically, the Bass diffusion model implies that the rate of purchases is equal to an initial value due to innovators plus a term that reflects the impact of the word of mouth influence process. In equation form:

$$S(t) = pm + [q - p]Y(t-1) - \frac{q}{m}(t-1)^2 \qquad (7.7)$$

where $S(t)$ is the sales in period t, m is the total number of potential buyers in the market, and $Y(t-1)$ is the total number of people who have bought by the end of period $t-1$. Interpreting Equation (7.7), we see that the market sales will first grow due to the effect of innovators (first term) and favorable word of mouth (second term) and then decline as the market saturates (third term). The decline results because as more people purchase a product in the market, fewer people are left to purchase the product in the future.

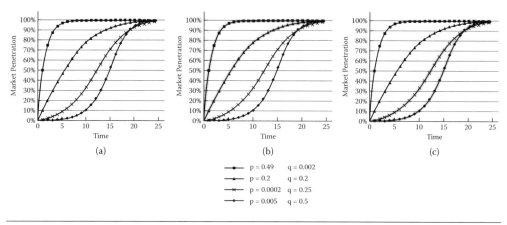

Figure 7.6 Bass diffusion curve for different p and q. (From Endnote 49.)

The adoption curve is S-shaped as shown in Figure 7.6. This figure demonstrates how the diffusion curve changes with different values for p, the coefficient of innovation, and q, the coefficient of imitation. In every period, some fraction of the potential market that has not yet purchased/adopted the product will adopt due to their propensity to try new things introduced to them through advertising. Notice that market penetration cannot be more than 100%, so this is where the curve levels out (see Figure 7.6a). The inflection point, or the point where the curve starts to turn up, is an indication that new adopters buy the product at a diminishing rate (i.e., sales rate slows down). This is seen by the decreasing value in $S(t)$.

Figure 7.6b demonstrates that the more innovators there are in the target market (high p), the quicker the adoption occurs. In every period, those who have bought will interact with those who have not bought. Some portion of these interactions results in adoption due to imitating others' purchases. For example, early adopters bought MP3 players because they were an easy way to store music. Later adopters (Bass's imitators) saw the early adopters (Bass's innovators) and went out and bought their own. MP3 players diffused more rapidly into the marketplace in comparison to microwaves, which surprisingly were very slow to diffuse. Microwaves were first introduced into the market in 1966, but weren't fully accepted by homeowners until the 1970s. Many people did not understand the usefulness of the microwave, and, hence, they were slow to adopt. Additionally, MP3 players were transportable, and microwaves were not, so imitation was slower. In Figure 7.6c, the curves where $p = 0.005$ and $p = 0.0002$ show slow adoption from the innovator, so it is said to have a slow takeoff.[50]

Estimating p and q

In using the Bass model for forecasting, it is important to determine appropriate values for m (total market size for your product), p (coefficient of innovation), and q (coefficient of imitation). Market size, m, can be determined through primary or secondary data analysis. Estimates of p and q are more challenging. There are three methods in which to obtain these estimates: (1) sales history, (2) using historical values of p and q, and (3) forecasting by analogy.

Using Sales History[51]   If historical sales at each time period, $Y(t-1)$, are known, then cumulative sales, $S(t)$, are easy to calculate. Rewriting Equation (7.8) into mathematical regression format, we obtain:

$$S(t) = a + bY(t-1) - cY(t-1)^2 \qquad (7.8)$$

where

$$m = \frac{-b - \sqrt{b^2 - 4ac}}{2c}; \; p = a/m; \; q = -m/c \qquad (7.9)$$

Coefficients a, b, and c can be easily determined using ordinary least squares (OLS) regression analysis as suggested by Bass in his seminal paper on diffusion.[52] Although this is beyond the scope of this chapter, in most cases nonlinear least squares (NLLS) regression model provides a better fit to Equation (7.8) than does OLS.

The downside of using historical sales is that Equation (7.8) cannot be calculated until there are several (typically more than 20) data points. This means that if sales data are only available monthly, it may take up to two years to determine the p and q estimations, which does not provide the manager with any foresight on how to plan for production, shipping, distribution, etc. We must use other estimates of p and q. One option is to use historical estimates of p and q from similar products that have diffused within the marketplace. If the product is an incremental innovation, the company can use the historical sales data from other products in the product line to calculate a p and q estimate. Similarly, marketing research firms that specialize in conducting Bass diffusion model forecasts will have historical estimates of p and q from similar products that can be used. Or, we can use estimates that have been determined from historical sales of common products.

Using Historical Estimates of p and q   In Table 7.4, we provide estimates of p and q as reported in various research studies. Van den Bulte[53] provides general estimates for durable products, non-durable (typically one-time buys, such as a watch), and industrial goods. It is important to note that these estimates are dependent upon a time period. For example, in Table 7.5, we see that Olfek[54] has given estimates of $p = 0.008$ and $q = 0.421$ for the cellular (mobile) telephone and Van den Bulte has provided estimates of $p = 0.226$ and $q = 0.635$. This is because the Olfek estimates are based on sales from 1986 to 1996 and Van den Bulte's are based on sales from around the world from 1950 to 1992. In both of these studies, the q value is greater than p, which is indicative of the "network effects" of mobile phones, so the results are not incongruent. It is important to understand how the researcher derived the estimates before using them in your own analysis.

Table 7.5 Historical Values for p and q

PRODUCT	YEARS	p	q
B&W TV	1946–1957	0.024	0.4
Blender	1958–1971	0.001	0.55
Cable television[a]	1981–1994	0.100	0.060
Calculator	1974–1986	0.0294	0.12
Calculator[a]	1973–1979	0.143	0.52
Can opener	1958–1971	0.0135	0.3
Cassette tape deck	1974–1990	0.0179	0.18
CD player[a]	1986–1996	0.055	0.378
Cellular telephone[a]	1986–1996	0.008	0.421
Cellular telephone[b]	1950–1992	0.226	0.635
Color TV	1954–1970	0.0001	0.66
Curling iron	1974–1982	0.0399	0.3
Deep fryer	1950–1961	0.0337	0.74
Digital watch	1974–1982	0.0204	0.41
Durables (average)[b]	1950–1992	0.016	0.409
Electric blanket	1946–1966	0.0029	0.26
Elec. clothes dryer	1947–1960	0.008	0.4
Electric toothbrush	1963–1971	0.0826	0.13
Fire extinguisher	1975–1982	0.0694	0.16
Food processor	1978–1984	0.0271	0.51
Hair setter	1968–1977	0.1305	0.35
Heating pad	1922–1931	0.0274	0.26
Home personal computer[a]	1982–1988	0.121	0.281
Industrial product[b]	1950–1992	1.058	1.149
Knife sharpener	1957–1969	0.0655	0.5
Lawn mower	1946–1961	0.0056	0.3
Microwave oven	1970–1989	0.0101	0.37
Nondurable product[b]	1950–1992	0.689	0.931
Power leaf blower (gas or electric)[a]	1986–1996	0.013	0.315
Refrigerator	1922–1931	0.0047	0.42
Refrigerator[a]	1926–1979	0.0250	0.126
Room A/C	1946–1960	0.0031	0.43
Slow cooker	1973–1982	0.0654	0.32
Styling dryer	1974–1984	0.0891	0.18
Trash compactor	1972–1982	0.0866	0.19
Turntable	1974–1990	0.0577	0.21
VCR	1976–1990	0.0012	0.53

Source: Endnote 55. With permission.

[a] Ofek (2009).[56]
[b] Van den Bulte (2005).[57]

Forecasting by Analogy   If historical sales data are not available, which they rarely are, the Bass diffusion model is ideally suited to **forecasting by analogy**, which may be necessary for really new or radical innovations where there is no sales history of similar products. For example, suppose we are forecasting sales for a new robotic office mail delivery system. If an existing product, such as a high-end professional photocopier,

Table 7.6 Bass Model Parameter Estimates Using Analogy

Product	p	q	Analogy Comparisons		Weighted Score	Weighted % Score
			Market Structure Characteristics[2]	Product Characteristics[2]		
Weightings for p & q			$w_a =$ 0.4	$w_b =$ 0.6		
Cassette Tape Deck[1]	0.0179	0.18	2	5	3.8	3.8/13= 0.292
Automobile Radio[1]	0.0161	0.41	3	5	4.2	4.2/13= 0.323
Cellular Phone[1]	0.008	0.421	8	3	5.0	5.0/13= 0.385
					13	1.000
Weighted Average for OnStar:	0.0135	0.347				

1. obtained from tables of p and q
2. managerial judgement

For example: Wt. numerical score for Cassette Tape Deck - (2*0.4) + (5*.6) = 3.8

New weighting for p = (0.0179*0.292) + (0.0161*0.323) + (0.008*0.385) = 0.0135

serves a similar population, has a similar price, and analogous customer benefit as the new product, then it can serve as an analogy for the new product. Using an *analogy* to associate a really new product to a set of existing products provides a means for calculating a relevant estimate of p and q. It has been recommended[58] that, in assessing the appropriateness of using a particular product for comparison, the following factors should be considered: environmental factors, market structure, buyer behavior, marketing mix strategies of the firm, and characteristics of the innovation. We seek products that have similar environmental factors or similar market structures or some relevant similarity to the new product.

With a radically new product, sometimes comparative products/services are not available. The diffusion model provides a method for using historical data from several products to come up with an estimated forecast. For example, in 1995 when OnStar, the automobile subscription-based communication system, was launching, there were no comparable services in the marketplace. When looking for existing products that have similar physical characteristics or similar market structures, car radios are an obvious choice since OnStar was used only in the car. A recent innovation that had similar technological characteristics to OnStar would be the portable CD player because it too provided quality digital sounds. It also is probably necessary to consider buyer behavior for subscription services that utilize satellite signals to carry information/content, such as cellular phones or satellite television (e.g., DIRECTV). Table 7.6 provides an example of how to use the coefficients p and q from these previous innovations to derive a set of p and q for a communication service, such as OnStar.

The first two columns of Table 7.6 show the estimates of p and q for the cassette tape deck, the car radio, and cellular telephone. Instead of taking the arithmetic mean, which would equally weight the contribution of each existing innovation, it is recommended to take a weighted average. A set of criteria, such as we previously discussed, can be used to weigh the factors of importance in the diffusion of the new product. For instance, the product characteristics of an OnStar system (e.g., its physical features, programming content) may be of more importance than the market structure for an OnStar system (e.g., subscription-based revenue model). As such, a weighting factor, w_a, of 0.4 could be given to a "market structure" factor, while a weighting factor, w_b, of 0.6 could be given to a "product characteristic" factor. Secondly, a score must be given to each existing innovation for its similarity to the satellite radio. For example, the cellular phone was given a score of 8 for market structure because a revenue model is subscription-based with an initial cash outlay for the hardware (phone), an approach that was expected to be used with satellite radio. Conversely, a score of 3 was given to the product characteristic of the cell phone because it is substantially different in form and function to satellite radio.[59] Thus, using weighted averages of these factor, values for p and q can be estimated for the satellite radio.

The original Bass model has been extended and applied in over 100 publications incorporating marketing-mix variables (price, promotions).[60] It is important to recognize that the use of this model is for the growth rate of a product class (or a market) not a specific product, which means that a company can forecast overall market sales, but not specific firm sales. For example, the diffusion of MP3 players could be estimated, but not the exact sales for iPods. The diffusion model also does not represent replacement after the product wears out. The model helps establish the initial growth rates for a category and, hence, the attractiveness to new product entry.

Regression to Estimate Purchase Probabilities[61]

As we progress through the Design phase from Building the Business Case stage out of the Development stage, there is a need for more accurate estimates of sales. After test product runs, actual purchase can be tracked for the new product through market testing. When sales data are available, the more complex regression logit analysis, an analytic technique to estimate purchase probabilities, can be used to translate preference measures into purchase predictions.

The first column of Table 7.7 shows the preferences from the model for four typical consumers. Compare consumer 1 to consumer 2. From the preference values, intuitively, we would expect that consumer 1 would be almost as likely to purchase Tylenol as the new product, while consumer 2 would be much more likely to purchase the new product. The basic idea behind logit analysis is a mathematical function that translates the preference values into purchase probabilities based on a theory of customer behavior. It has been shown that the following equation predicts actual probabilities as a function of observed consumer preferences.[62]

$$L_{cj} = \frac{e^{\beta p_{cj}}}{\sum_{j=1}^{J} e^{\beta p_{cj}}} \tag{7.10}$$

Table 7.7 Comparison of Rank Order Transformation to Logit Predictions for Four Typical Analgesics (Aspirin) Consumers

	PREFERENCE VALUE	PREDICTED PROBABILITY LOGIT
CONSUMER 1		
Bufferin	1.3	0.08
Excedrin	2.1	0.09
Tylenol	9.8	0.41
New product	9.9	0.42
CONSUMER 2		
Anacin	1.4	0.11
Private label	2.0	0.13
Tylenol	2.3	0.14
New product	9.9	0.62
CONSUMER 3		
Bayer	1.2	0.10
Bufferin	2.2	0.12
Tylenol	5.0	0.21
New product	9.9	0.57
CONSUMER 4		
Anacin	5.9	0.33
Tylenol	6.0	0.33
New product	6.1	0.34
Share of new product		**0.49**

Source: Ref. 63. With permission.

where L_{cj} is the likelihood that customer c purchases product j and let be the measured preference for new product j, and β be a parameter that is used to fit the model to the data. For example, if $\beta = 0.2$, then the probability of consumer 1 in Table 7.7 choosing "new product" would be given by:

$$L_{cj} = \frac{e^{.2*(9.9)}}{e^{.2*(1.3)} + e^{.2*(2.1)} + e^{.2*(9.8)} + e^{.2*(9.9)}} = 0.42 \qquad (7.11)$$

This says that there is a 42% probability that customer 1 will choose "new product" when given the choice set of Bufferin, Excedrin, Tylenol, and the new product. Thus, once we know the value of β for a product category and a particular preference measure, p_{cj}, we can estimate more accurately purchase probabilities. To estimate β, we use statistical analysis.

Estimation of Parameters

To estimate β, we need a data sample. We can measure preferences using research techniques, such as direct measures or conjoint analysis.[64] At this stage in the NPD process, measures that could be collected are (1) product choice in a simulated purchase environment, (2) reported last similar product chosen, or (3) reported frequency of purchase of all similar products. The first measure is appropriate when an actual product is ready for testing and stated choice correlates well to actual market choice. The other measures are used in early design or in product categories, such as services, industrial products, or consumer durables where the interpurchase interval makes simulated purchase infeasible.

With observed choice or frequency data and measured explanatory variables, we can use a statistical technique known as maximum likelihood techniques to estimate β. A number of computer programs are available to perform logit analyses including most popular statistical packages, such as SAS, SPSS, or R. To use the model to predict demand for a new idea, concept, or product, the analyst must:

1. use the preference model (direct measures, preference regression, conjoint analysis) to determine how preferences change as the result of design changes in the new product; and
2. use the logit model (equation (7.10)) to estimate the probabilities that each customer will actually try (or use) the new product if it is available and the consumer is aware of it.

In a study of preference models to project the impact on preference of a change in transportation service in Evanston, Illinois, a logit model based on the last mode of transportation chosen produced an estimate of β = 3.35[65] that can be used in Equation (7.10).

The advantages of logit analysis are that it is based on a realistic model of customer behavior and tries to explain as much behavior as is feasible. It is relatively easy to use, provides reasonably good estimates of demand, and provides explicit statistical measures that can be used to judge the usefulness, accuracy, and significance of the model. A disadvantage is that the logit model assumes the same preference-to-choice process for all customers and does not take into account complementarities or similarities among products.

Managerial Use of the Model

Purchase intentions and conjoint probability models give estimates of demand that can be used early in the design process. Logit analysis is used to refine these estimates as the product progresses through the design process. Since the logit model is based on the magnitudes or intensities of preference via a model of customer choice behavior, it has the potential to provide more accurate estimates of demand. This accuracy is important in making a commitment to complete the new product design and committing resources to initiate the testing phase of the NPD process (i.e., advance to the next stage).

This completes the basic discussion of purchase potential models. Together, the ATAR model, the Bass model, and the logit model give the new product manager a set of techniques to estimate the purchase potential of new product ideas. The intent or probability scales give a direct measure of the customer's beliefs about whether he or she will actually choose the new product. While not exact, the scales provide a good indication of behavior that is sensitive to effects that may not otherwise be captured. The disadvantage of direct scales is that they are limited to testing specific concepts or a relatively few changes to those concepts. The logit model provides a more sophisticated and potentially more accurate model that is sensitive to the intensity of the measured preferences. It is used later in the design process than the ATAR model and can update or confirm predictions made by earlier analysis.

It is sound practice to use multiple techniques to get the best estimates of purchase potential. When used appropriately, these combined models provide accurate estimates that are relevant to managerial design decisions.

Chapter Summary

Success in new products is linked to the ability of a new product to attract and retain customers. Whether the goals are profit, market share, or increased utilization, the key marketing input is an estimate of how many customers will purchase or use the new product. Previous chapters presented techniques to design a product that would be preferred by customers. In this chapter, we looked at a number of methods for forecasting. We looked in detail at two, ATAR and the Bass diffusion model. ATAR is generally used for CPGs and/or services forecasting, and the Bass model is more generally used for forecasting durable goods and technical products. Companies have many options available to them for forecasting methods, which can be as simple as qualitative expert judgments or more sophisticated, such as time series analysis. Poor accuracy of forecasting methods is often an issue. The mean error in forecasting has been found to be about 53%.[66] Although you might think this is high, firms continue to use all methods of forecasting in planning in an uncertain market. Some sense of the potential market size is needed to guide the firm's strategic planning.

Glossary

Affiliate marketing: Marketing practice in which a business rewards a partner company (the affiliate) for each visitor or customer brought to the sponsoring company's Web site through the affiliate's own marketing efforts.

Agent-based models: A computer-based simulation approach that models micro-level consumer choice in order to predict macro-level events (e.g., diffusion, adoption, sales). It allows various exogenous market influences to be added to the model to allow "what if" scenarios.

Assumption-based modeling: A technique that attempts to model the behavior of the relevant market environment by breaking the market down into observable factors. Then, by assuming values for these factors, forecasts are generated.

ATAR model: Awareness, Trial, Availability, and Repeat purchase. Model used to forecast long-run share of consumer goods.

Autoregressive moving average (ARMA)/autoaggressive integrated moving average (ARIMA): A set of advanced statistical approaches to forecasting, which incorporate key elements of both time series and regression model building, sometimes called Box–Jenkins models. Three basic activities (or stages) are considered: (1) identifying the model, (2) determining the model's parameters, and (3) testing/applying the model.

Bass diffusion model: Forecasting method that focuses on (a) "innovators" who buy first and then, through an influence process, (b) encourage others, the "imitators" to adopt. Model suggests that consumers will adopt a product based on a combination of how innovative they are in trying new products (the p coefficient) and how they are influenced by their peers (the q coefficient).

Causal/regression modeling: Techniques use statistical analysis to relate an influencing factor to predicted sales. The term *causal* is used very loosely because these models only represent correlations between variables and not true cause-and-effect relationships.

Concept testing: A survey asking consumers to evaluate their probability of buying a new product/service based on a concept description.

Consideration set: Subset of product/brands that consumers consider when making a purchase decision.

Customer/market analysis: New product development strategy in a *New Market* with *Current Technology*, which includes new use and new market products (market extensions). The purpose of this forecasting strategy would be to reduce risk by understanding the new market.

Decision trees: A probabilistic approach to forecasting where various contingencies and their associated probability of occurring are compared against each other. Commonly used in operations research and financial management to help identify a strategy with the greatest probability of reaching a goal.

Delphi method: A forecasting technique based on subjective expert opinion gathered through several structured anonymous rounds of data collection.

Diffusion models: Models that estimate the growth rate of a *durable product* by considering the impact of mass media and word-of-mouth influence on the consumer adoption process.

Evoked set: See Consideration set.

Executive opinion: A corporate top-down forecasting technique where the forecast is arrived at through the opinions, predictions, and experiential judgment made by informed executives and experts.

Expert systems: Typically, computer-based heuristics or rules for forecasting. These rules are determined by interviewing experts and then constructing numerous "if-then" statements for various scenarios.

Exponential smoothing: A set of techniques that develop forecasts by addressing the forecast components of level, trend, seasonality, and cycle. Whereas, in the simple moving average, the past observations are weighted equally, exponential smoothing assigns exponentially decreasing weights over time.

Extrapolation: A method for projecting out a forecast based on past sales. There are many ways of making the projection: linear, nonlinear, continuous, periodic, etc. Which extrapolation method to apply relies on *a prior knowledge* of the sales history.

Forecast by analogy: A forecasting technique that assumes that two different kinds of phenomena share the same model of behavior.

Intent transition model: A purchase intention model most useful early in the product design process when the new product is still a concept statement or a mathematical position on a perceptual map.

Judgment techniques: Forecasting models that utilize managerial expertise, judgment, and intuition into formal forecasts.

Judgmentally unaided recall: Used to estimate awareness, unaided recall is often easier for advertising managers to estimate than a consumer's consideration set.

Linear regression: Assesses the relation between one or more managerial variables and a dependent variable, strictly assuming that these relationships are linear in nature.

Logistic regression: Assesses the relation between one or more managerial variables and a binary outcome, such as purchasing versus not purchasing. You either have bought an iPad or you have not.

Market components testing: A market research process in which consumers evaluate the effectiveness of a marketing campaign.

Market testing: A market research process by which targeted customers evaluate the marketing plan and the new product in a marketing setting.

Moving average/weighted moving average: In financial applications, a simple moving average (SMA) is the (weighted) mean of the previous n data points. A moving average is commonly used with time series data to smooth out short-term fluctuations and highlight longer-term trends or cycles.

Neural networks: An advanced computer program that attempts to decipher patterns in a particular sales time series. They are based on an information processing paradigm that is inspired by the way the brain processes information.

Nonlinear regression: Assesses the relation between one or more managerial variables and a dependent variable (sales), strictly assuming that these relationships are nonlinear in nature.

Premarket testing: A market research procedure that uses syndicated data (as collected by marketing research firms) and primary consumer research to estimate the sales potential of new product initiatives through forecasting of purchase intent and preference share.

Probability scale: Categories representing an estimate of some probability of purchase.

Product life cycle: A new product strategy in a *Current Market* with *New Technology* for line extensions representing product technology uncertainty.

Product use testing[67]**:** A market research process by which customers (current and/ or potential) evaluate a product's functional characteristics and performance.

Quantitative techniques: Consists of time series, "causal"/regression modeling, and other quantitative forecasting methods. Time series models use data from past sales to make project estimates of future sales. Causal/regression models consider factors that might impact sales and use a mathematical model to quantify this relationship.

Relevant set: See Consideration set.

Sales analysis: A new product development strategy in a *current market* with *current product* technology is where the uncertainties of market and product technology are lowest and include situations with cost reductions and product improvements.

Sales force composite: A corporate bottoms-up forecasting method where frontline employees (i.e., sales personnel and customer service representatives) provide forecasts based on customer feedback. These forecasts are then aggregated across all employees to calculate a higher level forecast.

Sales forecasts: Reasonable estimate of company sales attainable in the given market under consideration with a proposed marketing mix: product, price, place, promotion, positioning.

Sales potential: Estimate of company sales reasonably attainable within a given market under a given set of conditions.

Scenario analysis: Various scenarios are envisioned to predict the future. Exploratory scenario analysis starts in the present and moves out to the future based on current trends. Normative scenario analysis leaps out to the future and works back to determine what path to take to achieve an expected goal.

Scenario analysis (what if): Strategy in *New Market* with *New Product Technology*, representing high market and product technology uncertainties akin to new-to-the-company (new category entries) and new-to-the-world products.

Review Questions

1. Why is forecasting important to the design of new products?
2. What is ATAR forecasting? What are its uses and its accuracy?
3. When should intent-to-purchase measures, such as ATAR, be used? When should the Bass model be used? When should logit regression be used for forecasting?

4. What are the advantages of making probabilistic forecasts? Why not simply assume customers will choose the product that preference regression or conjoint analysis says they should prefer?
5. How can Bass's model be used to forecast the dynamics of the diffusion process?
6. Why is it important to consider awareness and availability when forecasting sales? How do you estimate awareness and availability?
7. What is the difference between sales potential and sales?
8. What diagnostic information is produced with ATAR? How is this information used to improve the new product launch?

Assignment Questions

1. Voras Chemical Supply Company is launching a new brand of exotic dinners designed for picnics where insect pests pose a potential problem. They estimate product awareness to be approximately 99% because the product has received an unexpectedly high press coverage. One supermarket chain that sells 37% of picnic dinners in the target market will carry the product. The probabilities of purchase given awareness and availability include:

PERCENT OF POPULATION	PURCHASE PROBABILITY	INTENT MEASURE
10	90%	Definitely would purchase
40	20%	Might or might not purchase
50	0	Definitely would not purchase

If the target market consists of a half million families each buying one dinner per purchase occasion, what will be the expected number of consumers buying the product? Do you expect that this product will be successful in the long run? Why or why not? How do these numbers change if there is repeat purchase three times yearly?

2. Using the ATAR model of intent, calculate a best "guesstimate" forecast for one of the following fast moving consumer goods:
   a. A high protein organic breakfast cereal
   b. Strawberry-flavored Cheerios
   c. Teeth whitening gum
   d. Compostable paper dinnerware
   e. Red Bull Total Zero energy drink
3. Using forecasting by analogy, use the Bass model to calculate forecasts for:
   a. Electric vehicles
   b. White strips for teeth whitening

   c. George Foreman indoor cooking grill (http://www.georgeforemancooking. com/)

   d. Compostable plastic milk containers

   e. Tablet PC for gaming

   f. 3D television with subscription—only movie viewing

State and support any assumptions that you have made.

4. Draw a hypothetical diffusion curve for each of the products in question 2. Will they have a long takeoff curve or a short takeoff?

5. Based on these findings of Table 7.7, would you introduce a new analgesic? Why or why not? What positioning statement would you give it compared to its competition?

6. Using the Executive Opinion method for judgmental forecasting, with a team of classmates, come up with a three-year and five-year forecast for (a) conventional university classroom learning, (b) hybrid online–classroom learning, and (c) only online classroom learning.

Appendices

Appendix A

Additional Forecasting References

Armstrong, J. S. (ed.). 2001. *Principles of forecasting: Handbook for researchers and practitioners.* Norwell, MA: Kluwer Academic Publishers.

Shocker, A. D., and W. G. Hall. 1986. "Pretest market models: A critical evaluation." *Journal of Product Innovation Management* 3, 86–107.

Makridakis, S., S. C. Wheelwright, and R. J. Hyndman. 1998. *Forecasting methods for management,* 3rd ed. New York: John Wiley & Sons.

Armstrong, J. S., and F. Collopy. 1992. "Error measures for generalizing about forecasting methods empirical comparisons." *International Journal of Forecasting* 8, 69–80.

Meade, N., and T. Islam. 2006. "Modelling and forecasting the diffusion of innovation—A 25-year review." *International Journal of Forecasting* 22, 519–545.

Sources for Estimates of Bass Model p and q

Ofek, E. 2009. "Forecasting the adoption of a new product." *Harvard Business Press,* 9-505-062.

Van den Bulte, C. 2002. "Want to know how diffusion speed varies across countries and products? Try using a Bass model." *PDMA Visions* 26(4): 12–15.

Dolan, R. J., 1992. "Concept testing." *Harvard Business School* Background Note 590-063.

Lilien, G. L., A. Rangaswamy, and C. Van den Bulte. 2000. "Diffusion models: Managerial applications and software." ISBM Report 7-1999. University Park, PA: Institute for the Study of Business Markets.

Sultan, F., J. U. Farley, and D. R. Lehmann. 1990. "A meta-analysis of applications of diffusion models." *Journal of Marketing Research,* 70–77.

Appendix B

Bass Diffusion Model Values of p and q, and Estimated Market Size, m

PRODUCT/TECHNOLOGY	PERIOD OF ANALYSIS	p	q	m
AGRICULTURAL				
Tractors (thousands of units)	1921–1931	0.000	0.211	1324.0
Hybrid corn	1927–1939	0.000	0.798	100.0
Artificial insemination	1943–1953	0.000	0.567	56.9
Bale hay	1943–1955	0.006	0.583	80.3
MEDICAL EQUIPMENT				
Ultrasound imaging	1965–1977	0.001	0.510	89.2
Mammography	1965–1976	0.000	0.738	56.4
CT scanners (50–99 beds)	1980–1990	0.036	0.572	47.8
CT scanners (>100 beds)	1974–1985	0.034	0.254	100.0
PRODUCTION TECHNOLOGY				
Oxygen steel furnace (USA)	1955–1970	0.000	0.477	56.2
Oxygen steel furnace (France)	1961–1974	0.003	0.384	58.2
Oxygen steel furnace (Japan)	1959–1968	0.048	0.324	83.9
Steam (vs. sail) merchant ships (UK)	1815–1900	0.000	0.311	77.0
Plastic milk containers (1 gallon)	1964–1975	0.024	0.331	73.6
Plastic milk containers (half gallon)	1964–1973	0.040	0.630	4.4
Stores with retail scanners (FRG, units)	1980–1993	0.001	0.605	16702.0
Stores with retail scanners (Denmark, units)	1986–1993	0.076	0.540	2061.0
ELECTRICAL APPLIANCES				
Room air conditioning	1950–1963	0.016	0.304	24.2
Bed cover	1949–1962	0.002	0.177	64.2
Blender	1949–1960	0.023	0.199	10.3
Can opener	1961–1971	0.027	0.341	51.8
Electric coffee maker	1955–1965	0.001	0.302	72.8
Clothes dryer	1950–1960	0.009	0.514	18.2
Clothes washer	1923–1936	0.004	0.093	100.0
Coffee maker ADC	1974–1979	0.077	1.106	32.2
Curling iron	1974–1979	0.101	0.762	29.9
Dishwasher	1949–1974	0.000	0.189	57.4
Disposer	1950–1966	0.008	0.256	15.5
Fondue	1972–1979	0.166	0.440	4.6
Freezer	1949–1959	0.043	0.213	25.3
Fry pan	1957–1967	0.301	0.000	51.0
Hair dryer	1972–1979	0.055	0.399	51.6
Hot plates	1932–1942	0.095	0.143	18.2
Microwave oven	1972–1983	0.012	0.383	33.1
Mixer	1949–1959	0.000	0.145	83.0
Power leaf blower (gas or electric)	1986–1996	0.013	0.315	26.0
Range	1925–1935	0.071	0.000	10.2
Range, built-in	1957–1969	0.030	0.000	41.3

Refrigerator	1926–1940	0.015	0.290	69.5
Slow cooker	1974–1979	0.000	1.152	34.4
Steam iron	1950–1960	0.000	0.376	63.8
Toaster	1923–1933	0.039	0.262	46.2
CONSUMER ELECTRONICS				
Cable television	1981–1991	0.080	0.167	60.8
Calculators	1973–1979	0.143	0.520	100.0
Camcorder	1986–1996	0.044	0.304	30.5
CD player	1986–1996	0.055	0.378	29.6
Cellular telephone	1986–1996	0.008	0.421	45.1
Cordless telephone	1984–1994	0.000	0.438	54.0
Electric toothbrush	1991–1996	0.110	0.548	14.8
Home PC (millions of units)	1982–1988	0.121	0.281	25.8
Radio	1922–1933	0.028	0.422	100.0
Telephone answering device	1984–1994	0.019	0.481	63.4
Television, black and white	1949–1959	0.100	0.353	90.1
Television, color	1965–1975	0.058	0.168	97.1
VCR	1981–1991	0.011	0.832	67.5
Average			0.040	0.398

Note: 25th percentile; median, 75th percentile (p, q): (0.001, 0.021); (0.055 0.255); (0.365, 0.519); unless indicated, the model was estimated on penetration data collected in the United States. (From Endnote 68.)

Endnotes

1. Based on R. Jaroslovsky, "The iPad Isn't Just Fun and Games," *Bloomsberg BusinessWeek* (2010, April 12): 20. Online at: http://www.isuppli.com/mobile-and-wireless-communications/news/pages/ipad-sales-to-hit-7-million-in-2010-and-triple-by-2012.aspx (accessed June 14, 2013); C. Guglielmo, "iPad Sales May Be Twice Forecast," *The Boston Globe* (2010, April 5), B7. Online at: http://www.boston.com/business/technology/articles/2010/04/05/ipad_sales_may_be_twice_forecast/ (accessed June 14, 2013).
2. Jaroslovsky, "The iPad Isn't Just Fun and Games."
3. Guglielmo, "iPad Sales May Be Twice Forecast."
4. T. Nguyen, "5 Reasons Tablets Suck and You Won't Buy One," *Tom's Hardware*. Online at: http://www.tomshardware.com/news/tablet-islate-ipad-netbook-notebook,9929.html (accessed June 14, 2013).
5. Jaroslovsky, "The iPad Isn't Just Fun and Games."
6. Ibid.
7. Guglielmo, "iPad Sales May Be Twice Forecast."
8. F. Sideco, "iPad Sales to Hit 7 Million in 2010 and Triple by 2012," *iSuppli*. Online at: www.isuppli.com/mobile-and-wireless-communications/news/pages/ipad-sales-to-hit-7-million-in-2010-and-triple-by-2012.aspx (accessed June 14, 2013).
9. The section is based on Glen Urban and John R. Hauser, "Forecasting Sale Potential," in *Design and Marketing of New Products*, 2nd ed. (New York: Prentice Hall, 1980), Chap. 11.
10. Ken Kahn, "Approaches to New Product Forecasting," in *The PDMA Handbook of New Product Development*, 2nd ed., eds. K. Kahn, G. Castellion, and A. Griffin, p. 385 (Hoboken, NJ: John Wiley & Sons, 2005).
11. Ibid.
12. Kenneth B. Kahn, *New Product Forecasting: An Applied Approach* (Armonk, NY: ME Sharpe, 2006.)

13. Ya-lun Chou, *Statistical Analysis* (Eugene, OR: Holt International, 1975).
14. For more on exponential smoothing, see Spyros Makridakis, Steven C. Wheelwright, and Rob J. Hyndman, *Forecasting Methods and Applications* (Hoboken, NJ: John Wiley & Sons, 2008); John T. Mentzer and Carol C. Bienstock, *Sales Forecasting Management* (Beverley Hills, CA: Sage, 1998); Jon Scott Armstrong, ed., *Principles of Forecasting: A Handbook for Researchers and Practitioners,* 30 (New York: Springer, 2001.)
15. George E. P. Box, Gwilym M. Jenkins, and Gregory C. Reinsel, *Time Series Analysis: Forecasting and Control,* 734 (Hoboken, NJ: John Wiley & Sons, 2011.)
16. Frank Bass, "A New Product Growth Model for Consumer Durables," *Management Science* 15(5) (1969): 215–227.
17. Jeffrey Morrison, "How to Use Diffusion Models in New Product Forecasting," *Journal of Business Forecasting Methods and Systems* 15 (1996): 6–9.
18. V. Mahajan, E. Muller, and Y. Wind, *New-Product Diffusion Models* (Boston: Kluwer Academic Press, 2000).
19. Kahn, *New Product Forecasting*.
20. For more information on neural networks, see: R. J. Thieme, M. Song, and R. J. Calantone, "Artificial Neural Network Decision Support Systems for New Product Development Project Selection," *Journal of Marketing Research* 37(4) (2000): 499–507.
21. For more information on agent-based models, see Rosanna Garcia, "Uses of Agent-Based Modeling in Innovation/New Product Development Research," *Journal of Product Innovation Management* 22(5) (2005): 380–398.
22. Ken Kahn, "Approaches to New Product Forecasting," in *The PDMA Handbook of New Product Development,* 2nd ed., K. Kahn, G. Castellion, and A. Griffin (eds.), p. 385. (Hoboken, NJ: John Wiley & Sons, 2005).
23. Ibid.
24. Kahn, "New Product Forecasting."
25. W. J. Infosino, "Forecasting New Product Sales Likelihood of Purchase Ratings," *Marketing Science* 5 (1986): 372–390.
26. Based on L. F. Jamieson and F. M. Bass, "Adjusting Stated Purchase Intention Measures to Predict Trial Purchase of New Products: A Comparison of Models and Methods," *Journal of Marketing Research* 26 (1989): 336–345.
27. Ibid.
28. Albert C. Bemmaor, "Predicting Behavior from Intention to Buy Measures: The Parametric Case," *Journal of Marketing Research* 32 (1995): 176–191.
29. Urban and Hauser, *Design and Marketing of New Products*.
30. A. J. Silk and G. L. Urban, "Pre-Test Market Evaluation of New Product Goods: A Model and Measurement Methodology," *Journal of Marketing Research* 15 (1978): 171–191.
31. Spring Tide is the false brand.
32. "Fridge Packs Appear to Be Plus for Coke System," *Beverage Digest* (2003, March 28). Online at: http://www.beverage-digest.com/editorial/030328.php (accessed July 23, 2011).
33. Charles Merle Crawford and C. Anthony Di Benedetto, *New Products Management*, 9th ed. (New York: McGraw Hill, 2008), p. 418.
34. Urban and Hauser, *Design and Marketing of New Products*.
35. See A. J. Silk and U. K. Manohar, "Measuring Influence in Organizational Purchase Decisions," *Journal of Marketing Research* 19 (1982): 165–18; Dolan, "Concept Testing."
36. Urban and Hauser, *Design and Marketing of New Products*.
37. Crawford Di Benedetto, *New Products Management*.
38. Urban and Hauser, *Design and Marketing of New Products*, p. 319.
39. Based on Urban and Hauser, *Design and Marketing of New Products*.
40. Glen L. Urban, "PERCEPTOR: A Model for Product Positioning," *Management Science* 21(8) (1975): 858–871; Urban and Hauser, *Design and Marketing of New Products*.

41. Edward M. Tauber, "Forecasting Sales Prior to Test Market," *Journal of Marketing* (1977): 80–84.

42. This is meant to be an example and not accurate numbers for repeat purchase with any product.

43. F. T. Juster, "Consumer Buying Intentions and Purchase Probability: An Experiment in Survey Design," *Journal of the American Statistical Association* 61 (1966): 658–696.

44. B. C. Gan, D. Esslemont, and P. J. Gendall, "A Test on the Accuracy of the Juster Scale as a Predictor of Purchase Behaviour," *Market Research Center Report No. 45*, Massey University (1985); Dianne Day, Boon Gan, Philip Gendall, and Don Esslemont, "Predicting Purchase Behaviour," *Marketing Bulletin* 2(5) (1991): 18–30.

45. Day, Gan, Gendall, and Esslemont, Ibid.

46. The exception seems to be for automobiles. Both Juster, "Consumer Buying Intention and Purchase," and Day et al., "Predicting Purchase Behavior," showed good results using the Juster scale for the prediction of sales of cars.

47. Joseph A. Foster, Peter N. Golder, and Gerard J. Tellis, "Predicting Sales Takeoff for Whirlpool's New Personal Valet," *Marketing Science* 23(2) (2004): 182–185.

48. F. M. Bass, "A New Product Growth Model for Consumer Durables," *Management Science* 15 (1969).

49. Based on Elie Ofek, "Forecasting the Adoption of a New Product," *Harvard Business Press*, 9-505-062, (2009), p. 4.

50. Peter N. Golder, and Gerard J. Tellis, "Will It Ever Fly? Modeling the Takeoff of Really New Consumer Durables," *Marketing Science* 16(3) (1997): 256–270.

51. For a nice example of how to use nonlinear regression to determine estimates of p and q, see David Cramer's example for the forecast of a social media computer game. D. Cramer, "Forecasting the Market of a Social Game Start-Up," *Hollywood and Wall*, 2012. Online at: http://www.hollywoodandwall.com/wordpress/forecasting-the -market-of-a-social-game-start-up/

52. Bass, "A New Product Growth Model for Consumer Durables."

53. Christophe Van den Bulte also provides estimates for different geographical regions, see C. Van den Bulte, "Want to Know How Diffusion Speed Varies across Countries and Products? Try Using a Bass Model," *PDMA Visions* 26(4) (2002): 12–15.

54. Ofek, "Forecasting the Adoption of a New Product."

55. Unless otherwise noted, estimates of p and q are from Gary L. Lilien, Arvind Rangaswamy, and Christophe Van den Bulte, "Diffusion Models: Managerial Applications and Software," ISBM Report 7-1999 (University Park, PA: Institute for the Study of Business Markets, 2000).

56. Based on Ofek, "Forecasting the Adoption of a New Product."

57. Van den Bulte, "Want to Know How Diffusion Speed."

58. Robert J. Thomas, "Estimating Market Growth for New Products: An Analogical Diffusion Models Approach, *Journal of Product Innovation Management* 2(1) (1985) 45–55.

59. Ofek, "Forecasting the Adoption of a New Product."

60. See Mahajan, Muller, and Bass, "New Product Diffusion Models in Marketing," for a review of the literature. See also R. Chatterjee and J. Eliashberg, "The Innovation Diffusion Process in a Heterogeneous Population: A Micro-Modeling Approach," *Management Science* 36(9) (1990); and D. Horsky, "A Diffusion Model Incorporating

61. This section is based on Urban and Hauser, *Design and Marketing of New Products*, p. 286.

62. Daniel L. McFadden, "Conditional Logit Analysis of Qualitative Choice Behavior," *Frontiers in Econometrics* (1973): 105–142.

63. Based on Urban and Hauser, *Design and Marketing of New Products*, p. 286.

64. For details on the different methods to obtain individual preferences, see Urban and Hauser, *Design and Marketing of New Products*, Chap. 10.

65. As reported in Urban and Hauser, *Design and Marketing of New Products*, Chap. 11.

66. Kenneth B. Kahn, "An Exploratory Investigation of New Product Forecasting Practices," *Journal of Product Innovation Management* 19(2) (2002): 133–143.
67. Lilien, Rangaswamy, and Van den Bulte, "Diffusion Models."
68. Ibid.

PROACTIVE NEW PRODUCT DEVELOPMENT PROCESS

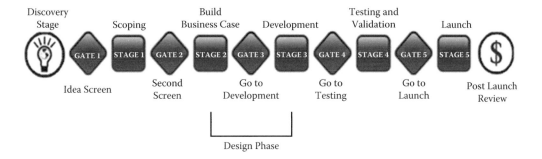

Learning Objectives

In this chapter, we will address the following:

1. How to translate customers' needs into winning product designs.
2. How to use the House of Quality to combine the voice of the customer (VOC) with voice of the engineer (VOE).
3. What is the role of marketing in engineering design/development?
4. What is the product platform? What is the product architecture?
5. How does Design Thinking alter the approach to traditional R&D?

Designing a Better TV Experience at TiVo

TiVo is rooted in a failed partnership between Time-Warner and Silicon Graphic Inc. An innovative new cable television experiment, called *Full Service Network*, was offered to 4,000 Orlando households. The service had some 500 television channels and a new kind of remote control allowing viewers to program their favorite shows. But the much-hyped Full Service Network never caught on, and reportedly cost Time-Warner $100 million.[1] Mike Ramsay and Jim Barton, co-founders of TiVo, had both been involved in work on the doomed market launch. The two realized that they could build a smarter system giving viewers control over their television programming, and their time, at a price the average consumer

could afford.[2] They went about creating the first mass-merchandised digital video recorder (DVR), which was eventually branded TiVo.

Ramsay and Barton assembled a team of designers to blend a service with an intelligent hardware device to create the world's first personal television service bringing freedom in television watching to the masses.[3] Early on in the design process, the remote control was an important issue. Ramsay and Barton knew they had to get it right. They brought on Paul Newby, a design engineer with experience from Caterpillar construction equipment. In June 1998, as the company was just starting up, Newby and a team of six designers were given 14 weeks to come up with a functioning remote control. In their design efforts, they relied on feedback from potential users on everything from the feel of the device in the hand to the best place for the batteries. Central to the process, Newby said, was producing prototypes "early, ugly, and often." Ugly products helped the design team understand what made a "good" remote.

Nonengineers on the TiVo staff were frequently solicited for feedback. This helped the designers refine the size and shape of the keys and the amount of space between them. TiVo remotes also were distributed to beta testers, consisting of technologically inexperienced friends and relatives of employees. Feedback led to improvement in button response time, fine-tuning for maximize comfort level, and best positioning of the keys.[4] In January 1999, the company unveiled its Personal Television Service at the Consumers Electronics Show. And, since that time, they have worked diligently to deliver products for the average consumer to enjoy and interact with television.

Introduction

In previous chapters, we looked at how the design stage is made up of three phases: Scoping, Building the Business Case, and Development. In Scoping, we discussed how the voice of the customer (VOC) is collected and used to generate a product definition, and in Building the Business Case, we developed a preliminary market plan with an early forecast. A new product design can only be handed off to development when we know who will buy (target), how it will sell (core benefit proposition (CBP)), why our offering is superior (positioning), what we will deliver to customers (features), and how to communicate and distribute the CBP (advertising, selling, channels, and technology). During the development phase, which is the focus of this chapter, the firm will invest considerable resources in designing and engineering the proposed product. Although most of the tasks are to be completed by R&D and engineering, the marketing department continues to play a major role in the new product development (NPD) process as customer needs are metamorphosed into design requirements. The VOC should continue to be heard as the product moves through development. To understand where the VOC fits into the product development phase, let us first look at what is meant by *design*.

Design

As we enter the development phase, we should now have (1) a carefully crafted CBP, (2) a set of features that substantiate a value-generating product, and (3) an indication of the potential to generate profits. Next, we need to translate the proposition and the set of features into detailed engineering and production designs. In order to engineer a product, we first need to generate specifications that will be used by the engineers to design the product. Product specifications are often described as **form, fit, function**, sometimes called **F3**. Definitions of F3 include:

Form: Visual parameters that are unique to a product, including its shape, size, dimensions, and mass. This is the look or physical package of the item. From a service perspective, form refers to the combination of "servicescape" and "physical evidence" that facilitate performance or communication of the service[5] (i.e., Ritz Carlton versus Best Western hotels). **Servicescape** is the setting in which a service occurs. It can include landscape, signage, parking, interior decor, music, air quality, temperature, and ambiance. **Physical evidence** in the servicescape refers to the tangibles, such as business cards, stationary, billing statements, reports, employee dress, uniforms, brochures, and Web pages used to present the service to consumers.

Fit: Refers to how the product physically interfaces or interconnects with part of another item or assembly. Tolerances are also part of fit, where tolerance refers to the allowable limit of variation in a physical dimension, manufactured object, or service, such as temperature, humidity, mean-time-to-failures, etc. From a service perspective, fit refers to how the **touchpoints**, where customers interact with the service, and the form (physical evidence and servicescape) come together.

Function: Function refers to the action that an item is designed to perform or a service is designed to provide. This defines the primary use of the product or the service. For example, in the TiVo digital video recorder (DVR), the function is a digital recorder of TV broadcasts, both a product and a service.

So how do consumers' needs metamorphose into product specifications that are characterized through the F3? As we saw in Chapter 4, after gathering VOC data, a new product team might identify 100 to 300 phrases that describe customer needs. For example, we saw in Table 4.6 a list of 18 of the 100+ customer needs for the Segway's balancing system, including 0° turn radius, minimal lean required, no special training, feels balanced, etc. After gathering these needs using the VOC method, the NPD team will then categorize these needs as *primary, secondary,* and *tertiary,* as shown in Table 8.1, which provides an example of a simplified hierarchy of needs for the Segway. At the **primary level** are the strategic benefits that characterize the CBP; at the **secondary level** are the tactical features that are associated with the perceptual dimensions of the CBP; and at the **tertiary** level are the operational needs that correspond to detailed engineering attributes of the product or service and that establish the design features. In Table 8.1, we have specified 3 primary (strategic) needs, 4 secondary (tactical) needs, and 18 tertiary (operational) needs. Notice how the complexity increases as one drills down to the final design level where products are specified with increasing operational detail. In typical applications, tertiary needs, the lowest level of the hierarchy, comprise a list of 100 to 300 need statements used in detailed product design. These hundred or more statements are grouped into 10 to

Table 8.1 Needs Hierarchy for a Segway

PRIMARY NEEDS	SECONDARY NEEDS	TERTIARY NEEDS
• Easy to use	• Easy to turn	• 0° radius turn
		• Minimum leaning
		• No training needed
		• Feels balanced when turning
		• Smooth turns, not jerky
• Smooth ride	• Shock absorbent	• No noticeable difference as terrain changes
		• No knee stress from standing
		• Automatic rider weight recognition/adjustment
		• Weather resilient
	• Wheels	• Radial tires
		• 200 hour usage
		• Less than 10 min. replacement
		• All-terrain
• Stylish body	• Easy to maintain	• Low/no maintenance
		• No greasing
		• Aluminum casing
		• No loose parts
		• Paintable surface for customization

Note: Hypothetical as composed by the authors.

30 secondary needs, and the secondary needs themselves are grouped into a relatively few (2–8) primary needs.

Top management strategists are most concerned with the primary needs as they relate to the CBP and the overall image of the product. The design team who implements and communicates the CBP focuses on the secondary needs, which fulfill the primary promises. At the most detailed tertiary level, the engineering team must "get all the little things right" to deliver the core benefits with the most effective design and in the most cost-efficient manner. The use of the hierarchy assures that the detailed design decisions are "on strategy" with respect to the customer benefits to be delivered.

Ulrich and Eppinger[6] demonstrate the steps for taking customer needs and developing them into a set of product specifications that can be used to design a new product (Figure 8.1). If marketing has identified the customers' needs using VOC, then marketing will need to work with engineering to translate these customer needs into product specifications. If the customer needs have been identified by engineering, specifications typically may come in the form of technical requirements that refer to the key design variables of the product, such as MIPS (Series 2) processors, MPEG-2 encoder/decoder chips, and high-capacity IDE/ATA hard drives, which are examples of TiVo specifications. Although a new product may initiate in R&D or engineering, it must still deliver on the CBP, therefore, these technical specifications must still align with the customers' needs.

In either case, **target specifications**, which represent the hopes and desires of the NPD team, must be generated. These target specifications are then used to generate several, not just one, product designs for consideration. Selected designs are typically discussed at design meetings or at a gate before final specifications are set for R&D consideration. Often the designs will be tested by engineering to determine their design feasibility. To set the **final specifications**, the team must frequently make hard trade-offs among different desirable characteristics of the product. The final specifications will incorporate the primary, secondary, and tertiary needs and should clearly delineate in detail *what* the product has to do. Product specifications do not tell the team *how* to address the customer needs, but provide clear guidelines on what the product must look like and how it needs to function in order to satisfy the customer needs.

As shown in Figure 8.1, it is rare that a successful product is finalized on the first iteration through the design process. Speed of refinement through the reiterations depends on the effective integration of marketing, R&D, engineering, production, and other functional areas of the organization. Establishing final specifications must

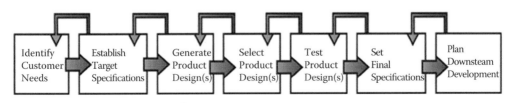

Figure 8.1   Product specification process. (From Endnote 7. With permission.)

be a cross-functional activity. Marketing ensures that the customer's voice is represented and that the communication plan provides a positioning consistent with the CBP; R&D ensures that the best technology is used so that the product is **parity plus** or "breakthrough"; engineering ensures that the physical product fulfills the CBP in a cost-effective manner; and production assures that the product can be built and the associated services delivered in a manner of consistent quality, low cost, and continuous improvement as outlined in the CBP. However, cross-function communication is often difficult as engineers talk objectively from a technical perspective and marketers often speak more subjectively. Fortunately, based on years of observation on cross-functional communications, researchers have found methods to ease knowledge transfer between the two disparate groups. We discuss these methods next.

Voice of the Engineer Blending with Voice of the Customer[8]

We clearly need a way to communicate customer needs uncovered by marketing to engineers so they can design products that meet customers' tertiary, secondary, and primary needs. Thus, descriptions of the physical product that are measurable and quantifiable and that relate to the customer needs must be identified in the language of the engineer. Before the industrial revolution, producers were close to their customers. Marketing, R&D, engineering, and production were integrated—in the same person. If a knight wanted armor, he talked directly to the armorer, who translated the knight's desires into a product. The two might discuss the material—plate, rather than chain armor—and details like fluted surfaces for greater bending strength. Then the armorer would design the entire process. For strength, he cooled the steel plates in the urine of a black goat (perhaps due to its acid content). For a production plan, he arose with the cock's crow to light the forge fire so that it would reach its critical temperature at midday. Today, clearly this type of communication is not feasible for each customer.

The challenge to an organization is to integrate the VOC into product development while retaining the creativity essential to innovative new product design. **Quality function deployment** (QFD), a component of the Six Sigma quality management,[9] is one method of merging the **voice of the engineer** (VOE) with the voice of the customer.[10,11] QFD itself is a set of processes that link customer needs all the way through to production requirements. Although the full QFD process is sometimes used, the first matrix of QFD, the **House of Quality (HOQ)**, is used most often by engineering managers. In the center of the house, a "relationship matrix" indicates how each engineering characteristic relates to each customer need. The basic structure of the HOQ consists of the short, accurate, relevant list of key customer needs identified in the fuzzy front end and structured into our primary, secondary, and tertiary needs (labeled 1 in Figure 8.2). These needs are related to product features (label 2), which are then evaluated as to how well they meet customer needs (label 3). Product features are "benchmarked" against competitors' features in their ability to meet customer needs (label 4, sometimes called *the porch*) and the HOQ is used to compare

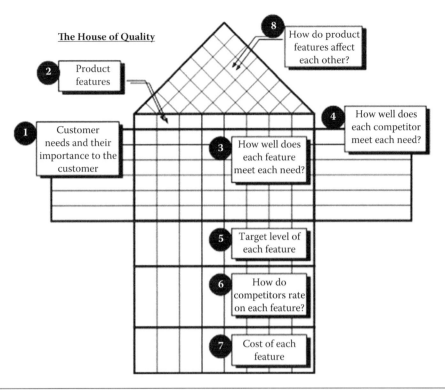

Figure 8.2   The House of Quality (HOQ). (From Endnote 12. With permission.)

how the features meet customer needs (labels 5 and 6). Finally, costs are compared to determine if the total product with its defined features meets customer needs more effectively and at lower costs than competitive products (label 7). The "roof" (label 8) is used to determine how tradeoffs between features should be made.[13]

Let's run through a hypothetical situation to illustrate how HOQ directs the engineering design. General Motors (GM) had an interest in building a car targeted toward the aging baby boomer market.[14] They noticed that existing doors are much more difficult to close from the outside than those on the competitors' cars and marketing data confirmed that ease of closing a car door was important to customers. Using the HOQ, it was identified which engineering characteristics affected this customer need: energy to close the door, peak closing force, and door seal resistance. Engineers judged the energy to close the door and the peak closing force as good candidates for improvement together because they are strongly, positively related to the customer's desire to close the door easily.

In the "roof of the house," label 8 in Figure 8.2, other engineering characteristics that might be affected by changing the door closing energy were identified. Door opening energy and peak closing force are positively related, but other characteristics (check force on level ground, door seals, window acoustic transmission, road noise reduction) are bound to be changed in the process and are negatively related. It is not an easy decision on which variables to change. However, with objective measures of competitors' doors, customer perceptions, and considering information on cost and technical difficulty, the design team decided that the benefits of the change outweigh the costs. A new design target was set for the door of the new car: 7.5 ft/lbs of energy.

Once the relationship matrix is complete, the new product team compares the customer perceptions to objective measures. For example, if the car door with the least energy to open is perceived by customers as the most difficult to open, then either the measures are faulty or the car suffers from an image problem that might be addressed with an improved marketing program. A change in the gear ratio on a car window may make the window motor smaller, but cause the window to go up more slowly. If the design team decides to enlarge or strengthen the motor, the door might be heavier causing it to be more difficult to open, especially on a slope.

The values in the relationship matrix are obtained by consensus of the interfunctional, new product team. In many cases, this judgment is influenced by market research studies measuring the impact of product features on perceptions or by experiments with alternative features or part characteristics. The roof matrix helps the team balance engineering changes to ensure that important customer benefits are not inadvertently adversely affected. Together the relationship matrix and the roof matrix indicate the complex interrelationships among various product changes.

The roof is a correlation matrix that links each engineering characteristic with every other engineering characteristic. For example, look now at the customer need "no road noise" and its relationship to acoustic transmission of the window. As seen in Table 8.2, larger relative importance numbers indicate that the customer values those features more than others. "Road noise" is only mildly important to customers, and its relationship to the characteristics of the window is not strong. "Easy to close from the outside" has a relative importance of 7, while "no road noise" might have a relative importance of a 2. Window design will help only so much to keep things quiet. Decreasing the acoustic transmission usually makes the window heavier. Examining the roof of the house, we see that more weight would have a negative impact on

Table 8.2 Customer Needs for a Car Door

NO.	PRIMARY NEED	TERTIARY NEED	IMPORTANCE
1	Good operation & use	Easy to close from the outside	7
2	Good operation & use	Stays open on a hill	5
3	Good operation & use	Easy to open from the outside	3
4	Good operation & use	Doesn't kick open	3
5	Good operation & use	Easy to close from inside	2
6	Isolation	Doesn't leak in rain	3
7	Isolation	No road noise	2
8	Isolation	Doesn't leak in car wash	3
9	Isolation	No wind noise	4
10	Isolation	Doesn't drop water when open	7
11	Isolation	Doesn't rattle	3
12	Isolation	Snow doesn't fall into car	2

characteristics (open–close energy, check forces, etc.) that, in turn, are strongly related to customer needs that are more important to the customer than quiet ("easy to close," "stays open on a hill"). If the customer was faced with two cars and all else were equal, the customer would choose the car with the easy-to-close door over the car with less road noise. Clearly these trade-offs are essential in designing the best product for customers. Thus, a decision is made to focus on the weight of the door instead of the acoustic transmission. This is how the voice of the engineer is blended with the voice of the customer.

An advantage of the HOQ and similar techniques is that it enhances communication among NPD team members by encouraging cross-functional departments to work as a team to produce a coordinated CBP, the physical product, and the marketing plan. Although many companies do not involve marketing in R&D activities, numerous studies have indicated the benefits of doing so.[15] This need for cross-communication also was documented in a study of 167 firms (109 marketing managers and 107 R&D managers), which found that the no. 1 barrier to integration across functions was lack of communication.[16] This study also found that R&D managers did not find marketing's input to be of the same level that marketing perceived it to be. However, marketing and R&D managers agreed that the five most important areas where integration is needed are: customer requirements, feedback on product performance, information on competitors, development of products according to market needs, and setting of new product goals and priorities. If NPD managers can align with the goals of the engineering/R&D departments, it is much easier for the two groups to resolve conflicts in the NPD process when they arise, which inevitably will occur. Open communication grows in importance as NPD teams become more dispersed and global.

Another advantage of the HOQ is related to speed to market. By various claims, QFD has reduced design time by 40% and design costs by 60% while maintaining or enhancing the quality of the design.[17] QFD achieves these goals by encouraging communication among functional groups early in the design effort and by assuring that the design is focused on the voice of the customer. More activity early in the design process means that fewer engineering changes need to be made as the product nears production and this means time to market is shorter.

The downside of HOQ is that strict adherence to the method can lead to overly complex charts and become extremely time-consuming, especially for products with many customer needs and engineering key characteristics. Ironically, complex projects that make QFD difficult to implement may be the very ones that benefit most from improved communication and coordination within the firm.

In summary, the HOQ is most useful after determining the primary customer needs, customer perceptions of competitive products with respect to those needs, and the CBP. In the QFD/HOQ process, R&D provides creative solutions to improve the engineering characteristics that relate to the tertiary customer needs, which fulfill the primary and secondary needs of the CBP. Engineering produces the final product design that is consistent with the CBP. Marketing's primary *input* to the HOQ includes identifying customer needs (fuzzy front end), measuring how products fulfill those needs,[18] and communicating to R&D the trade-offs among customer needs and among potential product features. From the *output* of the HOQ, marketing uses the defining product features to firm up the marketing program and looks to the relationship matrix to select highlights that establish the credibility of the CBP. Ultimately, the HOQ method translates customer priorities, as captured by a prioritized list of needs, into engineering and/or design priorities by identifying those product features that contribute the most to satisfying customers better than competitive offerings. The HOQ has proved effective in a variety of applications including consumer packaged goods, consumer durables, consumer services, business-to-business products, and business-to-business services.[19]

Generate Product Designs

We have now set target specifications and R&D has taken them to begin developing product designs. Starting with the core benefit proposition, the customers' needs list and the target product specifications are the required inputs to the product generation process. For example, in designing the TiVo, the CBP could have been identified as an "affordable, feature rich, intelligent, user-friendly digital video recorder that gives viewers control over their television programming and their time." Then a list of customer needs would be identified:

- Record to a set-top box TV broadcasts from antenna, cable, or direct broadcast satellite transmissions for future viewing
- Easy-to-use, ergonomically pleasing remote control
- Viewer ability to pause live television, and rewind and replay
- User programmable for custom recording
- Download programs to iPod or computer

R&D teams subsequently translate these customer needs into target product specifications, such as:

- 40 to 250 GB data storage supported with Linux operating system
- 30-button (about) "peanut-shaped" consumer controllable remote
- Analog-to-digital conversion receiver, HDTV receiver
- Easy menu options
- USB interface

Once the goals of design are understood, the problems can be decomposed into **subproblems**. As we previously noted, the TiVo design process included one subproblem of the remote, which was integral to the design as the set-top box had no buttons and all functions were to be controlled by the remote. Other subproblems for the TiVo could include hardware components of set-top box and software programming of set-top box.

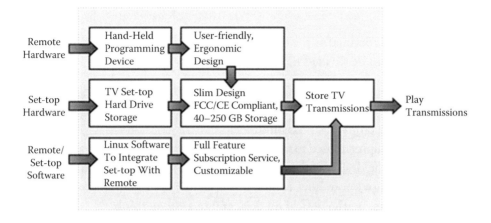

Figure 8.3 Functional decomposition of TiVo.

Ulrich and Eppinger[20] suggest dividing subproblems either by (a) functional decomposition, (b) sequencing by user actions decomposition, or (c) key customer segments. Figure 8.3 shows the functional decomposition of the TiVo into the three subproblems: remote hardware, set-top hardware, and remote/set-top software. This product decomposition helps us to identify the product architecture and the product platform.

Product Architecture and Platform in Product Design

When evaluating a product's functional decomposition, the product's platform and architecture must be a consideration, which are used to help a firm develop and manage a family of products based on a common set of assets. Reuse of components, processes, and design solutions leads to competitive advantages in process engineering and economies of scale. In its product planning stage, TiVo concentrated on perfecting its product platform for its remote controller across the different product offerings. The company believed that the ease of use of the remote could make or break marketplace acceptance of the TiVo DVR, so they wanted to "get it right."

Product Platform

Product platform is defined as the collection of assets that are shared by a set of products. These assets can be divided into four categories:

Components: The bill or material of a product, the fixtures and tools needed to make the product, the circuit designs, and the software programs burned into programmable chips or stored on disks.

Processes: The equipment used to make components or to assemble components into products and the design of the associated production process and supply chain.

Knowledge: Design know-how, technology applications and limitations, production techniques, mathematical models, and testing methods.[21]

People and relationships: Teams, relationships among team members, relationships between the team and the larger organization, and relationships with a network of suppliers.

Taken together, these shared assets constitute the product platform. Generally, products using the same platform share many if not most development and production assets. In contrast, although efforts to standardize parts across products lines may lead to the sharing of a modest set of components, shared components are generally not considered a product platform.[22] A platform is "a set of subsystems and interfaces that form a common structure from which a stream of derivative products can be efficiently developed and produced."[23] Honey Nut Clusters may share a couple of the ingredients of Honey Nut Cheerios, but they are not derivatives of each other due to completely different manufacturing processes, and, thus, don't share the same platform. (Both are products of General Mills.)

Can you identify the platform for Black and Decker's line of handheld tools, as shown in Figure 8.4? What are the commonalities in the core technology or design across the tools? What commonalities might be shared with competitors? A successful platform strategy should lead to innovation and generation of new revenue growth by leveraging existing

Figure 8.4   Black and Decker portable tools platform.

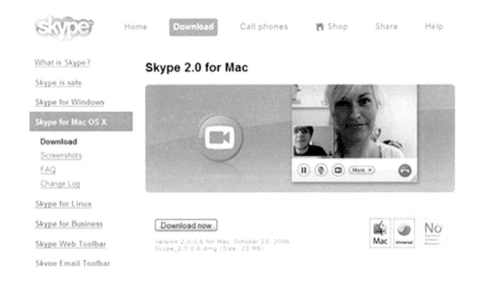

Figure 8.5   Skype's instant messaging platform.

brands, modules, and subsystem technologies into new product offerings. What new product ideas do you envision for Black and Decker that might use their existing platform?

Software platforms often can play a competitive role in the acceptance of a product/service in the marketplace. The support of 1080p and 1080i HD formats for digital cable viewing for TiVo helped to drive its sales. Another example is when Skype opened up its platform in 2005 to anyone who wanted to integrate Skype's voice-over-the-Internet and instant messaging services into their application. By opening up its platform to the Web, Skype quickly created the world's most popular instant messaging service (Figure 8.5). However, platform dominance is not required to be successful; Apple's strategy has been a "closed" platform, yet, its products sell well worldwide.

The platform approach is also a way to achieve successful **mass customization**, which is the manufacturing of products tailored-made to the preferences of the individual consumer. Chocomize allows customization of its chocolate bars based on a common platform for all of its products (Figure 8.6). This type of product strategy allows highly differentiated products to be delivered to the market without consuming excessive resources.

In 1987, Fuji, the Japanese camera film company, was first to market the QuickSnap 35 mm single-use in the U.S. market. Kodak, the American-based film company, did not have a comparable product of its own and was caught off guard to an exponentially growing market. "By the time Kodak introduced its first model almost a year later, Fuji had already developed a second model, the QuickSnap Flash. Yet, Kodak won market share back from Fuji. By 1994, Kodak had captured more than 70% of the U.S. market. The success of Kodak's response resulted in part from its strategy of developing many distinctively different models from a common platform. Because

Figure 8.6   Chocomize's mass customization strategy. (From Endnote 24. With permission.)

Kodak designed its four products to share components and process steps, it was able to develop its products faster and more cheaply. The different models appealed to different customer segments and gave Kodak twice as many products as Fuji, allowing it to capture precious retail space and garner substantial market share."[25] A strategic product platform that can easily address a wide range of user requirements in fast-changing technologies allows a firm to quickly address evolving market needs. Yet, this alone cannot keep a company afloat. As of early 2012, Kodak was making plans to declare bankruptcy, as it failed to compete against digital photography.

Product Architecture

The product platform will drive the product architecture. **Product architecture** is the structure that integrates product components and subsystems into a coherent entity to perform intended behaviors and functions.[26] It can be thought of as the structure of the product (similar to a building's architecture), whereas the product platform is the commonality between structures. It also reflects rationale and intention of the design, such as the products functions, methods of use, methods of maintenance, and manufacturing. Ulrich and Eppinger refer to the product architecture subsystems as *chunks*,[27] as shown in Figure 8.7. Architecture is not limited to products alone, but also pertains to service offerings. How do the chunks one encounters in shopping at Ikea, the furniture store, differ from shopping at Pier One Imports, a similar type of home furnishing store?

Establishing a product's architecture is a major development decision that most impacts a firm's ability to efficiently deliver high product variety.[28] However, as variety increases, component standardization (complexity), product performance, manufacturability, and product development management are significantly impacted.[29] Just as adding more rooms to a house will determine how easily it is to build and to navigate through after construction, increasing the number of chunks in a product increases the complexity of the product design and the difficulty in manufacturing. This becomes a trade-off between **distinctiveness** and **commonality** in a product's architecture, as shown in Figure 8.8. If a firm wants to strive for distinctiveness in a competitive market,

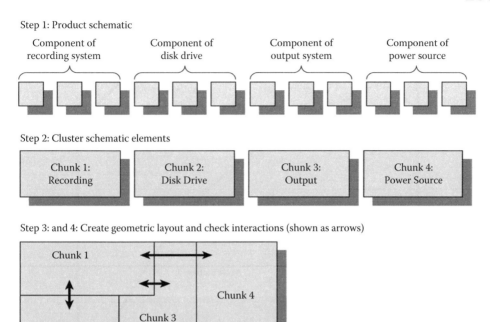

Figure 8.7 Product architecture's chunks. (From Endnote 30. With permission.)

it will operate at either Scenario A or Scenario D. If commonality is more important than distinctiveness, the firm will operate at Scenario B or Scenario C. Custom-made furniture is often seen as works of art and, thus, would be very distinctive. Dell's strength is in custom orders using a common product architecture. The Apple iPod series have very common parts, but the nano is distinct from the iTouch and the iPod. If you are not familiar with custom M&Ms, you can order them in your favorite color with your face on them; very distinctive from the same M&M covered-chocolate candy at your local convenience store. What other products can you think of that can be broken into architectural chunks? How does common chunks help MiniCooper market its cars?

Product platforms and product architectures are the "tactical" tools used in planning product design. Technology roadmaps, which we discuss next, help to outline a strategy for which platforms and architectures will most benefit the product portfolio.

Technology Roadmapping

Technology roadmapping[31] is a strategic and long-range planning technique that is widely used to support the development, communication, and implementation of technology and business strategies. "The approach provides a structured (and often graphical) means for exploring and communicating the relationships between evolving and developing markets, products, and technologies over time. It is proposed that the roadmapping technique can help companies survive in turbulent environments

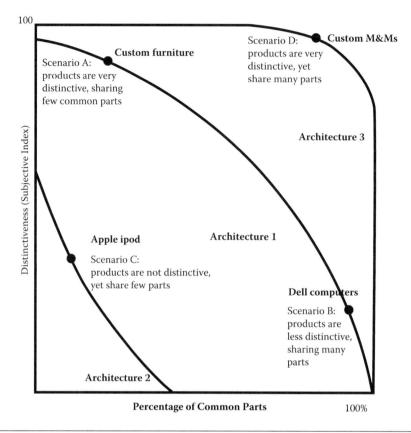

Figure 8.8 Trade-off between distinctiveness and commonality. (From Endnote 32. With permission.)

by providing a focus for scanning the environment and a means of tracking the performance of individual, including potentially disruptive, technologies."[33] Many companies use the technique to involve their suppliers with the long-term vision of the organization and the industry. It is closely related to other graphical planning approaches, such as PERT (program evaluation and review technique) and Gantt planning tools,[34] except that it takes a holistic viewpoint of technology evolution.

A number of different types of roadmaps have emerged. Several are shown in Figure 8.9 and explained below:

(a) *Product planning* (most common type of technology roadmap): Relates technology to manufactured products, often including multiple generations of products. Figure 8.9a shows a roadmap based on Philips Electronics, of The Netherlands, where the approach has been widely adopted.[35] This roadmap links planned technology and product developments.

(b) *Service/capability planning*: Relates organizational capabilities to technology and business strategies for service-based firms. Figure 8.9b shows a Royal Mail roadmap[36] used to investigate the impact of technology developments on mail delivery service.

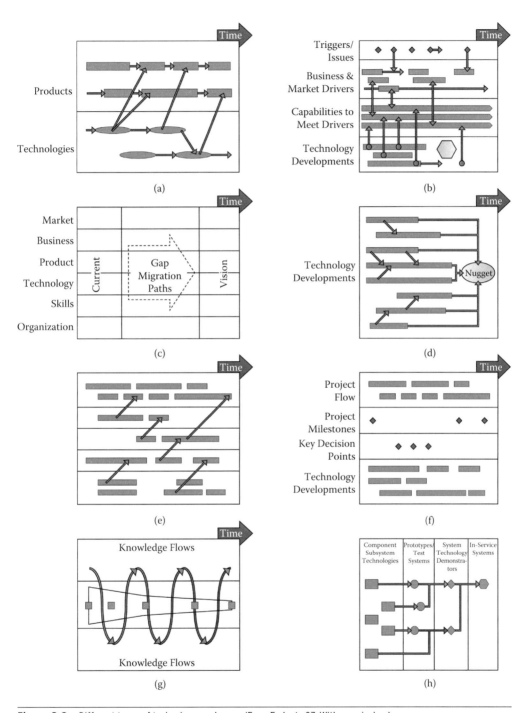

Figure 8.9   Different types of technology roadmaps. (From Endnote 37. With permission.)

(c) *Strategic planning*: Relates general strategic appraisal to different opportunities or threats at the business level. The roadmap in Figure 8.9c[38] focuses on the development of a vision of the future business, in terms of markets, business, products, technologies, skills, culture, etc. Any gaps identified are addressed with strategic planning.

(d) *Long-range planning*: Addresses a long-range planning horizon at the sector or national level, and "can act as a radar for the organization to identify potentially disruptive technologies and markets."[39] Figure. 8.9d is based on the U.S. Integrated Manufacturing Technology Roadmapping Initiative, which focuses on information systems, showing how technology developments are likely to converge on an industry "nugget," in this case, toward the "information-driven seamless enterprise."

(e) *Knowledge asset planning*: Aligns knowledge assets and knowledge management initiatives with business objectives to meet future market demands. Figure 8.9e shows an example developed by the Artificial Intelligence Applications Unit at the University of Edinburgh.[40]

(f) *Program planning*: Relates to project planning, for example, for R&D programs. The U.S. National Aeronautics and Space Administration (NASA) roadmap[41] of Figure 8.9f was used to explore how the universe and life within it has developed. It shows the relationships between technology development and program phases and milestones.

(g) *Process planning*: Focuses on management of knowledge within a process area, such as the fuzzy front end of NPD. Figure 8.9g shows the NPD knowledge flows incorporating both technical and commercial perspectives.

(h) *Integration planning*: Focuses on integration and/or evolution of technology, in terms of how different technologies combine within products and systems, or to form new technologies. Figure 8.9h, another type of NASA roadmap,[42] shows how technology feeds into test and demonstration systems to support scientific missions.

Attempts have been made to generalize the technology roadmap architecture so that firms can more easily coordinate company-specific roadmaps with their suppliers and customers viewpoints. In this universal map shown in Figure 8.10, the "top layers relate to the organizational purpose that is driving the roadmap (know-why). The bottom layers relate to the resources (particularly technological knowledge) that will be deployed to address the demand from the top layers of the roadmap (know-how). The middle layers of the roadmap are crucial, providing a bridging or delivery mechanism between the purpose and resources (know-what). Frequently, the middle layer focuses on product development, because this is the route through which technology is often deployed to meet market and customer needs. However, for other situations, services, capabilities, systems, risks, or opportunities may be more appropriate for the middle layer, to understand how technology can be delivered to provide benefits to the

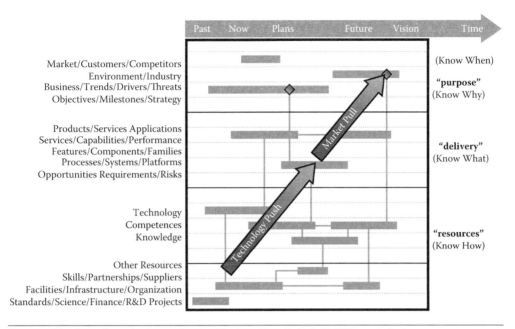

Figure 8.10 Generalize technology roadmap architecture. (From Endnote 43. With permission.)

organization and its stakeholders."[44] Most roadmaps include market pull and technology push dimensions, reflecting the balance between product and market development drivers.

Design Thinking and the NPD Process

In this section, we move from the traditional analytical approaches (product platforms, product architectures, and technology roadmapping) of product design to a more organic approach in product development. During the mid-2000s, there was a renaissance in **design thinking** as a way of approaching the product development process. "Design thinking is a human-centered approach to innovation that draws from the designer's toolkit to integrate the needs of people, the possibilities of technology, and the requirements for business success."[45] More simply, it is a creative method that uses the voice of the customer to identify innovative product design solutions to meet customer needs. GE Healthcare has used design thinking in revamping its MRI and CT appliances for its youngest patients who often required sedation because of the frightening-looking machine. "Instead of being slid into an ominous machine in a sterile hospital room, [GE] redesigned the entire MRI experience so that children enter an adventure, from a pirate ship to a camping trip—tent included. The result? A dramatic drop in sedation rates. Happier patients, caregivers, and families."[46]

Problem-Solving Approach or Process[47]

Design thinking can be regarded as a problem-solving method or, by some definitions, a process for the resolution of problems.[48] It is particularly useful for addressing so-called "wicked" or ill-defined problems. Even when the general direction of the problem may be clear, considerable time and effort may be needed on clarifying the requirements. Thus, in design thinking, a large part of the problem-solving activity is comprised of defining and shaping the problem. The problem resolution process is regarded as creative, fluid, and open, with the search for an improved future result. Products such as the Swiffer Carpet Flick, toothpaste caps for Crest, and emergency room flow at community hospitals were all created using the design thinking techniques of IDEO, the popular design company based in Palo Alto, California.

Process and Methods

Design thinking comprises the *design process* using *design methods*. The **design process** is the way in which the methods come together through a series of actions, events, or steps. Several different models of the design thinking process have been proposed (see Table 8.3 for examples), including a three-step simplified triangular process by Tim Brown from IDEO and a six-step process with feedback loops from Stanford's d.school and Hasso-Plattner-Institut D-School of Potsdam, Germany. Although there is no specific process that defines design thinking, the most popular approach comes from the d.school at Stanford University. We describe this process below. Design methods are all the techniques, rules, or ways of doing things that are employed by a design discipline. Some of the methods used in design thinking originate from traditional **human–computer interaction** (HCI) methods and **user-centered design** (UCD) methods, while others are more specific to designers, or borrowed from creativity training.

The design thinking process is similar to the stage-gate process. It consists of a number of stages, typically between three and seven, and can be linear or circular, returning to the starting point and beginning a new iteration, as shown in Figure 8.11.[49] Design thinking has a mindset that is used throughout the process (Figure 8.12). For a virtual crash course on design thinking, see the d.school site (http://dschool.stanford.edu/dgift/).

Process Stages   The Design Thinking process as established by the Stanford d.school program consists of six stages: Understand, Observe, Point of View, Ideate, Prototype, and Test. As we have seen in the stage-gate process, the steps are not necessarily sequential and should involve feedback loops. Although *test* is the last stage shown in the process, it is actually one of the most important stages. Only by testing a product design with the end user can it be determined whether the product solves a problem.

Table 8.3 Different Design Thinking Processes

PROTOTYPICAL STAGES	HERBERT SIMON (1969)[50]	IDEO TOOLKIT[51]	TIM BROWN/IDEO (2008)[52]	D.SCHOOL STANFORD/D-SCHOOL (HPI)[53]	D.SCHOOL BOOTCAMP BOOTLEG (HPI)—MODES[54]	BAECK & GREMETT[55] (2011)	MARK DZIERSK[56] (FAST COMPANY)
Understand the Problem	Define	Discovery	Inspiration	Understand	Empathize: Observe, engage, immerse	Define the problem	Define the problem
Observe Users	Research			Observe		Look for inspiration	
Interpret the Results		Interpretation		Point of View	Define (Problem statement)		
Generate Ideas	Ideation	Ideation	Ideation	Ideate	Ideate	Ideate multiple ideas	Create and consider many options
Prototype, Experiment	Prototype	Experimentation	Implementation	Prototype	Prototype	Generate prototypes	Refine selected directions; repeat steps as needed
Test, Implement, Improve	Objectives/Choose, Implement, Learn	Evolution		Test	Test (includes *refine* and *improve* solutions)	Solicit user feedback	Pick the winner

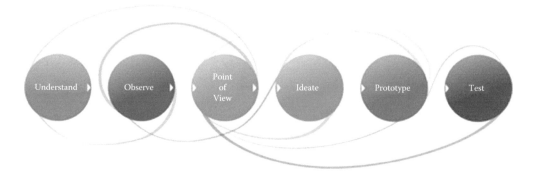

Figure 8.11   Design thinking process. (From Endnote 57. With permission.)

In testing, the end user's point of view is acknowledged and ideation begins again to make product improvements. A summary of the stages includes:

- *Understand*: Get an initial understanding of the problem.
- *Observe*: Observe users, visit them in their (work) environment, observe physical spaces and places.
- *Point of View*: Interpret the empirical findings by combining insights with a needs-focused approach.
- *Ideate*: Engage in brainstorming sessions to generate as many ideas as possible (expand the solution space).
- *Prototype*: Build "quick and dirty" prototypes and share them with other people (narrow down the solution space again, experimental phase).
- *Test*: Test, implement, and refine the design (narrow down the solution space again; solution-driven phase).

Focus on Human Values   Show Don't Tell   Create Clarity from Complexity   Get Experimental and Experiential

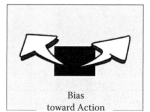
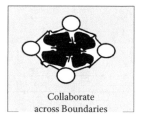

Be Mindful of Process   Bias toward Action   Collaborate across Boundaries

Figure 8.12   d.school mindsets. (From Endnote 54.)

A Creativity Approach   There is much more creativity involved in the process than is evident in the simple diagram in Figure 8.11. Unlike analytical thinking, which is associated with the "breaking down" of product ideas, such as we have seen with product platforms, product architecture, and technical roadmaps, design thinking is a creative process based on the "building up" of ideas. Analytical approaches reportedly focus on narrowing the design choices, while design thinking focuses on going broad, at least during the early stages of the process.[58] Similar to other ideation strategies we have looked at in this book, in design thinking, designers do not make any early judgments about the quality of ideas. This minimizes the fear of failure and maximizes input and participation in the ideation (brainstorming) and prototype phases. "Wild ideas" are encouraged in the earlier process stages, since this style of thinking is believed to lead to creative solutions that would not have emerged otherwise. The motto here is "everyone is a designer." Despite being driven by creativity, it is a multidisciplinary approach. The Hasso-Plattner-Institut (HPI) D-School leaders "believe that innovation happens when strong multidisciplinary groups come together and build a common collaborative culture to explore their different perspectives."[59] In their experience, "Design Thinking is the glue that holds different types of disciplines together and makes the projects successful."[60]

As a methodology or style of thinking, design thinking combines *empathy* for the context of a problem, *creativity* in the generation of insights and solutions, and *rationality* and feedback to analyze and fit solutions to the context. All this helps derive a solution that meets user needs and at the same time generates revenue, that is, drives business success. The *core attributes* of design thinking, together with their descriptions, are summarized in Table 8.4.

A User-Centered Approach That Brings Design into the Business World   According to the design thinking experts Baeck and Gremett,[61] design thinking is a more creative and user-centered approach to problem solving than traditional design methods. They point out that "Design Thinking defies the obvious and instead embraces a more experimental approach." The heart of the method is in understanding the customer— all ideas and subsequent work stem from knowing the customer. In this vein, design thinking has been characterized as a discipline in which the designer's sensibility and methods match people's needs, by applying what is technically feasible and by contemplating what a viable business strategy can convert into customer value and market opportunity.

Methods

In the course of the design thinking process, a wide variety of methods can be employed with no strict rules as to which method to choose. Some of the methods are typical of the way designers work; others are based on ones used in human-centered interface (HCI)/user-centered design (UCD), or have been borrowed from creativity

Table 8.4   Core Attributes of Design Thinking

ATTRIBUTE	DESCRIPTION	DETAILS
Ambiguity	Being comfortable when things are unclear or when you don't know the answer	Design Thinking addresses wicked problems = ill-defined and tricky problems
Collaborative	Working together across disciplines	People design in interdisciplinary teams
Constructive	Creating new ideas based on old ideas, which also can be the most successful ideas	Design Thinking is a solution-based approach that looks for an improved future result
Curiosity	Being interested in things you don't understand or perceiving things with fresh eyes	Considerable time and effort is spent on clarifying the requirements; a large part of the problem-solving activity, then, consists of problem definition and problem shaping
Empathy	Seeing and understanding things from your customers' point of view	The focus is on user needs (problem context)
Holistic	Looking at the bigger context for the customer	Design Thinking attempts to meet user needs and also drive business success
Iterative	A cyclical process where improvements are made to a solution or idea regardless of the phase	The Design Thinking process is typically nonsequential and may include feedback loops and cycles
Nonjudgmental	Creating ideas with no judgment toward the idea creator or the idea	Particularly in the brainstorming phase, there are no early judgments
Open mindset	Embracing design thinking as an approach for any problem regardless of industry or scope	The method encourages "outside the box thinking" ("wild ideas"); it defies the obvious and embraces a more experimental approach

Source: Endnote 62.

training. In Table 8.5, we provide a list of methods that are used and promoted by the Stanford d.school[63] to illustrate the types of approaches that can be used in the design thinking process. We contrasted these methods with exemplary traditional HCI/UCD methods.

Role of Marketing in Design

We have examined analytic approaches to product design, including product platforms, product architectures, and technology roadmapping. We also looked at the more creative approach, design thinking. From these analyses, it may not be obvious the role that marketing should take in the new product design process. MIT Professors Urban and Hauser identify six primary roles that marketing should take in this stage of the NPD process:[64]

> 1. A New Product is both a physical entity and a psychological positioning.
> A product must perform from an engineering viewpoint, but customers will not buy the product if they do not perceive that it delivers the benefits they need. For example, the right amount of fluoride is important in toothpaste, but it is not enough. The toothpaste must be seen as decay preventing, tasting great, preventing of gum disease, etc. Procter &

Table 8.5 Selected Methods of the Design Thinking Process Contrasted with Traditional HCI/UCD Methods

STAGE (MODE)	METHODS FROM D.SCHOOL BOOTCAMP BOOTLEG (2010)	TRADITIONAL HCI/UCD/ETHNOGRAPHY METHODS
• Understand (Background Knowledge)[65]	• Interview experts • Conduct research (domain, literature, market)	• Interview experts • Conduct research (domain, literature, market) • Define problem statement
• Observe Users (Empathy)	• Assume a beginner's mindset • Question everything • What? How? Why? • User camera study • Interview (for empathy) • Extreme users • Analogous empathy • Story share-and-capture • Bodystorming[66]	• Observe users (may include "think aloud" protocols) • Conduct site visits (contextual inquiry) (users doing their tasks at their work places) • Interview users, send questionnaires
• Point-of-view[67] (Synthesis, Interpret the results)	• Space saturate and group • Empathy map • Journey map • Composite character profile • Powers of ten • 2 x 2 matrix • Why–how ladder • Point-of-view madlib • Point-of-view analogy • Point-of-view want ad • Critical reading checklist • Design principles • "How might we" questions	• Contextual inquiry models (including affinity diagrams, etc.) • Personas, roles • Use cases, scenarios, user stories, day-in-a-life scenarios, etc.
• Ideate (Generate ideas)	• Powers of 10 • Stoke; game playing • Brainstorming (+ selection) • Bodystorming • Impose constraints	• Brainstorming • User days, focus groups
• Prototype (Experiment)	• Bodystorming • Impose constraints • Prototype for empathy • Prototype to test	• Prototyping • Low fidelity: Wireframes, paper prototypes, simple HTML prototypes • High fidelity: More or less functional prototypes
• Test and Improve	• Testing with users • Prototype (in 30 minutes or less) to decide • Identify a variable • User-driven prototyping • Wizard of Oz prototyping • Feedback capture grid	• User tests in the lab (may include think aloud protocols) • Remote user tests • User tests in the field • Informal user tests (e.g., with colleagues, friends, etc.) • KPI studies
• Other	• Storytelling • Shooting video (+ editing) • I wish, I like, what if	• Storytelling • Card sorting

Gamble's focused marketing on the ingredient, fluoride, in Crest toothpaste for reducing tooth decay, but within the image that using Crest was being a good parent. Their "Look Mom, No Cavities" advertising strategy showed peer group approval for the parents who had their children use Crest. Similarly, new tartar control toothpastes and rinses must be perceived as effective before consumers will buy them and the benefits of tooth hygiene must be of value to consumers. Although vitamins are perceived as a way of reducing stress in Japan, nutrition is the chief benefit in Europe and the United States. Understanding the link from customers' values to engineering variables is critical for the NPD managers to deliver on good product designs that meet worldwide opportunities.

2. The Design Process is iterative.

   Design is not accomplished in one step. As we saw with design thinking, evaluation, refinement, and learning take place in an iterative fashion. Early in the process, the design team concentrates on developing a good basic appeal and image of the product. Later they concentrate on the specific physical and psychological features that achieve the chosen positioning. Finally, they select the full marketing strategy. As the product proceeds through this process from idea to concept to prototype to production, it is continually refined and improved. At all times, the team must be willing to backtrack and reassess the market, position, product, and technology.

3. New Product Design is integrative.

   All functions, marketing, engineering, R&D, and production, must cooperate if the new product is to be a success. R&D provides the technology, but technology is good only when it works. Engineering must assure that the CBP is delivered when the customer uses the product under field conditions. Marketing can identify customer wants and needs, but this information must be integrated with engineering to select the most profitable CBP. Production factors also must be integrated early in the design process. Integration leads to better quality products, less expensive products, and products that are brought to market faster.

4. Both prediction and understanding are necessary.

   If prediction were the only goal of marketing research in the design process, the strategic summary of the customer's input would be unnecessary. Because the design process is iterative and new product teams are fallible, obtaining the voice of the customer throughout the development process provides valuable information with which to improve the product and marketing strategy and with which to refine the CBP. If a new product or service is not "right" the first time, VOC research can tell the new product team why and how to get it "right."

5. Marketing should guide trade-off decisions between time-to-market and completeness.

   Marketing should provide input at each gate. It is always possible to spend more time and more funds on building the perfect product. Marketing should assist in identifying the "best" design, which should be the one that most efficiently supplies the required input at the specific point in the decision process. In the concept development phase, money spent on predictive accuracy might better be invested in improving the product features and/or the positioning. All uncertainty cannot be removed so spending too much time on design could be a mistake, but not having sufficient customer response data could lead to an inferior design and a poor fit of market needs and product features. These problems need to be balanced so that the time-to-market can be short and the quality of the product high.

6. New product development is both an art and a science.

   The design process blends managerial judgment with analytical and creative techniques. Each aspect, managerial judgment, quantitative analysis, and creative input, is important for developing a winning new product. No one aspect can stand alone. Blending these aspects is an art, but once learned, it is a powerful art. The effective new product team combines managerial prowess, analytical approaches, and creative thinking throughout the NPD process.

Chapter Summary

In this chapter, we examined the Development stage. We saw how the Voice of the Engineer merged with the Voice of the Customer to develop a product that delivers on the Core Benefit Proposition of the new product/service. This stage stresses understanding the technical, production, and marketing capabilities of the organization to develop a new product based on customer needs. Form, fit, and function are basic product design characteristics with which the new product manager must be concerned. Another goal of the new product manager is to ensure the CBP is incorporated into the product platform as well as the product architecture. The effort put forth in the early stages will be reflected in the quality of the output of the product because engineers will be focused on the "right things." The importance of cross-functional teams in the development process is critical in delivering the best product/service solution to customers.

The technology roadmap can provide a holistic viewpoint on where the product platform and product architecture fits into the overall firm or industry technology strategy. We also saw how companies are beginning to use design thinking to drive the creative development process using the d.school mindset. This mindset assures that the customers' needs and wants are integrated into creative product designs.

Designing a superior innovative product is perhaps the most creative and challenging part of the new product development process.

Glossary

Chunks: Modules that make up a new product's architecture

Commonality: Common parts across a product line or similar products.

Core benefit proposition: A short statement of the key benefits a product fulfills; it forms the keystone upon which all elements of the marketing strategy for a new product/service are built.

Design methods: All the techniques, rules, or ways of doing things that are employed by design thinking discipline.

Design process: The way in which the design methods come together through a series of actions, events, or steps.

Design thinking: A human-centered approach to innovation that draws from the designer's toolkit to integrate the needs of people, the possibilities of technology, and the requirements for business success.[68] It is a creative method that uses the voice of the customer to identify innovative product design solutions to customer needs.

Distinctiveness: Unique parts across a product line or similar products.

Final specifications: A list of the primary, secondary, and tertiary needs used to identify a new product/service design.

Fit: How the product physically interfaces or interconnects with part of another item or assembly, or the touchpoints where customers interact with service providers.

Form: Visual parameters that are unique to a product, including its shape, size, dimensions, and mass; the servicescape for a new service.

Function: The action that an item is designed to perform or a service is designed to provide.

House of Quality: A diagram, resembling a house, used for defining the relationship between customer needs and wants (VOC) and the firm/product design (VOE). It is used to identify trades-offs between customer needs and design capabilities compared to competitive offerings.

Human–computer interaction (HCI): The study, planning, and design of the interaction between people and computers. It is often regarded as the intersection of computer science, behavioral sciences, and design.

Integration planning: Focuses on integration and/or evolution of technology, in terms of how different technologies combine within products and systems, or form new technologies.

Knowledge asset planning: Aligns knowledge assets and knowledge management initiatives with business objectives to meet future market demands.

Long-range planning: Addresses a long-range planning horizon at the sector or national level, acting as a radar for the organization to identify potentially disruptive technologies and markets.

Mass customization[69]: Production of personalized or custom-tailored goods or services to meet consumers' diverse and changing needs at near mass production prices. Enabled by technologies, such as computerization, Internet, product modularization, and lean production, it portends the ultimate stage in market segmentation where every customer can have exactly what he or she wants.

Observe: A step in Design Thinking, which focuses on observing users, visiting them in their (work) environment, observing physical spaces and places.

Parity plus product: A product that has all the same features, functions, and benefits of a competitive product, but adds on additional attributes to increase the product's value.

Physical evidence: Refers to the tangibles like business cards, stationary, billing statements, reports, employee dress, uniforms, brochures, and Web pages used to present a service to consumers.

Point of view: A step in Design Thinking, which focuses on interpreting the empirical findings by combining insights with a needs-focused approach.

Primary level: The strategic benefits that characterize the core benefit proposition (CBP) of a new product.

Process planning: A type of technical roadmap that focuses on management of knowledge within a process area, such as the fuzzy front end of NPD.

Product architecture: (1) The structure that integrates components and subsystems of a product into a coherent mechanism that performs intended behavior and functions.[70] (2) The scheme in which the functional elements of the product are arranged into physical chunks.[71]

Product planning: Most common type of technology roadmap, which relates technology to manufactured products, often including multiple generations of products.

Product platform: (1) The collection of assets that are shared by a set of products. These assets might include components, knowledge, and production processes.[72] (2) Elements of technologies and concepts that are shared by multiple products in order to maximize the utility for responding to the wide range of requirements including cost, user choices, environmental adaptation, manufacturing, product development, operational compatibility, and services.[70]

Program planning: A type of technical roadmap that relates to project planning, for example, for R&D programs.

Prototype: A step in Design Thinking that focuses on building "quick and dirty" prototypes and sharing them with other people (narrow down the solution space again, experimental phase).

Quality function deployment (QFD): A set of processes that link customer needs all the way through to production requirements; one method of merging the voice of the engineer with the voice of the customer.

"Roof" matrix: (See label 8 in Figure 8.2) The matrix that links each engineering characteristic with every other engineering characteristic.

Secondary level: The tactical features that are associated with the perceptual dimensions of the core benefit proposition.

Service/capability planning: Relates organizational capabilities to technology and business strategies for service-based firms.

Servicescape: The setting in which a service occurs; includes the mode being established through customer touchpoints and the atmosphere created within the place of business.

Strategic planning: Relates general strategic appraisal to different opportunities or threats at the business level.

Subproblems: The decomposition of product design broken into smaller more manageable components.

Target specifications: Product specs that satisfy the core benefit proposition of a new product.

Tertiary level: The operational needs that correspond to detailed engineering attributes of the product or service and that establish the design features.

Test: A step in Design Thinking that focuses on testing, implementing, and refining the design (narrow down the solution space again; solution-driven phase).

Touchpoints: Where customers interact with service providers.

Understand: A step in Design Thinking that focuses on getting an initial understanding of the end user's problem.

User-centered design: A design process in which the needs, wants, and limitations of end users of a product are integrated into each stage of the design process.

Voice of the customer (VOC): Strives to record in the customer's own words the benefits of a product or service; it is a description of the customer's needs within a specific product category.[73] Voice of the customer (VOC) studies should be a combination of qualitative and quantitative research.

Voice of the engineer: Input from the R&D or engineering team on the technical feasibility to design a product or service that adheres to the voice of the customer and the core benefit proposition.

Review Questions

1. A key goal of the design process is the core benefit proposition (CBP). What is a CBP? Why is it important? What should a CBP contain?

2. How does the voice of the engineer merge with the voice of the customer in the House of Quality? What information is provided that helps the new product team refine the new product ideas?

3. What are the types of technology roadmaps that can be used in the new product development process? How would a firm use a roadmap to integrate into its suppliers strategic goals?

4. What advantages are there to establishing a product platform? A product architecture?

5. How do the steps in d.school approach to Design Thinking differ from the more traditional human-centered interaction and user-centered design approaches? How does the d. mindset alter the design perspective?

6. What role should marketing play in the product development process?

Assignment Questions

1. What are the product platforms for each of the following products? What is the product architecture?
   a. Kodak Single Use Flash camera
   b. Schick personal razors
   c. MiniCooper vehicle
   d. Apple's iPhone

2. Take the d.school virtual crash course in Design Thinking at: http://dschool. stanford.edu/dgift/. Explain how design thinking may change the design for:
   a. A digital camera
   b. A personal razor
   c. A high miles-per-gallon vehicle
   d. A mobile phone

3. Using one of the examples in the Appendix, take a product and explain the tradeoff that must be considered during a House of Quality evaluation.

4. Schedule a meeting with an R&D/engineering manager at the company you work for or use a social media site, such as Linkedin, to find one that may be willing to talk with you. Discuss with the manager how his/her company incorporates voice of the engineer with the voice of the customer. If the company does not, find ways that it might be able to do so.

5. Create a technology roadmap for one of the following industries:
   a. Athletic shoes
   b. E-readers
   c. Organic foods

    d. Electric vehicles

    e. Mobile banking

    f. Bottled water

    g. The industry you work in

6. One of the methods in design thinking is "composite character profiling." Create a profile for the user of a library, a museum, or a concert hall. You can find more details about this method and other design thinking methods at http://dschool.stanford.edu/wp-content/uploads/2011/03/BootcampBootleg2010v2SLIM.pdf (accessed June 20, 2013). How did this process fostered new product ideas?

Appendix

Additional References for Design Thinking

Aline Baeck and Peter Gremett. 2011. Design thinking. In *UX best practices—How to achieve more impact with user experience,* eds. H. Degen and X. Yuan. New York: McGraw-Hill Osborne Media.

d.school (Hasso Plattner Institute of Design at Stanford).

d.school Bootcamp Bootleg. 2010. Online at: http://dschool.stanford.edu/blog/2010/12/17/2010-bootcamp-bootleg-is-here/; and d.school Bootcamp Bootleg. 2009. Online at: http://dschool.stanford.edu/blog/2009/12/16/the-bootcamp-bootleg-is-here/

Hasso Plattner, Christoph Meinel, and Ulrich Weinberg. 2009. *Design thinking.* Potsdam, Germany: Hasso Plattner Institute.

HPI School of Design Thinking. Hasso-Plattner-Institut, Potsdam, Germany. (German)

Tim Brown. 2008. Design thinking. *Harvard Business Review.*

Tim Brown. 2008–2011. Design thinking—Thoughts by Tim Brown (blog; IDEO), http://designthinking.ideo.com/

IDEO: DesignThinkingforEducators (free toolkit; IDEO). Online at: http://www.designthinkingforeducators.com/

Mark Dziersk. 2006. Design thinking…What is That? (fastcompany.com).

Helen Walters. 2012. The seven deadly sins that choke out Innovation (fastcompany.com).

Helen Walters. 2012. Design Thinking Isn't a Miracle Cure, But Here's How It Helps (fastcompany.com).

Studio Noshoku: Design Thinking (blog page with a collection of citations on Design Thinking).

Critiques

Peter Merholz (2009) (ex-Adaptive Path; HBR Blog network, *Harvard Business Review*): Why Design Thinking Won't Save You.

Bruce Nussbaum (2012): Design Thinking Is a Failed Experiment. So What's Next?

Dan Saffer (2012): How to Lie with Design Thinking (Interaction 2012, Dublin, Ireland; vimeo; 8:21 min).

Endnotes

1. Michael Ramsay and James Barton Biography, Advamag, Inc. Online at: http://www.notablebiographies.com/news/Ow-Sh/Ramsay-Michael-and-Barton-James.html (accessed June 2, 2012).
2. Online at: http://www.tivo.com/abouttivo/jobs/historyoftivo/ (accessed June 2, 2012).
3. Ibid.
4. J. Atwood, "Coding Horror: The Tivo Remote." Online at: http://www.codinghorror.com/blog/archives/000008.html (accessed June 2, 2012).
5. Bernard H. Booms and Mary Jo Bitner, "Marketing Strategies and Organization Structures for Service Firms," *Marketing of Services* (1981): 47–51.
6. K. T. Ulrich and S. D. Eppinger, *Product Design and Development*, 4th ed. (New York: McGraw-Hill/Irwin, 2007).
7. Ibid, p. 74.
8. Glen L Urban and John R. Hauser, *Design and Marketing of New Products* (Upper Saddle River, NJ: Prentice Hall, 1993).
9. Forrest W. Breyfogle, III, *Implementing Six Sigma: Smarter Solutions Using Statistical Methods* (Hoboken, NJ: John Wiley & Sons, 2003).
10. For more on QFD, see: Joseph P. Ficalora and Louis Cohen, *Quality Function Deployment and Six Sigma: A QFD Handbook* (Hoboken, NJ: Prentice Hall, 2009); Amy Tessler, Norm Wada, and Robert L. Klein, "QFD at PG&E," in *Transactions from The Fifth Symposium on Quality Function Deployment*, June, 1993; Xiao-Xiang Shen, Kay C. Tan, and Mien Xie, "An Integrated Approach to Innovative Product Development Using Kano's Model and QFD," *European Journal of Innovation Management* 3(2) (2000): 91–99; Abbie J. Griffin, "Evaluating QFD's Use in US Firms as a Process for Developing Products," *Journal of Product Innovation Management* 9(3) (1992).
11. For a nice example of the components of QFD, see: http://www.webducate.net/qfd/qfd.html and for downloadable excel worksheets see http://www.qfdonline.com/templates/qfd-and-house-of-quality-templates/ (accessed June 18, 2013).
12. Ely Dahan and John R. Hauser, "Product Development—Managing a Dispersed Process," in *Handbook of Marketing*, eds. Barton A. Weitz and Robin Wensley (London: SAGE Publications Ltd., 2002).
13. Ibid.
14. Urban and Hauser, *Design and Marketing of New Products*.
15. For a review of the literature in this area, see A. Griffin and J. R. Hauser, "Integrating R&D and Marketing: A Review and Analysis of the Literature," *Journal of Product Innovation Management* 13(3) (1996, May): 191–215.
16. Raj Gupta and David Wilemon, "R & D and Marketing Dialogue in High-Tech Firms," *Industrial Marketing Management* 14(4) (1985): 289–300.
17. Hauser and Clausing, "The House of Quality."

18. For example, Paul E. Green, Donald S. Tull, and Gerald Albaum, *Research for Marketing Decisions* (Englewood Cliffs, NJ: Prentice-Hall, 1988); Gilbert A. Churchill, *Marketing Research: Methodological Foundations* (Fort Worth, TX: Dryden Press, 1999); William L. Moore and Edgar A. Pessemier, *Product Planning and Management: Designing and Delivering Value* (New York: McGraw-Hill, 1993).

19. J. R. Hauser, "How Puritan Bennett Used the House of Quality," *Sloan Management Review* 34(3) (1993): 61–70. See also Griffin, "Evaluating QFD's."

20. Ulrich and Eppinger, *Product Design and Development*.

21. David Robertson and Karl Ulrich, "Planning for Product Platforms," *Sloan Management Review* 39(4) (1998).

22. Ibid.

23. Online at: http://npdbook.com/tools-and-definitions/definitions/ (accessed June 19, 2013).

24. Online at: http://www.miniusa.com/#/shop/motoringGear-m (accessed June 12, 2012).

25. Robertson and Ulrich. "Planning for Product Platforms." Online at: http://sloanreview.mit.edu/article/planning-for-product-platforms/ (accessed June 18, 2013).

26. Online at: http://bauhaus.id.iit.edu/index.php?id=1163 (accessed June 2, 2012).

27. See Ulrich and Eppinger, *Product Design and Development*, for more information on project/concept selections.

28. Ulrich and Eppinger, *Product Design and Development*.

29. Ibid.

30. Charles Merle Crawford and C. Anthony Di Benedetto, *New Products Management* (New York: McGraw-Hill Irwin, 2011); based on Ulrich and Eppinger, *Product Design and Development*.

31. This section is based on Robert Phaal, Clare J.P. Farrukh, and David R. Probert, "Technology Roadmapping—A Planning Framework for Evolution and Revolution," *Technological Forecasting and Social Change* 71(1) (2004): 5–26.

32. Robertson and Ulrich, "Planning for Product Platforms."

33. Ibid.

34. J. S. Martinich, *Production and Operations Management—An Applied Modern Approach* (New York: John Wiley & Sons, 1997).

35. P. Groenveld, "Roadmapping Integrates Business and Technology," *Research Technology Management* 40(5) (1997): 48–55.

36. R. Brown and R. Phaal, "The Use of Technology Roadmaps as a Tool to Manage Technology Developments and Maximise the Value of Research Activity," paper presented at the IMechE Mail Technology Conference (MTC), Brighton, U.K., April 24–25, 2001

37. Phaal, Farrukh, and Probert, "Technology Roadmapping."

38. Phaal, Farrukh, and Probert, "Technology Roadmapping," p. 12.

39. Ibid.

40. A. Macintosh, I. Filby, and A. Tate, "Knowledge Asset Roadmaps," paper presented at the Proceedings of the 2nd International Conference on Practical Aspects of Knowledge Management, Basel, Switzerland, October 29–30, 1998.

41. NASA, Origins technology roadmap. 1997. Online at: http://origins.jpl.nasa.gov/ (accessed June 19, 2013).

42. Ibid.

43. Phaal, Farrukh, and Probert, "Technology Roadmapping," p. 12.

44. Ibid.

45. Tim Brown, IDEO President and CEO, "Design Thinking," *Harvard Business Review* (2008).

46. Paula Felps, "Is Design Thinking Missing in Your Corporate DNA?" *Insigniam Quarterly*. Online at: http://insigniamquarterly.com/2013/03/13/is-design-thinking-missing-in-your-corporate-dna/ (accessed June 18, 2013).

47. This section is co-written by Gerd Waloszek, SAP AG, "SAP User Experience," 2012, based on his summary review of design thinking. Online at: http://www.sapdesignguild.org/community/design/design_thinking.asp (accessed June 18, 2013).

48. See the Appendix at the end of the chapter for a list of additional resources on Design Thinking.

49. d.school Bootcamp Bootlegs from 2009, 2010. Online at: http://dschool.stanford.edu/wp-content/uploads/2009/12/bootcampbootleg20091.pdf and http://dschool.stanford.edu/wp-content/uploads/2011/03/BootcampBootleg2010v2SLIM.pdf (accessed June 19, 2013).

50. Online at: http://www.designthinkingforeducators.com/ (accessed June 19, 2013).

51. Herbert Simon, *The Sciences of the Artificial* (Cambridge, MA: MIT Press, 1969).

52. Brown, "Design Thinking."

53. d.school at Stanford University, Palo Alto, California; D-School at Hasso-Plattner-Institut (HPI) in Potsdam, Germany.

54. Online at: http://dschool.stanford.edu/wp-content/uploads/2011/03/BootcampBootleg2010v2SLIM.pdf (accessed June 19, 2013).

55. Aline Baeck and Peter Gremett, "Design Thinking," in *UX Best Practices—How to Achieve More Impact with User Experience,* eds. Helmut Degen and Xiaowei Yuan (New York: McGraw-Hill Osborne Media, 2011).

56. Mark Dziersk, Design Thinking … What is That? 2006. Online from Fast Company: http://www.fastcompany.com/919258/design-thinking-what (accessed June 19, 2013).

57. Online at: www.designthinkingblog.com/wp-content/uploads/2009/10/Design-thinking-process.png (accessed June 18, 2013).

58. Baeck and Gremett, "Design Thinking."

59. Online at: http://www.hpi.uni-potsdam.de/d_school/designthinking/core_elements.html ?L=1 (accessed July 9, 2013).

60. Ibid.

61. Ibid.

62. Based on Baeck and Gremett, "Design Thinking."

63. Online at: http://dschool.stanford.edu/wp-content/uploads/2011/03/BootcampBootleg2010v2SLIM.pdf (accessed June 19, 2013).

64. Urban and Hauser, *Design and Marketing of New Products.*

65. Adapted from: The k12 Lab Wiki— Steps in a Design Thinking Process. Online at: https://dschool.stanford.edu/groups/k12/wiki/17cff/ (accessed July 5, 2013).

66. Bodystorming is a "user experience design" (UXD) technique often used to design physical spaces such as the interior design of a shop; considered analogous to "brainstorming," but instead, actively moving the body through the design. See Antti Oulasvirta, Esko Kurvinen, and Tomi Kankainen, "Understanding Contexts by Being There: Case Studies in Bodystorming," *Personal and Ubiquitous Computing* 7(2) (2003): 125–134.

67. Also known as *Define* mode.

68. Brown, "Design Thinking."

69. Online at: http://www.businessdictionary.com/definition/mass-customization.html (accessed June 20, 2013).

70. Adriano B. Galvao and Keiichi Sato, "Affordances in Product Architecture: Linking Technical Functions and Users' Tasks." Proceedings of 17th international conference on Design Theory and Methodology (2005) 1–11.

71. Ulrich and Eppinger, *Product Design and Development.*

72. Ibid.

73. Abbie Griffin and John R. Hauser, "The Voice of the Customer," *Marketing Science* 12(1) (1993): 1–27.

9

PRODUCT/MARKET TESTING

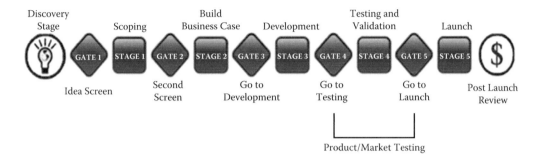

Discovery Stage — Idea Screen — GATE 1 — STAGE 1 — Scoping — Second Screen — GATE 2 — STAGE 2 — Build Business Case — Go to Development — GATE 3 — STAGE 3 — Development — Go to Testing — GATE 4 — STAGE 4 — Testing and Validation — Go to Launch — GATE 5 — STAGE 5 — Launch — Post Launch Review — $

Product/Market Testing

Learning Objectives

In this chapter, we will address the following questions:

1. What are the different types of prelaunch tests that a firm can conduct?
2. How do product use and market tests reduce risk?
3. Why conduct a product use test? Why not?
4. What is the difference between a beta test and an alpha test?
5. How does a marketing test differ from a marketing components test?
6. A controlled sale is? A pseudosale is? A full sale is? How do the goals for each differ?

Product Testing the MetaCork, a Better Wine Closure[1]

As a wine connoisseur, retired engineering professor Bill Gardner had his share of broken and crumbled corks while opening a bottle of wine. He thought to himself, "Why does opening a bottle of wine have to be so tricky?"[2] In 1996, Gardner patented a technology for an alternative wine closure, which did not require a corkscrew. He named it the MetaCork. "The MetaCork consists of an outer hard plastic capsule with a threaded interior surface, a matching plastic threaded cap and a natural cork or synthetic closure fitted with an anchor pin. A simple, effortless twist, followed by a few turns, and the cork is eased out of the bottle, thanks to the triple-helix threads, mated to ones on the bottle, that are longer and smoother than those on a screw cap." By spinning the capsule back into place, the bottle provides a neat package that doubles as a drip-resistant pourer. The plastic cap can be screwed on to the bottle top to form a leak proof bottle reseal.

Initial responses by first-time users were positive. People really liked how easy it was to open a bottle of wine. A concept test and a preference test promised big success for Gardner's MetaCork. In a press release, Gardner Technologies enthusiastically reported that it was confident in its go-to-market strategy. "Since we have focused from the outset on improving the wine experience, we knew that we had to fully understand all the dynamics facing a consumer at the moment of purchase and the moment of use," said CEO William Borghetti.[3] A national study conducted by a third-party marketing research vendor showed strong support for MetaCork; 73% of the survey group said that they would purchase MetaCork sealed wines over traditional packaging. Respondents loved the fact that wine was easier to open, was resealable, and did not require a corkscrew. Furthermore, 83% of the respondents described their image of wineries using MetaCork as innovative, more progressive, and more in tune with consumers' needs.

Gardner Technologies next conducted beta testing with its MetaCork. Products were sent home with employees to test with friends and families. The word started to come back that the wine closure was easy to open and easy to close, but two different problems occurred. It wasn't obvious to consumers that (1) the capsule needed to be screwed back into place over the threaded wine necks to act as the drip-resistant pouring feature (note the capsule in the right photo needs to be put back on the bottle) and, (2) the drip-resistant capsule had to be removed before resealing and only then would the screw cap create a leak proof seal. Instead, many people snapped the cap on to the capsule, which it did easily, but this did not give a tight seal. The testers would then lay the bottle of wine down in the refrigerator to have it leak. Beta testing showed that the MetaCork was too complicated to use for the average wine consumer.

Without adequate funding to redesign and retool the MetaCork, the company decided to refocus launch efforts on business markets, such as hotels, airlines, and restaurants, that would not typically reseal the bottles (thereby really only getting

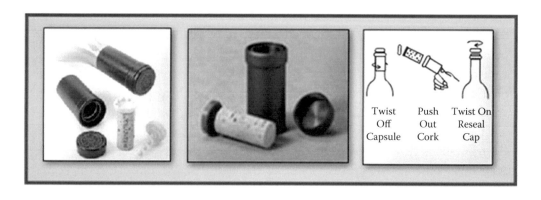

Twist Off Capsule   Push Out Cork   Twist On Reseal Cap

METACORK™- [*noun*]

ETYMOLOGY: [Latin from *Meta* - higher, transcending, more highly organized or specialized form of + *cork*]

a. a revolutionary new opener that uncorks a bottle of wine without a corkscrew and includes a convenient reseal cap and a drip-resistant pouring spout **b.** an opener designed to help waitstaff and consumers open wine gracefully and effortlessly **c.** a new innovation that gives one wine brand a significant competitive advantage over another **d.** a product designed to bring new consumers to wine and improve the wine experience for all wine drinkers

half of the value of the innovation). Gardner Technologies signed a contract with a large winery that supplied wine to restaurants. The winery ultimately broke the contract, resulting in an early demise for MetaCork and Gardner Technologies.

Introduction

At this stage, the new product development (NPD) team has successfully developed a new product. Your organization has made major investments, both monetarily and in regards to time. The team began with just the fuzzy idea of a new product and has arrived triumphantly at the stage of an actual product and a well-formulated marketing strategy. The team has spent a year or more pursuing research and design of the product and the conjecture of potential sales looks very positive. The organization is putting a lot of pressure and emphasis on embarking on a full-scale launch. You should not give in to this kind of pressure. Although your models from the design phase portray an accurate account of the market response, these models are not based in reality. Things can still go wrong and often do. Every company has experienced its failures. Not even Coca-Cola is immune from product launch failures as evident from OK Soda, Surge, and New Coke[4]—all Coca-Cola bombs in the marketplace. Instead of moving right from development to a full-scale launch, a better managerial policy is to test the overall strategy and the marketing mix components while the consequences of failure are small. Many of the same lessons can be learned for $200,000—or even $50,000—with product testing, rather than $50 million it can take to launch a new product nationwide. You also will have the added advantage of further polishing the product during the testing.

In this chapter, we will discuss Stage 4 (Testing and Validation). Once the product has been engineered, product use testing and market testing are the next step in the stage-gate process. It is rare that the product design entering R&D comes through the

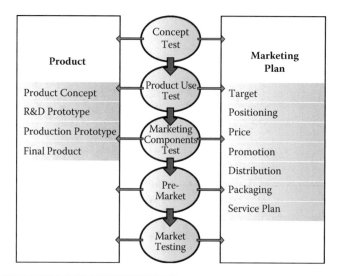

Figure 9.1 Role of testing in new product development.

process unchanged, so it must be verified that the resulting product design still meets customers' desires. The purpose of the **Testing and Validation** stage is to provide final validation of the entire project—the combination of the final product, its production, and its marketing plan.

There are several different types of tests a firm may undertake. How to interpret the test results and how to use the tests to reduce risk and maximize benefit will be discussed in this chapter. We consider the tradeoffs among cost, time, and benefit to the organization and indicate what testing methods are appropriate for various industries. The different types of tests are shown in Figure 9.1. The **concept test**, which we have already studied extensively, informs both the product design and the marketing plan. The **product use testing** after initial product development informs refinements in the final product design and helps to set a pricing strategy. The **market components test** (ad copy testing, pricing testing, distribution channels testing) is used to build the marketing plan. In **premarket testing**, the product, its production, and the launch plan are tested together under prelaunch conditions. **Market testing** is the dress rehearsal for the marketing mix—products from a pilot production run, packaging, price, advertising, and promotion all come together. It is a last chance to catch any needed improvements. The results of market testing are then used to finalize launch plans. We provide a summary of the different types of tests below, and then provide details on how to conduct these tests.

> *Product Use Testing*: A process where current and/or potential customers evaluate a product's functional characteristics and performance under the conditions in which it is meant to be used, whether the conditions are real or simulated. The purpose of product use testing is to prove the product's functionality.

Market Components Testing: A process in which consumers evaluate the effectiveness of a marketing campaign. Although advertising testing is the most popular, testing of price points and distribution channels sometimes occurs.

Premarket Testing: A procedure that uses syndicated data (as collected by marketing research firms) and primary consumer research to estimate the sales potential of new product initiatives through forecasting of purchase intent and preference share. The product and the marketing program are tested together for the first time in premarket testing.

Market Testing: A process by which targeted customers evaluate the marketing plan and the new product in a marketing setting. The purpose of the market testing is to evaluate the marketability and financial viability of a new product launch. A market test can be thought of as a dress rehearsal prior to full launch.

Reducing Risk[5]

New products can always be assessed from the viewpoint of its given benefits and risks. There is bound to be some risk when introducing a new product to the market. Gardner Technologies certainly was aware of the risks of introducing a new wine closure to a very traditional industry with a mature 3,000-year history. When assessing risk, it is helpful to refer to a "decision frontier" as the smallest benefit that must be gained from any given level of risk. You also may look at it as the most risk you are willing to take for a given anticipated benefit. You will find an illustrated example of the decision frontier in Figure 9.2. The diagram depicts an amalgamation of other product and marketing solutions. All of the possible products are below the curve of the decision frontier. Suppose we select the starred (*) product for attention. Market testing at its core is a means of experimentation. This experimentation hypothetically is refining the product so that it makes the cut of the decision frontier. For example, if the product undergoes an advertising test that helps to clear initial uncertainty and ends up being more noteworthy than expected, that might boost the product along the dashed line. Unfortunately, the risk/benefit may still be below par. The following steps can be seen in the pretest market

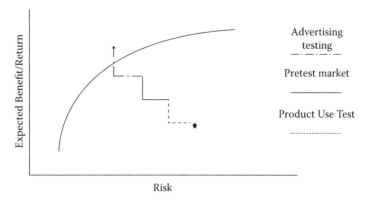

Figure 9.2 Role of risk reduction in testing. (From Endnote 6. With permission.)

(solid line) and the test market (double-dashed line). Ultimately, the objective is to cross the decision frontier with as little cost as possible. In order to achieve this, we must reduce risk and increase the anticipated benefit gained. Gardner Technologies took this type of sequential strategy first with an in-house alpha test, then a beta test with family and friends, followed by a small market test with a vested partner.

Many companies are reluctant to undergo product testing when they are so close to launch. Testing can cause delays and give competitors a time advantage, so care must be taken to avoid unnecessary delays. Product testing will delay revenue flow as well and possibly give competitors a chance to more quickly introduce their own products. Another downside is that costs increase as we approach the decision frontier. However, the costs of a failed product can be very high, so the risk of not testing must be weighed against delaying the launch to test the product. The testing strategy should balance the uncertainty and the cost by careful use of a sequential strategy. At low costs, these pretest market tests can eliminate many failures without revealing the product to competitors. But, the testing procedure must produce actionable diagnostics to improve the product or it is not worth the time or expense. Apple Inc. is one company that rarely conducts any type of product or market testing.

Product Use Testing

A well-designed product fulfills the core benefit proposition (CBP). Back in the design phase, we conceptualized the product by means of perceptual mapping in order to distinguish how we would manage the CBP. Through conjoint analyses, we were able to pinpoint the essence of what consumers perceived as the actualization of the CBP. House of Quality methods were proposed for micro design aspects to meet product feature requirements.

In the Testing and Validation stage prior to production, we must now determine if the product performs as planned and if its formulation can be improved. **Physical product testing**, sometimes called *field testing* or *user testing*, determines whether the production-ready product delivers the CBP and also generates diagnostic information to improve the product and/or reduce costs.

Product use tests provided Gardner Technologies, inventor of the MetaCork, with valuable customer insights into these types of questions:

- Does the product perform as intended or does it need design changes?
- Is the MetaCork easier to use compared to competitors' products?
- How will consumers react to a "new" type of wine closure, which eliminates the traditional corkscrew?
- What features/benefits drive are most appealing about the MetaCork?

Table 9.1 New Product Launch Gone Awry

PRODUCT LAUNCH OOPS
• A dog food package discolored on store shelves.
• The packaging for a prebrewed coffee drink melted when microwaved.
• The antenna reception on a mobile phone was hindered depending on how the phone was held.
• Chocolate flavored French fries never were a big hit with kids.
• A biodegradable bag of a popular snack was so noisy sales plummeted.
• A laptop computer weighed in at 16 lbs., making it not very portable.
• In cold weather, baby food separated into a clear liquid and sludge.
• The electric cord in a powered microscope produced shocks.
• A laptop battery that would catch on fire.
• A new 24-hr medical service resulted in a doctor being sent to a home of a patient who had died the previous year.
• Automated telephone service centers led to customer complaints and loss of sales.

Source: From Endnote 7.

- Is there an opportunity to extend the product line with different variations of the MetaCork?
- Does the MetaCork really make opening a wine bottle easier? Or is this just the opinion of the designers?

A goal of the new product manager is to test the product or service performance under actual consumer use situations. Table 9.1 lists a variety of things that have gone wrong in products launches. These product/service defects are embarrassing and costly. In fact, with the legal product liability that exists in many countries, defects or bad design can create huge losses. Design faults must be eliminated by careful testing There are several different types of product tests a company can undertake: preuse reactions, alpha testing, and beta testing, which are discussed next.

Preuse Reactions

Preuse reactions, also called preference tests, are commonly done with a print or online survey that queries users on measures of satisfaction and/or preferences for a product/service's features. It is a test to get a reading of likely market acceptance of the product and provides clues to minor design improvements that can make the product better. Questions that get respondents to express preferences and rank alternatives can yield quantifiable information on what drives customer decision making. Figure 9.3 shows a preference survey for podcasts. Issues to be concerned about when conducting preference tests include:

- Just as when conducting the concept test, it is important not to *oversell* the product in the preference test.
- Price sensitivity cannot really be measured at this stage because customers cannot judge the quality of the final product.

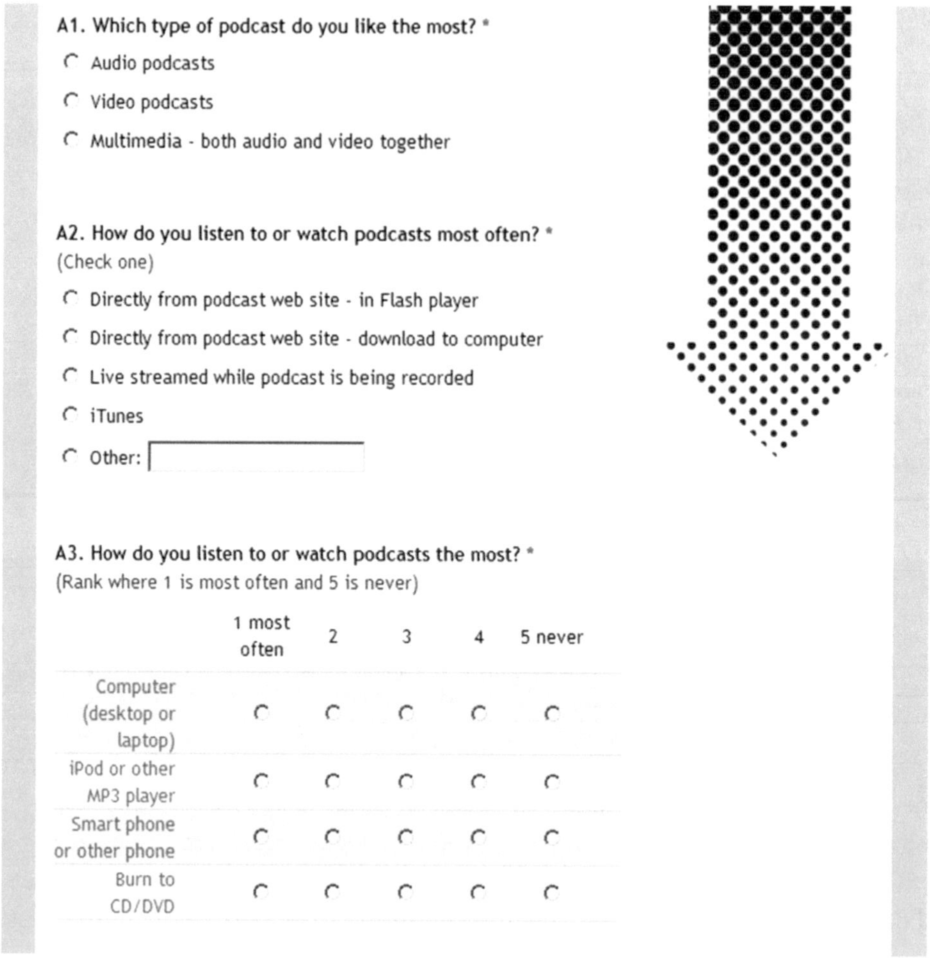

A1. Which type of podcast do you like the most? *

    ○ Audio podcasts

    ○ Video podcasts

    ○ Multimedia - both audio and video together

A2. How do you listen to or watch podcasts most often? *
(Check one)

    ○ Directly from podcast web site - in Flash player

    ○ Directly from podcast web site - download to computer

    ○ Live streamed while podcast is being recorded

    ○ iTunes

    ○ Other: [_____]

A3. How do you listen to or watch podcasts the most? *
(Rank where 1 is most often and 5 is never)

	1 most often	2	3	4	5 never
Computer (desktop or laptop)	○	○	○	○	○
iPod or other MP3 player	○	○	○	○	○
Smart phone or other phone	○	○	○	○	○
Burn to CD/DVD	○	○	○	○	○

Figure 9.3 Preference survey for podcasts.

- Preference tests measure intention, and don't always translate into what customers will actually do. When products are offered for free, stated intentions are typically high. For this reason "freemium" pricing for online services, where basic content is free and premium content costs extra, may be one way of testing actual behaviors.

Alpha Testing

In the alpha test,[8] "*does it work?*" knowledge is gathered. This preproduction product testing sets out to determine and eliminate design defects or insufficiencies. Alpha testing typically occurs in a lab setting or within the firm, although in some cases it may be conducted with lead users in closely monitored situations. The alpha test should not be performed by the same people who are doing the development work. Because this is the first "flight" for the new product, basic questions of fit and function should be evaluated. As the testing is conducted in-house, selecting unbiased personnel to test the product in order to gain objective insights is important. Of course, the MetaCork easily passed alpha testing with flying colors.

Beta Testing

Beta testing is short-term use tests at consumer sites. The goal of beta tests is to find out if the product works as expected by testing all functions in a breadth of field situations to find system faults that are more likely to show in actual use than in the firm's more controlled in-house alpha tests. It is critical that real customers perform this evaluation, not the firm developing the product or a contracted testing company.[9] During beta testing, the MetaCork started to show signs of possible use problems.

Computer hardware and software firms often conduct beta tests to determine if the product is free of bugs. In this industry, beta testing really means: "This is a new product, use at your own risk. There is no obligation on either parties' side, yours or ours, to acknowledge issues with the product." Companies then monitor user forums and blogs to identify problems users are having with the product. For example, Google's Gmail was launched as an invitation-only beta release on April 1, 2004. It was opened to the general public in February 2007 still in beta status. The service was upgraded from beta status to general release in 2009.[10] Google Scholar, an academic-based Web site was still in its beta version as of April 2011 after having been launched in 2004[11] (see Figure 9.4 for Google's beta notification on its Web sites). That's a seven-year beta. Launching of beta versions of software is sometime a preemptive competitive strategy to be first to market with a new software program. It is sometimes used as well as a marketing tool, which gives early adopters the feeling of being a part of an exclusive group.

Typically, beta tests are conducted in short time periods of days or weeks. However, the test period could be longer with extended user trials, especially for industrial or other complex products. In February 2008, General Motors announced a three-month test drive of the Equinox Fuel Cell vehicle, which is shown in Figure 9.5. A 24-hour test drive would not have given customers enough insights into the unique issues driving a fuel cell powered vehicle might entail.

Figure 9.4 Google's example of beta sites.

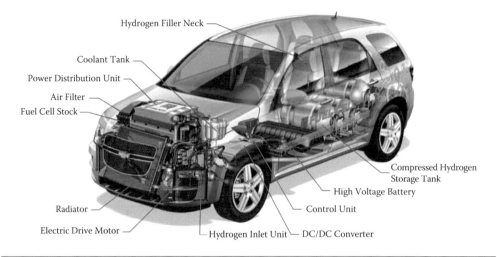

Figure 9.5   General Motor's concept car: Equinox fuel cell vehicle. (From Endnote 12. With permission.)

These extended beta trials are typically called field trials. Care should be taken when selecting customers to conduct these types of trials. Some issues to be concerned with include:

- Have a written agreement with the customer in advance about details of the trial, such as the length of the trial, the cost of the trial, and what happens in the event of a product malfunction.
- Have personnel readily available or onsite, especially for industrial products, in the event of problems.
- Regularly check in with the customer to gain their insights both verbally and with written reports. Know in advance what type of feedback you are looking for from the customer.
- Establish expectations of what happens when the trial is over. Is the test product for sale or will you expect return of the product?
- If leading-edge user analysis is used in the design of the product, it should be carried over to the testing stage because the lead users are the most demanding in terms of use requirements. They will cooperate in testing the product because they gain advantages by early use and users will not only point out problems, but also help solve them.
- Ensure there is a confidentially agreement in place, so that early testers won't share overly negative results with other users or with competition.

Conducting Product Use Tests

There are three commonly used procedures for conducting customer tests: **single product evaluation**, **blind comparisons**, and **experimental variations**. In each procedure, it is important that customer attitudes and experience be understood in the evaluation process.

Single Product Evaluation[13]   The best way to go about customer testing is to engage the customers in evaluation of the product, seeking their approval. These evaluations are typically polled on a scale called the *hedonic scale* or overall scale of liking. This scale ranges between five and seven points and is an accurate indicator of the customer's preference toward the product. Other evaluations may be compiled based on different features of the product, giving further insight as to how customers react to a new product on a particular aspect of the CBP. For example, a company may ask customers to try their new shampoo formulation. Or, a tablet computer might be distributed to university professors for efficacy testing. Engineers from different fields may be asked to give their opinion on the reliability of a CAD (computer-aided design) program.

Blind Tests   In the event of a blind test,[14] a new product is posited against other existing products in the market. The identity of the company is not revealed during the test in order to ensure that the physical response is not directly linked to brand perceptions. For example, if a new coffee is being tested on the CBP premise of "mildness," a comparison can be made between two brands—one will likely be more mild than the other. For the most part, brand names and market positioning affect the outcome of testing, so the blind test can sometimes be switched out for testing with hypothetical labels and other graphics.

Experimental Variations[15]   Often the firm will develop more than one design and need to determine which design best meets the needs of the customer. For example, in a study of paper towels, consumers were given a rack that contained three different towels. They were asked to use them sequentially. Across the sample, the towels were varied in weight, adhesive content, and plastic reinforcement. Consumers rated the towels on overall preference: softness, absorbency, and durability. Based on these results, a best combination of weight, adhesive content, and reinforcement was selected to fulfill the CBP of "soft" and "absorbent" while maintaining a competitive cost and acceptable strength.

Issues in Product Use Tests

There is evidence that customers are influenced by more than the physical characteristics of a product. One study found that when consumers tasted and rated labeled beers, they rated the brand of beer they drank most often significantly higher than other beers.[16] However, when they tasted and rated the same beers without labels, there were no significant differences between the ratings of brands. The label and its psychological associations affected the taste evaluation. In another study,[17] all of the beer was physically the same and only the price level was changed. The taste ratings (undrinkable, poor, fair, good, and very pleasant) were significantly related to price, with the higher-priced beer receiving better ratings. These studies indicate that although physical ingredients are important, past experience and perceptions may affect consumer evaluations.

Psychosocial cues also have an impact on consumer evaluations of durable products. The same physical car may be evaluated differently if it is manufactured by Volkswagen rather than Ford or Mazda. Consumer perceptions of the reliability of a mobile phone can depend upon whether the manufacturer is Apple, Nokia, or Blackberry. Even with industrial products, cues can be important. An air-conditioning system must not only be reliable, but appear and sound reliable. The right faceplate for an electronic component helps communicate that it is state-of-the-art. Design work should have identified the major cues, but testing should assure that they operate as expected in building customer attitudes.

Summary of Product Use Testing Procedures

New products are unlikely to flourish in the market if they are not perceived to be of high quality, preferably of higher quality than complementary products. New product managers must carefully consider the engineering design and how these materials and designs reflect in the quality of the product and consumer acceptance of the new design. Through product testing, the product manager is supplied with information necessary for informing such decisions. Coupling laboratory testing with expert evaluation finds a well-balanced view of how the product will be perceived and received by consumers. The tests should determine the best utility-versus-cost trade-offs and result in an optimal quality product.

At the end of the day, there is no optimal method that will be sufficient for every product. The product manager must be sensitive to the strengths and weaknesses of every available process and must have the acumen to discern which combination of tests will provide the most accurate and helpful information for product design and creation.

Market Components Testing

After the product has been tested, the NPD team also will want to test the components of the marketing mix to determine if they adequately support the innovative new product about to be launched. As we see in Figure 9.1, the results of the product test will feed into market component testing. Before a product is ready to be tested in the market, advertising copy and budget, personal selling approaches, distribution variables, prices, promotion, packaging, branding, and pre- and postsales service need to be specified. Although our goal is to set these variables at the "best" levels, it is difficult to know exactly what best means because at the design stage there is still too much uncertainty with respect to customer and competitive response. Thus, a manager sets an initial marketing mix that is consistent with the CBP, determines initial sales forecasts based on best estimates of customer response to marketing efforts, and establishes best estimates of competitive reaction. As more information becomes available during the testing phase of new product development, this initial marketing mix is refined for greater strategic advantage and profitability. Testing of the marketing mix is often overlooked by companies. However, there are times when a company will want to test out components of the marketing mix prior to committing large resources to an advertising or brand building campaign. In this section, the focus is on the 3Ps of the marketing mix (promotion (primarily advertising), price, and place) since we have covered the 4th P, product, in detail already.

Testing Advertising

Advertising is a key component in the new product marketing mix. Good copy creates awareness and relays the CBP in a way that consumers want to act upon their feelings. Advertising testing allows the best ad that relays the product's CBP to be selected, assesses the effectiveness of the ad, and generates ideas on how to improve the ad. Most firms rely on managerial judgment for selecting advertising copy. Unfortunately, many times individual discernment is no better than a randomized selection. For example, in a study of 24 print advertisements where market results were known, managers' judgments had almost no correlation with the market results.[18] For large or complex campaigns, the question is not whether to test, but rather which technique should be used and how many advertisements to test. We begin by examining the criteria for evaluating advertising copy.

Criteria for Evaluating Advertising Copy[19]   Advertising is very productive in yielding advantageous sales, but it must accomplish several transitional objectives before a sale as well. A consumer response hierarchy, as shown in Figure 9.6, indicates areas to determine what is effective about an advertisement and what must be improved.

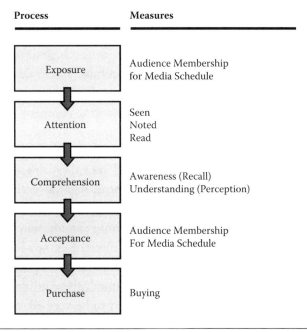

Figure 9.6   A model for the consumer response hierarchy to advertising copy. (From Endnote 20. With permission.)

Table 9.2 Selected Copy Testing Approaches

APPROACH	MEASURED BY	PRIMARY MEASURE
On-Air Testing	Comprehension	Day after Recall
Theater Testing	Comprehension and Acceptance	Immediate Recall, Choice of Prize
Trailer Testing	Acceptance	Coupon Redemption in Store
Simulated Buying	Comprehension and Acceptance	Buying in "Retail Store"
In-Market Test	Purchase	Panel Records of Buying

Source: Endnote 21. With permission.

Exposure is essential for any consumer feedback and is evaluated by a barometer of whether a consumer has seen a particular clip of media. **Attention** is determined by whether or not a potential consumer will absorb the content of the ad during exposure. **Comprehension** ties together awareness and understanding. Awareness is determined by whether or not a potential consumer can recall the product name or the premise of the advertisement, with or without assistance. This assessment of awareness can take place at any time: immediately, the following day, or much later. **Acceptance** shows how comprehension of the product and the ad message turns into product preference. **Preference** is measured by consumers' preferred choice in "intent to buy" or an actual purchase in a simulated buying situation.

There are numerous methods for testing advertising copy; some major methods are shown in Table 9.2.[21] Various organizations may place different emphasis on the criteria for evaluating advertising copy, but each organization should select the method that is best for its needs and resources. In more recent years, the Internet has played a large role in testing of advertisements because of video and voice streaming capabilities. For example, a large consumer products manufacturing company wanted to test several radio spots for a new product introduction. Prior to launching one of the spots, they had the MarketingVision Research (www.mv-research.com/) company conduct a quick Internet-based survey with potential product buyers to evaluate the various ads. Several different radio ads were streamed to customers for their responses. The winning radio spot from the test was used by the sponsoring company in a very successful campaign.

It is also important to measure why certain effects occur within the consumer response hierarchy. Likability, believability, and meaningfulness are three important data sets that have been found to be relevant in regards to the response hierarchy. **Likability** is found to improve the rate of acceptance, but there also is evidence that advertisements that generate a negative response can ultimately produce positive results. Perhaps you remember the famous line from the campy lifeline commercial: "Help! I've fallen and I can't get up." Or the annoying: "Can you hear me now?" ad nauseum from Verizon mobile commercials. Granted, these advertisements are generally disliked by the public, but they have proved to be very effective campaigns for the companies. However, the effectiveness of nonlikability must be looked at very cautiously. The studies have shown that if attention can be garnered for a presumably

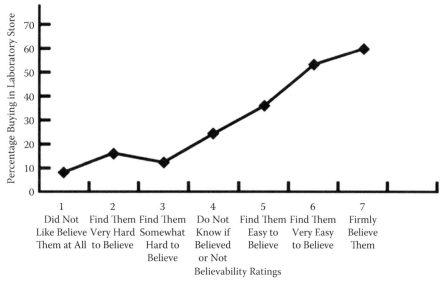

Average Trial of 40 New Brands versus Believability after Ad Exposure

(a)

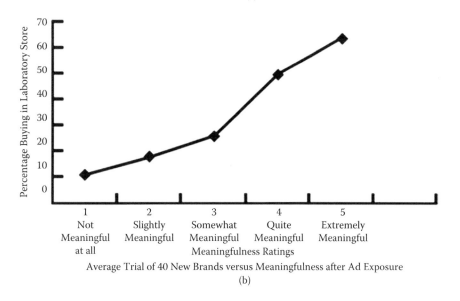

Average Trial of 40 New Brands versus Meaningfulness after Ad Exposure

(b)

Figure 9.7 Average trial of 40 new brands versus believability and meaningfulness. (From Endnote 22. With permission.)

likable ad, then the amount of consumers who will try the product is much higher for likability ratings of 6 or 7, and not so good for ratings under 5.

Believability and **meaningfulness** are often found to have a high correlation with trial once a product receives attention. Figure 9.7a illustrates the interaction between high believability ratings and high trial. However, intermediate values of 4 and 5 show a correlation with consumer reaction. This reaction can be attributed to what is commonly known as "curious disbelief," which means that the consumer is not sure whether they believe the claims the company has made, and, therefore, will try the

product to see if the claims are true. *Meaningfulness* can be defined as personal importance. When someone sees the ad, they might have a personal reaction depending
on their situation, wants, or needs. Higher levels of meaningfulness are definitively
associated with higher levels of trial, as indicated in Figure 9.7b. As an example, in
1998, Procter & Gamble initially launched Febreze targeted to smokers. However,
most smokers do not mind the smell of smoke, so the product had little meaning to
them. Additionally, many consumers did not believe that the product could actually
eliminate odors instead of just masking odors. P&G had to change their marketing
approach repositioning the product for families with small children and even college students, and they also addressed the believability issue on their Web site. Good
advertisements will get more response if they are liked, believed, and perceived as
personally relevant.

Testing Price

In the design process, some measures of price response are needed. **Willingness to
pay** (WTP) denotes the maximum price a buyer is willing to pay for a given quantity
of a good. It can be measured with different techniques, including:[23]

- Conjoint studies, which captures WTP for a product presented as a bundle
  of attributes
- Contingent evaluation,[24] which captures WTP for a product through a what-
  if (contingency) survey
- Market transaction scanner data, which captures price paid at point of purchase
- Simulated test markets, which captures price paid in simulated markets
- Vickrey auctions,[25] which captures prices paid in an auction environment

Early in the product development stage, **conjoint studies** or **contingent valuation
studies** are used more frequently because they can provide information on valuation of
specific attributes, and they do not require a finished product to test WTP.

Conjoint Study   Conjoint is a market research technique in which consumers choose
from different sets of product attributes. These choices are trade-offs between two or
more features or benefits of a product. **Conjoint analysis** is a method for evaluating
different pricing points.[26] For example, consumer WTP for recyclable packaging can
be assessed during a conjoint study, and these results can be used to determine the
value of designing a product with environmentally friendly packaging. See Table 9.3
for attributes and levels for a cream cheese product. Figure 9.8 shows an example of
a conjoint study using these attributes and levels for cream cheese product offerings.
Adding price as an attribute in a conjoint study provides WTP information.[27]

Contingent Valuation   Contingent valuation is a survey-based technique for the valuation of a product/service or even nontangible items, such as environmental preservation.

Table 9.3 Attributes/Levels for Conjoint Study of Packaged Cream Cheese

ATTRIBUTE	PACKAGING MATERIAL	PACKAGING MATERIAL RECYCLABLE	PRICE	FAT CONTENT	RECLOSEABLE
Level 1	Plastic	Yes	$2.30	30%	Yes
Level 2	Cardboard	No	$1.99	15%	No
Level 3	Alufoil		$1.60	6%	
Level 4	Glass			0%	

Source: Endnote 28. With permission.

Typically, the survey asks how much money an individual would be willing to pay (or willing to accept) to maintain the existence of (or be compensated for the loss of) a product/service/concept (see Figure 9.9 for an environmental example). Contingent valuation is often referred to as a **stated preference** (intent to buy) model, because survey respondents state their preferences, but are not required to act upon that preference. Like conjoint studies, contingent valuation captures a utility (valuation) for a product or its attributes. Open-ended contingent valuation requires respondents to directly state their WTP, whereas closed-ended contingent valuation asks consumers to make repeated choices of whether they would buy a good at various prices.

Transaction Data **Transaction data**, such as a scanner, and **simulated test market** data[29] are good procedures for eliciting WTP at the point of purchase. Actual purchases are observed under realistic marketing-mix conditions to reveal a preference by consumers for a product at a set price. These are called *revealed preference* studies as consumers make actual purchases. They differ from conjoint and contingent valuation studies, which are stated preferences where a consumer only states their intent to purchase and is not required to actually make the purchase. The downside of transaction

Choice	A	B	C
Packaging Material	AluFoil	Plastic	AluFoil
Recyclable	No	Yes	Yes
Price	$1.99	$2.30	$2.30
Fat Content	15%	0%	6%
Reclosable	No	Yes	No
Choose One	◎	◎	◎

Figure 9.8 Example of conjoint including price.

Contingent Valuation Study on Environmental Issues

In the Four Corners Region of Colorado, rivers provide 2,465 river miles of critical habitat for nine species of fish that are listed as threatened or endangered. Continued protection of these areas required habitat improvements, such as fish passageways, as well as bypass releases of water from dams to imitate natural water flows needed by fish. A contingent valuation survey was used to estimate the economic value for preserving the critical habitat. A questionnaire was developed to ask local residents if they were willing to pay for the improvements needed. The question was posed as:

Suppose a proposal to establish a Four Corners Region Threatened and Endangered Fish Trust Fund was on the ballot in the next nationwide election. How would you vote on this proposal? Remember, by law, the funds could only be used to improve habitat for fish. If the Four Corners Region Threatened and Endangered Fish Trust Fund was the only issue on the next ballot and it would cost your household $_____ every year, would you vote in favor of it? (Please circle one.)

YES/NO

The dollar amount blank in the above illustration, was filled in with one of 14 amounts ranging from $1–$3 to $350, which were randomly assigned to survey respondents.

Figure 9.9   Example of contingent valuation study for price. (From Endnote 30.)

data is that they can only ascertain that a buyer's willingness to pay is at least as high as the posted price and a nonbuyer is not willing to pay the posted price. The highest possible price a consumer is willing to pay is not revealed.

Auctions   Auctions can be one way of ascertaining the highest possible WTP. **Vickery auctions,**[31] named after William Vickery who did early research on sealed bid auctions, provide another way of observing revealed prices. In a Vickrey auction, bidders submit written bids without knowing the bid of the other people in the auction. The highest bidder wins, but the price paid is the second-highest bid. Bidders have an incentive to reveal their WTP truthfully because they don't know what others are bidding and they must buy the goods if their bid wins the auction. In this way, WTP is revealed through the auction transaction purchase.

A downside of WTP studies is that if subjects believe that their answers will be used to set long-term market prices, they have an incentive to understate their WTP. If they believe that their responses will influence whether a new product is introduced or not, they may overstate their WTP to ensure launch.

Testing Distribution Options

We have looked at methods for testing advertising and price. We now turn to distribution. The most effective channel must be chosen from existing distributors, agents, retailers, franchisees, direct marketing, and other new avenues, as shown in Figure 9.10. Sometimes, decisions about how to maintain sufficient levels of stock, minimum order quantities, delivery methods, delivery frequency, and warehouse locations may require evaluation. Many times a company's distribution network will already be established so testing for the best channel may not be required. However, sometimes a new channel will be more efficient in a new product launch. Specialized

Figure 9.10 Distribution alternatives for new products.

M&Ms are sold only online because of the personalization of each order (http://www.mymms.com/). This product could not be easily sold in a retail environment. It is easy to test the success of a product in an online environment; it is also very cost-efficient.

Another recent method of testing out a retail distribution strategy is by using a **pop-up store.** A pop-up store opens for a few months, weeks, or even a few days in a retail location. This type of ephemeral store can test how well a new product sells in retail environments. Walmart has been using the pop-up store strategy as a way to learn what customers want and to see how all of its innovations are meshing together.[32] Table 9.4 lists some of the issues that should be considered when making distribution channel choices.

The sales team also can provide support with testing distribution options. When sales reps are the primary sales channel, testing of personal selling presentations can be useful. In industrial products, these presentations are as important as advertising is to consumer products sales. Alternative presentations could be made to stress particular benefit claims, use various demonstrations, or be based on direct competitive comparisons. Such a laboratory testing procedure also could test industrial advertising in conjunction with the sales presentation. By exposing one subsample to ads and sales presentations and another to only the sales presentation, data would indicate the effect of advertising in improving sales productivity.

Table 9.4 Some Issues in Testing Channels of Distribution

- How complex is the product? Does it need detailed explanations on how to use? If not, online testing may be feasible.
- How strong is the relationship with current distributors? Strong relationships may allow testing the success of an existing channel.
- Is there a need to build up relationships with new distributors? Testing the channel with a desirable new product may be a good way to develop a relationship.
- Is the product highly competitive? Sales reps/agents who represent competitors should be avoided.

Summary of Marketing Components Testing

Just as a new product should be tested in the marketplace, so too should the other three "Ps"—promotion, price, and place (distribution). Often firms will overlook testing the marketing mix if the focus is on an innovative product. It is up to the new product manager to make sure this does not happen. Finding a price that maximizes sales can be crucial to the success of the product. Using stated preference methods (conjoint or contingent evaluation) is efficient and can be completed at a relative low cost compared to the benefits obtained from gathering consumer reactions to prices. There are a variety of options a firm can use to test distribution strengths. Ephemeral locations, such as online stores and pop-up stores, are becoming popular because of their relatively low cost and ease of setting up a test site.

Premarket Testing

In regards to testing, so far in the NPD process, we (1) have conducted product-use testing to determine how well the designed product meets the needs of the buyer and (2) have tested various components of the marketing mix for conveying the CBP. We now combine the product with the marketing program in simulated market environments. This is the dress rehearsal before full launch. The input to a pre-market analysis is the physical product, advertising copy, packaging, price, distribution strategy, service policy, and advertising and promotion budgets. We focus here is on three primary pre-market testing methods: *pseudo sales, trial and repeat* and *controlled sales*. Other methods of pre-market testing include: j*udgment and past product experience*,[33,34] *attitude-change models*,[35] *convergent approach*,[36] and *information acceleration*.[37] Table 9.5 lists a number of these alternative approaches. Each approach has its relative strengths and weaknesses; managers may wish to combine two or more approaches to enhance accuracy.

The premarket test must be done early enough before the launch so that major expenditures can be stopped if the analysis indicates the product is likely to be a failure. In packaged goods, premarket tests should be undertaken before the test market (12–18 months before launch) and before national launch. For consumer durables there are two significant points - before commitment to production capacity (e.g. in autos 36–48 months before launch) and before committing to specific levels of launch expenditure (e.g. 12 months before launch). In industrial products, the investment in production is the milestone (often one-two years before introduction). In services, pre-marketing testing should occur before large capital expenditures (e.g. new computers for credit services) or full launch.

Pseudosale

A pseudosale is a mock sale; a buying environment is simulated, but for a few select buyers. There are two types of pseudosales: simulated test markets (STMs) and

Table 9.5 Approaches to Premarket Analysis

PSEUDOSALE	TRIAL/REPEAT MEASUREMENT	CONTROLLED SALE	OTHER METHODS
• Simulated Test Market • Speculative Test Market	• Home Delivery • Laboratory Measurement	• Informal Selling • Direct Marketing • Minimarkets • Scanner Market Testing	• Judgment and Past Product Experience • Attitude-Change Models • Convergent Approach • Information Acceleration[38]

speculative sales. STMs are used with **fast moving consumer goods** (FMCGs). Speculative sales are typically used with industrial products or large ticket item services, such as corporate cloud computing capabilities.

Simulated Test Market (STM) STM is the predominant method used by consumer packaged goods manufacturers to evaluate new products, line extensions, and other new business opportunities prior to large-scale in-market introduction. One method of conducting an STM has been through a shopping mall "intercept," where the potential buyer may be asked to watch a series of advertisements, one of them about the product being tested, although it is not identified as such. The buyer would then be asked to *shop* at a mock store and make a purchase choosing between different competing products and the firm's new product. Buyers also will have an opportunity to make repeat purchases at a later time. The downside of mall intercepts is that sample populations are biased with young, female, suburban, middle-income, frequent shoppers.

The goal of the STM is to get estimates of trial probabilities and repeat purchasing probabilities in order to make forecasting estimates using ATAR (awareness–trial–availability–repeat, discussed previously in other chapters). Simulated test markets can be conducted quickly to assess not only a new product, but also its marketing mix. If STM results are strongly positive, a company may decide to proceed with the NPD process without further testing. When results are poor, the product needs to go through a thorough evaluation for redesign or may need to be killed altogether.

A variety of marketing research firms offer different types of simulated test markets; for example, ASSESSOR by M/A/R/C, BASES by Nielsen, DESIGNOR by Ipsos, and Launch eValuate by TNS-Research International. The simulated test market was initially focused on forecasting for consumer packaged goods (CPG), but more recently has expanded to services (e.g., financial, insurance), durables (e.g., automotive), pharmaceuticals, healthcare, and other industries. Thus, it's important that the vendor specialize in the industry of interest. STM companies use a combination of historical data on previous product launches combined with current STM results to make forecasts about a new product launch. STM market researchers claim that "90% of their forecasts are within 20% of actual volume and over half are within 10%."[39] Figure 9.11 outlines ASSESSOR's capabilities.

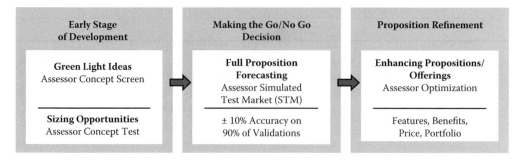

Figure 9.11 ASSESSOR capabilities by M/A/R/C research. (From Endnote 40. With permission.)

Despite their popularity, STMs have limitations and downsides. STMs were initially proposed in 1978 by Silk and Urban at MIT in 1978.[41] Thus, these tests are based on marketing plans that are largely driven by advertising and other mass media promotional vehicles, which are not the only means of marketing communication today. Mass media-driven campaigns are no longer the norm. The Internet with YouTube, bloggers, and media-rich content are now an integrated part of marketing campaigns. In addition, in-store factors, such as packaging, shelf-configuration, in-store promotions, and other **category management** issues, drive consumer purchasing behavior.

Firms are beginning to make use of the Internet and virtual reality in simulated tests to incorporate new media marketing techniques in the simulated test market. Using virtual reality has the advantage of being relatively inexpensive and highly flexible. It is easy to create a large number of simulated surroundings for several different products with several different forms, sizes, or colors. Second Life, the online virtual world, has been a forum for virtual market testing. "Second Life's real potential may be that of an experimentation platform," says Eric Klopfer, a professor at MIT.[42] However, virtual reality too comes with its caveats. According to Greg McMahon, the senior vice president of Synovate,[43] "The average response rate to a Web survey is less than 1%, making it even less likely to get a reliable read on the potential of a new product."[44] Additionally, Second Life has not become mainstream for many consumers so its reach is even more limited. A screenshot from Second Life is shown in Figure 9.12.

Speculative Sale The second type of pseudosale is the speculative sale. Industrial goods manufacturers, consumer durable manufacturers, and other business-to-business companies will use the speculative sale technique. In this type of pseudosale, salespeople make a sales call with full use of marketing collateral, including brochures, price sheets, presentations, and, if relevant, samples, even though the product/service is not ready for market. Because the speculative sales are typically conducted with existing customers, it is possible for marketing teams to use the opportunity to get at the voice of the customer as well as test the marketing approach. After the presentation, sales representatives can ask outright, "Would you be willing to buy this product?" in order to record customer feedback. Another goal of the speculative sale is

Figure 9.12   Second life as a simulated market test platform.

to get an understanding of willingness to pay for the product by asking: "What would you expect to pay for this product?" Because customers are not told they are being sold **vaporware** (products/services that are still in R&D stage and, thus, not ready for launch), speculative sales work when:

1. the commercial/industrial firm has a close relationship with its key buyers;
2. the product is highly technical and/or has a high capital expenditure, requiring multiple sales calls;
3. little risk is involved in introducing the product to potential consumers;
4. customers are not likely to discuss the new product with competitors;
5. the manufacturer is searching for the killer application (killer app) for the new product; and
6. the product is new-to-the world and introducing it to the mainstream marketplace may be complex.

Trial/Repeat Measurement[45]

The next types of premarket tests are trial/repeat measurements. Accurate forecasts of long-run sales can be made if the percentage of consumers who will try the product (cumulative trial) and the percentage of those who will become repeat users (cumulative repeat) can be obtained. Trial/repeat pretests are one way of collecting this type of data. Two types are **home delivery measures** and **laboratory measures**. The advantage of this approach is that it is based on direct observation of consumer response to the new product. The disadvantage is these are still mock tests, not actual purchasing environments.

Home Delivery Measures[46] One approach to forecasting with a trial/repeat model is done in a panel (i.e., 1,000 households), which is visited each week by a salesperson. The advantage of this approach is that it is based on a consumer panel that is familiar with these types of campaigns, so it is a fairly good approximation to the actual product launch. The direct measures of trial are obtained based on exposure to real advertising via direct mail, salesperson pitches, or even TV ad placement just to the participant's home. Realistic repeat purchasing through salesperson visits enhances forecasting accuracy. One disadvantage of this approach is that the panel must be run long enough to get good estimates, probably longer than other pretest market approaches. Therefore, costs are relatively high.

Laboratory Measurement In a laboratory approach, consumers are recruited, exposed to advertisements (television or print), and given the opportunity to buy in a simulated retail store. After buying in the simulated store, the consumer takes the product home and repeat measures are taken with a call-back interview. The success of the laboratory measurement depends on the ability of the approach either to minimize the bias of such a laboratory simulation or to develop procedures to correct for any bias.

Controlled Sales[47]

The third type of premarket test is a controlled sale, which is a product launch limited to a very few select customers or distribution outlets. The intent is to test the market in a real environment, but tightly controlling the launch variables. There are four types of controlled sales market testing: **informal selling**, **direct marketing**, **minimarkets**, and **scanner market testing**.

Informal Selling **Informal selling** is business-direct-to-consumer testing of a new product. For example, a few top ranking salespeople are given the product (or product description) with the appropriate selling materials and are told to get orders. This type of market testing is often done at tradeshows for industrial products. Another method of informal selling is launching the product on the Internet directly to customers.

Direct Marketing **Direct marketing** includes the sale by the manufacturer to the end users via mail, telephone, fax, and/or Inter- and intranets. The advantages are that it's easy to keep new product ideas secret if only a select group of people can view the offering, feedback is quick, and costs are low, especially with the ease of producing customizable advertising collateral. Again, distribution partners are typically not part of this type of test.

Minimarkets Distribution channel partners make this type of testing viable. Good relationships with retail partners allow companies to market test new products in a limited number of stores. MetaCork did this with CostPlus/World Market, which sells wines in many of its markets. End aisle displays were set up with the Amusant brand of wines using the MetaCork. Initial interest was positive, but repeat sales were poor. This **minimarket test** added to Gardner Technologies redirecting their focus away from the consumer toward third-party sales (hotels, restaurants, etc.). IRI, Information Resources Inc., is the world leading provider of minimarket testing.

Scanner Market Testing **Scanner market testing** is also a service provided by IRI. They describe their product, IRI BehaviorScan Rx™, as an ad testing service that "delivers different TV advertisements in the same market, and has the ability to link TV viewing to consumer and doctor prescription drug activity. This in-market, split cable design solution provides the most accurate measurement of the impact of direct-to-customer (DTC) advertising on prescription activity."[48] Other companies have similar offerings.

Summary of Premarket Testing

The goal of the premarket test is to determine where problems may arise prior to a full market launch. The goal of a pseudosale and trial/repeat measurements is to get estimates of trial purchasing and repeat purchasing in order to make forecasting estimates typically using ATAR. Products that don't receive trial and repeat may need to be taken back into R&D for improvements. The goal of controlled tests is to examine how different markets react to the same marketing program. Figure 9.13 provides a summary of when the different types of premarket testing are best used.

Market Testing

There are two methods of conducting full-sale tests: **test markets** and **rollouts**. In these types of tests, the finished product, with the complete marketing mix, is tested in the market with the intent that if these mini launches are success, full launch will soon follow.

Test Markets

In a **test market**, one or more metropolitan markets are chosen for a full dress rehearsal. Test markets allow products to be tested in real settings. Information about sales, consumer usage/abuse, prices, reseller reactions, publicity, distribution issues, competitive reactions, and production issues can all come to light in a test market. The downside is that it has the ability to slow down full launches and to show your hand to the competition. Some of the issues that must be determined for conducting a test market are: which markets to enter, for how long, and if there should be a control market site that also is monitored for comparison.

	Industrial		Consumer		
	Goods	Services	Packaged	Durables	Services
Pseudosale					
Speculative Sale	✓	✓	✗	✓	✓
Simulated Test Marketing	✗	✗	✓	✓	✓
Trial/Repeat Measurements					
Home Delivery	✗	✗	✓	✓	✓
Laboratory Measurement	✓	✓	✓	✓	✓
Controlled Sale					
Informal Selling	✓	✓	✗	✓	✓
Direct Marketing	✓	✗	✗	✓	✗
Minimarketing	✓	✓	✓	✓	✓

Figure 9.13 Summary of best uses for pseudo and controlled sales. (From Endnote 49. With permission.)

The test market's goal is to determine if the planned launch strategy is sound. However, a test market can determine that it is best not to launch at all. For example, with much hype, Pepsi Kona was test marketed in Philadelphia in the late 1990s/ early 2000s. Based on that test market, the decision was made not to conduct a national launch. Coca-Cola also tried a coffee-flavored drink, Coca-Cola BlāK (see Figure 9.14). It was first tested in France in 2005 before being launched in the United States and Canada in early 2006 using a glass bottle. In 2007, Coca-Cola announced

Figure 9.14 Competing coffee-flavored colas.

its discontinuance of the product in the United States, although it was still available in France and Central Europe using an aluminum bottle. Would you consider Coca-Cola BlāK a successful product launch or not?

Rollouts

Rollouts are when the product is introduced into one market and then continues to be introduced to more markets. They have become the preferred method of conducting test markets. Failures in one market can be fixed before launching in the next market. There are a number of methods for rollouts including geographical segmentation, industry segmentation, and channel segmentation. Rollouts are actual full product launches, and, thus, signal to competitors that total market launch is forthcoming. A rollout, however, can show poor consumer acceptance, such as with Zima, which was a carbonated, alcoholic beverage distributed by the Coors Brewing Company. A similar problem is too much enthusiasm for a product, such as the Nintendo Wii when it was first launched. Shortages led to slower sales.

Geographical Segmentation One of the most important strategy issues that must be addressed when planning a rollout is which market is to be targeted first. The choice can significantly impact future decisions. Companies can choose from single national rollouts, where one U.S. city is targeted and then launch is rolled out over time to other cities. Or, it can conduct international rollouts starting with one country and then moving to others. A lead country strategy has the product being introduced in one country before launching in the next country. This is the strategy utilized with Coca-Cola BlāK.

Industry Segmentation Selecting different business segments is another strategy for rollouts. However, industrial (business-to-business) segmentation is not always easy because of changing industry dynamics, heterogeneity in firm types all in the same industry, dependency between supplier–firm relationships, and complexity of B2B consumer needs/requirements. Thus, there are numerous ways to segment industrial markets, including **two-stage market segmentation**,[50] **nested approach**,[51] and **bottom up segmentation**.[52] Segmentation should identify the main reasons the buyers buy. A two-stage segmentation is one of the most common methods. It takes a macrosegmentation coupled with microsegmentation approach where the micro level is a homogeneous group of buyers within a defined macro segment.[53] At a macro level, company size, geographic location, standard industry classification (SIC) code, purchasing situation, decision-making stage, institution type (i.e., bank, hospital, transportation, etc), and benefit segmentation are used to determine market segments. Micro considerations include buying decision criteria (i.e., quality, price, technical support), supplier relationship (i.e., closed/open vendor lists, RFP (request for proposal) requirements, who the decision makers are (i.e., engineering, CEO, procurement center), critical needs of buyer (core to business or peripheral), and attitudes toward the buyer.

Channel Segmentation   Another way of selecting the segment is through different channels. As previously mentioned, the Internet is one channel that companies will often use, where they test the market first through online buyers. Another strategy is to start with small national stores (Brookstone's) and then move to larger chains (Macy's). Some new products can only be found with discount chains, such as Sam's Club, before moving over to Walmart. For industrial products, in-house sales personnel may be the initial sellers and then the products will be rolled out to manufacturer's representatives over time.

Market Testing for Durable Consumer Goods and Industrial Products/Services

Market testing for industrial products and services is a bit more complicated due to their high cost and customer involvement. Some obvious directions for possible application are to pursue the use of information acceleration methods for services and the integration of lead user analysis with flow models for industrial and high technology products.

Information Acceleration   Information acceleration[54] was introduced as a design tool to be used before production investments are made. Information acceleration (IA) uses the power of the Internet to place consumers in a virtual buying environment. With IA, the survey respondent is in control of what they view; for example, they can virtually test the product, discuss with a salesperson the product, join blog forums to learn more about the products, view advertisements, among other things. IA is also a good way of market testing durable products and industrial goods/services.

This is particularly useful if a new industrial product is revolutionary. Mock technical articles and advertisements could represent real articles that would be written as if the product was on the market. Word-of-mouth communication could be created by filming lead user comments on the technology and the (hypothetical) new products that would be produced. Sales brochures could be created based on the core benefit proposition and the price and service to be delivered in the market. The product use experience could be even be simulated by prototypes. For example, new composites for auto bodies could be supplied from the lab at the specification level of a new production process. With these stimuli, the future scenario could be represented and measures taken of willingness to buy, preference versus competition, and attribute ratings. These measures would then be input to the analysis in determining the future success of the product.

Services are probably easier to test with an information accelerator than industrial products because it is often easier to simulate the service on a computer. For example, a comprehensive personal investment, legal, and banking service could be simulated by a trained team of people backed by a workstation and Wi-Fi at the user's home.

Summary of Market Testing

Market testing is the final dress rehearsal before final product launch. We examined two market testing methods for fast-moving consumer products: test markets and rollouts. Rollouts may occur through geographical, industry, or channel segmentation. Testing of durables and industrial goods/services can be completed using information acceleration, which is a method of simulating the purchase environment on a computer.

Chapter Summary

There are numerous advantages to testing a product/service and its associated marketing mix prior to a full launch. Testing and validation is the final stage where potential problems can be fixed before considerable expenses are made on launch. Product use testing, such as preuse reactions and alpha and beta testing, can be used to flush out any product issues. Market components testing can be used to test the 3Ps (price, place, promotion) each separately. Premarket test models provide a low cost and rapid method to test the combined product, advertising, price, promotion, selling, and distribution plan. This "full proposition" analysis is sufficiently accurate to identify most winners and eliminate most losers. They provide an effective way to control the risks of failure and supply actionable managerial diagnostics to improve the product. Market tests are dress rehearsals where the final product and all its components are tested in the market prior to a full launch.

Glossary

Acceptance: A step in the "consumer response hierarchy to advertising copy" which reflects the translation of comprehension of a new product into preference for that product.

Attention: A step in the "consumer response hierarchy to advertising copy" which is measured by indicators of whether the consumer will see/listen to an advertisement and note its content during exposure.

Attitude-change models: The process by which consumers alter their beliefs or behaviors toward a product or service.

Believability: A measure in the consumer response hierarchy that determines how believable a new product/service is in delivering on the promised value.

Blind comparisons: Procedure used for conducting customer tests where the respondent does not know the brands of two products that are being tested.

Bottom-up segmentation: Industrial segmentation approach that searches for similarities in individual firms to determine a homogeneous segment of buyers.

Category management: Marketing strategy in which a full line of products (instead of the individual products or brands) is managed as a strategic business unit (SBU).

Comprehension: A step in the "consumer response hierarchy to advertising copy" which measures awareness and understanding of a new product/service.

Concept test: Quantitative survey asking consumers to evaluate their probability of purchasing a new product/service based on a concept description.

Conjoint analysis: Also called *stated preference analysis*, a statistical technique that requires research participants to make a series of choices between two or more sets of product attributes that represent a potential new product. This method forces respondents to make trade-offs between preferred combinations of attributes and less preferred combinations. Analysis of these trade-offs can reveal the relative importance of component attributes. Conjoint analysis can give an estimate of "share of preference" in simulated markets.

Controlled sale: A controlled sale is a product launch limited to a very few select customers or distribution outlets. The intent is to test the market in a real environment but tightly controlling the launch variables. There are four types of controlled sales market testing: **informal selling**, **direct marketing**, **mini-markets**, and **scanner market testing**.

Convergent approach: Using more than one type of market test to determine if the two or more methods converge to the same outcome. It provides greater accuracy and confidence in resulting forecasts and strategic planning.

Direct marketing: Includes the sale by the manufacturer to the end users via mail, telephone, fax, and/or Inter- and intranets.

Experimental variations: Prelaunch testing where consumers are asked to evaluate one or more experimental designs of the product/service to determine their preferences.

Exposure: A step in the "consumer response hierarchy to advertising copy" which is a prerequisite for any consumer response and is measured by indicators of whether a consumer has been exposed to a particular media insertion.

Fast-moving consumer goods (FMCGs): Frequently purchased essential or nonessential goods, such as packaged food, household cleaners, clothing, etc.

Field testing: See physical product testing.

Informal selling: Business-direct-to-consumer testing of a new product. For example, a few top ranking salespeople are given the product (or product description) with the appropriate selling materials and are told to get orders.

Information acceleration: Information acceleration uses the power of the Internet to simulate information to a potential buyer of a new product. The IA technique places consumers in a virtual buying environment.

Judgment and past product experience: A premarket testing method in which past experiences with similar products are considered.

Likability: A measure in the consumer response hierarchy that determines how likable a new product/service is to potential consumers.

Market components test: Testing the effectiveness of the 3Ps (place, price, promotion) used in the marketing mix of a new product launch.

Market testing: Is the dress rehearsal for the marketing mix—products from a pilot production run, packaging, price, advertising, and promotion all come together.

Meaningfulness: A measure in the consumer response hierarchy that determines how meaningful a new product/service is to potential customers.

Minimarkets: A type of controlled sale where distribution channel partner makes a product available on a limited basis.

Nested segmentation approach: Hierarchical industry segmentation that looks at demographics, operating variables, purchasing approaches, situational factors, and buyers' personality nested within each other.

Physical product testing: (Sometimes called **field testing** or **user testing**) Determines whether the production-ready product delivers the CBP and generates diagnostic information to improve the product and/or reduce costs.

Preference: A measure of consumers' preferred choice in "intent to buy" or an actual purchase in a simulated buying situation.

Premarket testing: The product, its production, and the launch plan are tested together under prelaunch conditions.

Preuse reactions: Also called *preference tests*, are commonly done with a print or online survey that queries users on measures of satisfaction and/or preferences for a product/service's features.

Product use testing: A process by which customers (current and/or potential customers) evaluate a product's functional characteristics and performance. The purpose of product use testing is to test the product's functionality.

Pseudosales: Premarket testing method where the product is sold in simulated markets.

Revealed preference: Actual purchases made by consumers under realistic marketing-mix conditions to "reveal" a preference by consumers for a product at a set price. This is the opposite of *stated preferences* where consumers state their intent, but it not required to follow through on it.

Rollouts: Method of conducting full-sale market tests. One segment is identified within which to sell the product first, and if the launch is successful, the product will rollout to other segments.

Scanner market testing: An ad testing service that delivers different TV advertisements in the same market, and has the ability to link TV viewing to consumer activity.

Single product evaluation: When conducting market tests on only a single product version, because there is no alternative reference product, it is often difficult to interpret the results.

Stated preference: Conjoint studies and contingent valuation are measures of what consumer say they will do, not their actual behavior.

Test markets: A method of conducting full-sale tests where one or more metropolitan markets are chosen for a full dress rehearsal. Test markets allow products to be tested in real settings.

Testing and Validation stage: Provides final validation of the entire project with the combination of the final product, its production, and its marketing plan.

Transaction data: Store scanner and simulated test market data, which are good at eliciting willingness to pay at the point of purchase.

Trial and repeat: Premarket testing method where potential consumers are provided with an opportunity to buy a product and then given the option to repeat the purchase.

Two-stage market segmentation: Industrial marketing approach broken down into micro-level homogeneous groups set within micro-level segments.

User testing: See physical product testing.

Vickery auctions[55]: Using sealed bid auctions to observe "revealed" prices.

Review Questions

1. What are the advantages and disadvantages of premarket tests? How accurate are pretest market models? Is this accurate enough?

2. Why should a manager accept less than a 100% probability of ultimate success when using pretest markets to screen new products?

3. What diagnostic information should a manager expect from:
   a. Product use testing?
   b. Market components testing?
   c. Premarket test analyses?
   d. Market testing?

4. How would you use pretest market analysis to test the effect of a price change and/or a price-off promotion?

5. What is a beta test versus an alpha test? Who would participate in each type of test?

6. For industrial goods, what is the type of testing that would be most beneficial? For a consumer package good (CPG)?

7. What are some of the ways of conducting a rollout?

Assignment Questions

1. Develop a preuse product reaction survey for the MetaCork. What type of questions would you ask?
2. Explain how a new line of clothing can use the Internet to test a proposed advertising campaign.
3. FAGE wanted to introduce to the market a new microbiotic Greek yogurt with real fruit. It is uncertain how to price the new product. Suggest some test pricing options they should consider.
4. Tesla, the electric vehicle company, is planning to market test the latest design in the European marketplace. Suggest a rollout strategy for the company.
5. Apple has developed a revolutionary tablet. It wants to move across the decision frontier to a less risky position. Describe a market test strategy for them. What are the benefits and weaknesses of your proposed strategy?

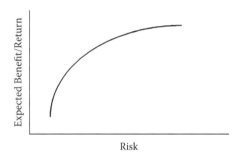

6. Take one of the B2B industries noted below and come up with a segmentation strategy for identifying new product opportunities.
   a. Cloud computing
   b. Electronic waste disposal
   c. Shipping
   d. Military defense
   e. Education

Endnotes

1. Based on http://www.businesswire.com/news/home/20030804005656/en/Gardner-Technologies-Unveils-Revolutionary-Wine-Opener-MetaCork (accessed June 22, 2013); http://www.fetzer.com/assets/press/Metacork_080403.pdf (accessed April 5, 2008); Gerald D. Boyd, "The Revolutionary MetaCork," *Management of Process, Technology and People* 29(6) (Nov/Dec 2003): 60-63.
2. Online at: http://www.fetzer.com/assets/press/Metacork_080403.pdf (accessed April, 2, 2008).
3. Ibid.
4. For those unfamiliar with the New Coke product failure, read more at: http://en.wikipedia.org/wiki/New_coke (accessed July 2, 2013).

5. Based on Glen L. Urban and John R. Hauser, *Design and Marketing of New Products* (Upper Saddle River, NJ: Prentice Hall, 1980).

6. Ibid.

7. Ibid.

8. Abbie Griffin, Stephen Somermeyer, and Paul Belliveau (eds.), *The PDMA Toolbook 3 for New Product Development* (Hoboken, NJ: John Wiley & Sons, 2007).

9. Ibid.

10. "Google Gets the Message, Launches Gmail," April 1, 2004. Online at: http://www.google.com/press/pressrel/gmail.html (accessed April 21, 2010); Matthew Glotzbach, director, Product Management, "Google Apps Is Out of Beta (Yes, Really)," *Google Enterprise*, July 2009. *Official Google Blog*. Online at: http://googleblog.blogspot.com/2009/07/google-apps-is-out-of-beta-yes-really.html (accessed June 19, 2010); Keith Coleman, "Gmail Leaves Beta, Launches "Back to Beta" Labs Feature," July 2009. Online at: http://google-blog.blogspot.com/2009/07/google-apps-is-out-of-beta-yes-really.html (accessed July 7, 2009).

11. "Anurag Acharya Helped Google's Scholarly Leap." Online at: http://www.indolink.com/SciTech/fr010305-075445.php (accessed May 11, 2011).

12. Online at: http://revocars.com/wp-content/uploads/2010/04/FuelCellEquinox.jpg (accessed June 24, 2013). Chevy Equinox is a registered trademark of General Motors.

13. Based on Urban and Hauser, *Design and Marketing of New Products*, Chap. 13.

14. Ibid.

15. Ibid.

16. Ralph I. Allison and Kenneth P. Uhl, "Influence of Beer Brand Identification on Taste Perception," *Journal of Marketing Research* (1964): 36–39.

17. J. Douglas McConnell, "The Price-Quality Relationship in an Experimental Setting," *Journal of Marketing Research* (1968): 300–303.

18. Urban and Hauser, *Design and Marketing of New Products*.

19. Ibid.

20. Ibid.

21. Ibid., p. 367.

22. Ibid., p. 368.

23. Other methods for measuring willingness to pay are available to the marketing researcher; for a review, see: Franziska Voelckner, "An Empirical Comparison of Methods for Measuring Consumers' Willingness to Pay," *Marketing Letters* 17(2) (2006).

24. Robert Cameron Mitchell and Richard Thomas Carson, *Using Surveys to Value Public Goods: The Contingent Valuation Method* (Oxon, U.K.: RFF Press, 1989).

25. David Lucking-Reiley, "Vickrey Auctions in Practice: From Nineteenth-Century Philately to Twenty-First-Century e-Commerce," *The Journal of Economic Perspectives* 14(3) (2000): 183–192.

26. Online at: http://www.businessdictionary.com/definition/conjoint-analysis.html#ixzz2WyvOv1iJ (accessed June 22, 2013).

27. For details, see Shlomo Kalish and Paul Nelson, "A Comparison of Ranking, Rating and Reservation Price Measurement in Conjoint Analysis," *Marketing Letters* 2(4) (1991): 327–335; Bryan K. Orme, "*Getting Started with Conjoint Analysis: Strategies for Product Design and Pricing Research*," (Madison, WI: Research Publishers LLC, 2006).

28. Based on John Thøgersen, "The Demand for Environmentally Friendly Packaging in Germany," MAPP working paper, 30, Handelshøjskolen Aarhus Universitet, Denmark, 1996.

29. Alvin J. Silk and Glen L. Urban, "Pretest-Market Evaluation of New Packaged Goods: A Model and Measurement Methodology," *Journal of Marketing Research* (1978): 171–191.

30. Dennis M. King and Marisa Mazzotta, "Ecosystem Evaluation." Website: HYPERLINK "http://www.ecosystemvaluation.org/contingent_valuation.htm" http://www.ecosystem-valuation.org/contingent_valuation.htm (last accessed January 10, 2014).

31. For more on Vickrey auctions, see Klaus Wertenbroch and Bernd Skiera, "Measuring Consumers' Willingness to Pay at the Point of Purchase," *Journal of Marketing Research* (2002): 228–241.

32. Online at: http://www.thestreet.com/story/11307719/1/wal-mart-discovers-new-purpose-for-pop-up-stores.html (accessed June 24, 2013).

33. Urban and Hauser, *Design and Marketing of New Products.*

34. Henry J. Claycamp and Lucien E. Liddy, "Prediction of New Product Performance: An Analytical Approach," *Journal of Marketing Research* (1969): 414–420.; Peter J. Lenk and Ambar G. Rao, "New Models from Old: Forecasting Product Adoption by Hierarchical Bayes Procedures," *Marketing Science* 9(1) (1990): 42–53.

35. Silk and Urban, "Pretest-Market Evaluation of New Packaged Goods."

36. Urban and Hauser, *Design and Marketing of New Products.*

37. Glen L. Urban, John R. Hauser, William J. Qualls, Bruce D. Weinberg, Jonathan D. Bohlmann, and Roberta A. Chicos, "Information Acceleration: Validation and Lessons from the Field," *Journal of Marketing Research* (1997): 143–153.

38. We introduced information acceleration as a method for presenting products in a concept test using virtual reality; see Urban, et al., "Information Acceleration."

39. Online at: http://researchexpertise.blogspot.com/ (accessed June 24, 2013).

40. Online at: http://www.marcresearch.com/methodologies/assessor.php (accessed April 12, 2011).

41. Silk and Urban, "Pre-Test-Market Evaluation of New Packaged Goods."

42. Steve Mollman For CNN, "Second Life's 2nd Value: Testing Ideas," September 23, 2007. Online at: http://edition.cnn.com/2007/BUSINESS/09/16/second.life/ (accessed June 24, 2013).

43. Synovate was sold to Ipsos in 2011.

44. Online at: http://researchexpertise.blogspot.com/(accessed June 24, 2013).

45. Urban and Hauser, *Design and Marketing of New Products.*

46. J. H. Parfitt and B. J. K. Collins, "Use of Consumer Panels for Brand-Share Prediction," *Journal of Marketing Research* (1968): 131–145.

47. Charles Merle Crawford and C. Anthony Di Benedetto, *New Products Management*, 9th ed. (New York: McGraw-Hill/Irwin, 2008).

48. Online at: http://us.infores.com/page/healthcare/IRI_BehaviorScan_Rx (accessed May 2008).

49. Based on Crawford and Di Benedetto, *New Products Management.*

50. Yoram Wind and Richard N. Cardozo, "Industrial Market Segmentation," *Industrial Marketing Management* 3(3) (1974): 153–165.

51. Thomas V. Bonoma and Benson P. Shapiro, *Segmenting the Industrial Market* (Lexington, MA: Lexington Books, 1983).

52. P. Kotler, *Marketing Management* (Englewood Cliffs: Prentice-Hall, 1984).

53. Fredrick Webster, *Industrial Marketing Strategy*, 3rd ed. (New York: John Wiley & Sons, 1991).

54. Urban, Hauser, Qualls, Weinberg, Bohlmann, and Chicos, "Information Acceleration."

55. For more on Vickrey auctions, see Klaus Wertenbroch and Bernd Skiera, "Measuring Consumer Willingness to Pay at the Point of Purchase," *Journal of Marketing Research*, 39, May (2002): 228–241.

INTO THE MARKET

Launch

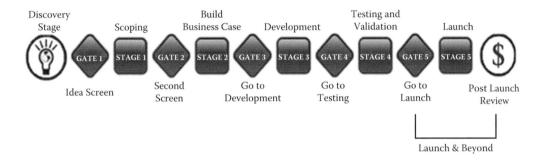

Learning Objectives

In this chapter, we will address the following questions:

1. What are the tasks involved in prelaunch strategizing? Prelaunch tactics? What is the difference between the two?
2. What are the issues to consider in developing a brand strategy for a new product/service?
3. What are the different pricing strategies that are typically used with new product launches?
4. Why should a firm monitor launch activities?
5. When should a firm kill off a seemingly failing product? When should it try to revamp it?

Nightmares for the Dreamliner Boeing 787

There had not been so much excitement in the aerospace industry regarding a commercial airplane model in many years. The Boeing 787 Dreamliner was reported to be "the future of flying,"[1] a "dream" for both carriers and customers. It was introduced with the anticipation of higher revenues for air carriers, greater flying range, more cargo volume, faster speed, lower emission and noise fees, higher reliability, and ease of repair.[2] The Dreamliner was said to be the first plane that put the passenger first with bigger windows, more room for carry-on baggage, and better cabin pressurization to make long flights more bearable.[3] Boeing's Web site described the new model: "The 787 Dreamliner increases revenue potential through significantly better performance, improved fuel efficiency, and lower operating and maintenance costs. Passengers will enjoy a superior flight experience with the 787—an experience that brings back the magic of flight."[4] All Nippon Airways (ANA) became the first customer to order the 787 because it would allow them to open new routes to cities not previously served, such as Denver, Moscow, and New Delhi.[5] Delivery was expected in late 2008.[6]

However, delays to product launch were announced in October 2007 due to numerous problems including foreign and domestic supply chain inadequacies.[7] Boeing had experienced difficulty in getting the right parts from its suppliers on time. In an effort to gain more control over the supply chain, Boeing announced in 2008 that it planned to buy Vought Aircraft Industries and its interest in Global Aeronautica,[8] a fuselage provider.

Throughout 2008, Boeing announced delays with initial deliveries expected in the third quarter of 2009. This delivery date was optimistic. On October 26, 2011, the 787 flew its first commercial flight from Narita International Airport (Tokyo) to Hong Kong for ANA. But the problems continued. In February of 2013, fears of a fire led to an emergency landing and evacuation of an ANA domestic flight.[9] Subsequently, air authorities in Japan, America, and India grounded the jet. A British newspaper reported, "Boeing's headache over its Dreamliner jets worsened today ... Authorities around the world have grounded the lightweight planes following problems with the model and Japan's transport ministry launched an extensive probe into the 787s after two incidents last month involving Dreamliners operated by All Nippon Airways."[10] The grounding of the Dreamliner was deemed by many to be more than just an embarrassment to Boeing, it caused big problems to airlines and their passengers with flight rescheduling, loss of revenues, and, more importantly, customer confidence in the dream. Speaking at a news conference on February 28, 2013, Raymond Conner, head of the commercial aircraft division at Boeing, said the incidents that led to the grounding of the entire fleet of Dreamliner 787 planes were "deeply regretful. On behalf of the Boeing Company and the 170,000 people which I represent today, I want first to apologize for the fact that we've had two incidents with our two very precious customers, ANA and JAL," he told reporters in Tokyo.[11] As this case shows, even well established companies such as Boeing can experience serious problems at launch.

Introduction

By the time a new product reaches the launch stage, significant resources have been expended and the company will be seeing the "light at the end of the tunnel" with a soon-to-be revenue stream. Although successive testing and development has reduced the risks, the stakes are now higher than ever; it is not the time to relax the rigor of the stage-gate process. Launch activities can be very expensive and require substantial commitments from all levels of the organization from production to sales to customer service representatives. Launch can easily be mismanaged in the handoff from the new product development team to the marketing team, as summarized in Table 10.1. In this chapter, we will examine both prelaunch and postlaunch considerations for the new product development (NPD) manager so as to minimize potential missteps such as those taken by Boeing. The benefits of a pragmatic, heavily monitored, and flexible launch program for new products has been well documented.[12]

Table 10.1   Issues in Product Launch

Frequently seen issues related to product launch failures include:

1. A failure to establish clear market windows for the introduction of products.
2. Product champions not assigned to lead important launches or, once assigned, do not keep close enough tabs on the progress of the launch.
3. Launch plans not synchronized with the business case or not included in the marketing plans.
4. Salesforce and channel organizations do not have the capacity to sell the product, and, in many cases, do not have their compensation plans adjusted to encourage the sale of new products/services.
5. Sales and marketing collateral is incomplete, inaccurate, or late.
6. Sales training is not carried out in a timely basis or the training is not sufficient.
7. Operational systems and infrastructure elements within the business are not ready to support launch, either because they are brought into the process too late, or not sufficiently staffed.
8. Launch metrics are missing, incomplete, or ignored.
9. Product teams are reluctant to kill the product midlaunch, even if that obviously is the right thing to do.

Source: Endnote 13. With permission.

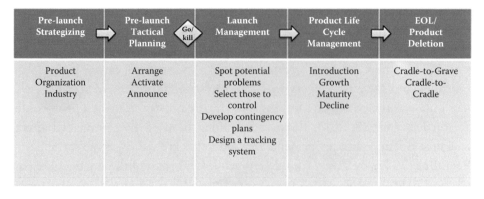

Figure 10.1 Launch process and postlaunch monitoring.

The **launch process** can be broken into five components, as shown in Figure 10.1. Prelaunch strategizing and prelaunch tactical planning are about making the launch decisions. Postlaunch requires diligent management to make sure that carefully laid plans succeed as planned. As the product moves through the product life cycle, the product/service should continue to be monitored and the marketing plan adjusted as needed. The product manager will one day have to decide when the product should be deleted and then a decision must be made on what strategy to take to move it though to its end-of-life. We look at each step in detail.

Prelaunch Strategizing and Tactics

Launch decisions are standardly conceptualized within two major categories: strategic and tactical.[14] Strategic launch decisions typically are made early in the stage-gate process and tactical launch decisions in the later stages. In comparison to tactical launch decisions, strategic launch involves a substantively greater resource commitment and is more difficult to alter once a trajectory has been selected.[15] Tactical decisions can be modified and normally will be changed even after launch as market conditions change. For example, Coke Zero, which is targeted at young men, was launched in Australia with online promotions using a fake grass-movement to reduce negativity in the world. The campaign was assailed by consumer advocates as misleading and a counter Web site emerged questioning the ethics of Coke's activities.[16] Coca-Cola removed the campaign from distribution and relaunched using conventional methods.

In general, strategic launch issues occur prior to and during the development stage and tactical issues occur postdevelopment,[17] although these tasks can take place anywhere in the stage-gate process. There are three primary strategic issues: **product concerns**, **organizational concerns**, and **industry concerns** as outlined in Figure 10.2. There is one primary issue for tactical issues, which is **market planning**. We discuss each of these next.

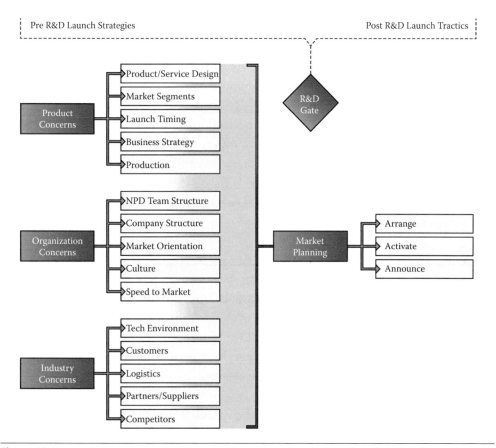

Figure 10.2 Launch considerations. (From Endnote 18.)

Strategic Launch

Product Concerns

Product concerns in strategic launch planning involve making sure that the product/ service is physically ready to launch. The firm's business strategy (online, retail, service-oriented, etc.) will have dictated what products and services will be designed. The market segments that have been identified by marketing will determine how many different product/service design models will need to be produced. Questions the product manager should be asking include:

- How does the new product align with business strategy?
- Is the launch strategy consistent with previous product launches?
- Has the right target market been identified?
- When should the launch be announced?
- Should there be a preannouncement to generate interest or desire for the new product or service?
- Have quality control standards been implemented and the production facility tuned to consistently produce the desired product specifications?

We address these types of product concerns next.

Business Strategy   It seems obvious that the firm's business strategy should be set prior to launch. This may not always be the case; the business model may actually change after launch. Zaarly, a U.S. Web-based company, was initially set up as a local customer-to-customer exchange site where you could post anything you wanted to buy (as opposed to sites where you post what you want to sell). About eight months after launch, it evolved into a commerce site where small local businesses posted services they could provide. It's become an "Etsy"[19] for services (Figure 10.3). Companies should closely monitor launch activities and remain flexible to change strategic course if new opportunities warrant redirection.

Market Segments   Segmentation involves the three-step STP process of segmentation–targeting–positioning. Understanding the segments in which the company will target is important for knowing because it will drive the different models/variations of a product or service that will be made available at launch. Lenovo, the computer manufacturer, created the ThinkPad tablet for business users and the IdeaPad tablets for home and personal use. The two products have very different uses and, thus, different approaches to the launching of new versions.

Product/Service Design   The design of the product or the service significantly impacts tactical launch designs. Does the product require customer service support for installation or troubleshooting? Will the service require a new customer–company interface? Nest, the Learning Thermostat, allows control of your home's temperature from your laptop, smartphone, or tablet when connected to Wi-Fi. Within days of launching in the United States, the company was receiving feedback from customers via the Internet. Because of the high involvement with customers, launch was initially limited to domestic markets with international launch

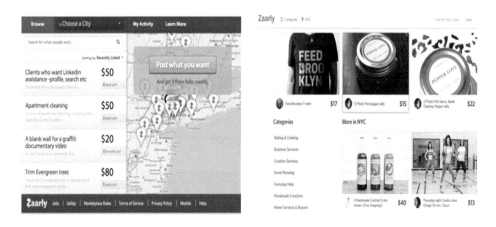

Figure 10.3   Morphing of zaarly.com from buyer-driven requests to small storefront offerings.

following almost a year later. Customer feedback was used to develop the second generation Nest Thermostat.

Launch Timing   Launch timing is about mitigating risks of being too early out of the gate and not being ready for market demand or being too late out of the gate and missing the market opportunity. Timing is often based on external events. For example, many in the electronics industry will be sure to launch new products at the Consumer Electronics Show. At the 2013 CES Show, Sony announced its Sony Xperia Z smartphone while Samsung announced its Galaxy SII plus smartphone. We discuss launch timing in detail below because it's an important issue for new product managers.

Production   Coordination with production is extremely important in launch planning and sometimes can be overlooked. Marketing and production teams should be working collaboratively during launch to ensure that products are available for the launch date. Advertisements or promotions campaigns introducing the new product or service should coincide with sufficient inventory. Contingency plans should be in place for both highly positive consumer reaction and highly negative reactions.[20] Joint planning of the teams has the potential to result in greatly increased profits.

Another important production question to consider is whether the product will be made locally or globally. Local product may allow speeding a product to market for glocal (local and global) product modifications. Producing in Asia may allow lower prices. LG's Optimus G Pro announced its new phone at the Mobile World Congress in 2013. The company debuted the device in South Korea in early 2013 before launching in North America, Japan, and other markets in the second quarter of 2013. LG could manufacturer products in any of these countries or all of the countries depending on how it expects to meet market demands.

Organizational Concerns

Organizational concerns focus on aligning the company's goals and overall strategy with the new product development process. The company structure can greatly impact which divisions are involved in product/service design and where launch directives originate—top-down or bottom-up. The organizational culture can have a strong impact on speed to market activities and the formality of the NPD process undertaken. And, of course, the NPD process will dictate the NPD team structure that will be involved with launch activities. Examples of organizational concerns are noted below.

Company Structure   The company structure can drive launch strategies within an organization. Some firms are termed *mechanistic*, which means that they have multiple layers of hierarchy, control is concentrated in the C-suites, and there is strong adherence to rules and policies within the organization.[21] *Organic* firms are much less formalized and allow flow of knowledge and new ideas throughout the organization without regard to hierarchies. However, even mechanistic companies can be successful at new products. Former CEO of Apple, Steve Jobs, was credited with overseeing the development of the iMac, iTunes, iPod, iPhone, and iPad.[22] He was also a prominent figure in the launching of Apple products and services until his death in 2011.[23] After his death, there was speculation that no one could possibly fill his shoes when it came to driving and launching Apple's new creations.

Culture   Risk-taking is important in the new product development process and acceptance of occasional failures can actually help with the success of new products.[24] It has been found that "mechanistic" organizations can stifle innovation because of "arrogance, short-term mentality, risk adversity, a not-invented-here syndrome," and a Wall Street "only shareholders matter" mentality.[25] Firms open to risk-taking in the innovation process show more creativity, which manifests itself in innovative products and, thus, higher firm performance. Risk-taking includes an affinity of venturing into the unknown or committing resources to uncertain projects.[26] Netflix, the online video rental/streaming company has had missteps in new service launches. In September 2011, Netflix CEO Reed Hastings announced that the DVD section of Netflix would be split off and renamed Qwikster. A million customers responded by leaving Netflix. By October 10, 2011, Hastings announced the cancellation of the planned Qwikster service and said that the DVD-by-mail service would remain a part of Netflix. Despite the negative backlash, it has been said that the company's willingness to take risks fostered its growth. *Forbes* magazine called Netflix the "turn-around story of 2012," due to its 50% market valuation that year.[27]

Speed to Market   There is pressure to move products through the development cycle at ever quicker rates because of windows of opportunity that remain open for increasingly shorter durations in the global competitive market.[28] It has been reported that Fortune 500 companies such as AT&T, Honda, Xerox, Hallmark, and Chrysler have cut development cycle time up to 50% without hurting their new product success rate.[29] Some methods utilized for reducing development cycle time include parallel or concurrent processing of NPD process steps, cross-functional teams that include marketing and R&D team members, reducing product complexity, sufficient allocation of resources in the front end of the project, and rewarding high R&D performance.

NPD Team Structure   The structure of the cross-functional team can take many forms and often will be dependent upon the company structure. Teams will differ on which functional groups are represented in the team, who the team reports to, who heads up the team, and even where the team members are located. Teams are typically structured along the continuum of **degree of projectization**,[30] which is the extent to which participants see themselves committed to or independent from the project. A functional team has the least projectization and project teams have the most degree of projectization:[31]

Functional: The project is divided into segments and assigned to relevant functional areas (marketing, finance, sales, development, production, etc.) and/or groups within functional areas. The team members report to their own functional managers and the team to top management.

Functional matrix: A project manager with limited authority is designated to coordinate the project across different functional areas and or groups. The functional managers retain responsibility and authority for their specific segments of the project. The team reports to the functional managers as well as top management.

Balanced matrix: A project manager is assigned to oversee the project and shares the responsibility and authority for completing the project with the functional manager. Project and functional managers jointly direct many workflow segments and jointly approve many decisions.

Project matrix: A project manager is assigned to oversee the project and has primary responsibility and authority for completing the project. Functional managers assign personnel as needed and provide technical expertise.

Project (venture/skunkworks) team: A project manager is put in charge of a project team composed of a core group of personnel from several functional areas and/or groups, assigned on a full-time basis. The functional managers or top management typically have no formal involvement. **Skunkworks** describe a team given a high degree of autonomy that is unhampered by bureaucracy, tasked with working on innovative, usually new-to-the-world projects. Project teams may be situated in buildings outside of main firm location.

One company that uses different team structures depending on the product or service is Google. Project-oriented teams are used at Google's X Lab where skunkworks projects are developed. One is Project Glass, the augmented reality, head-mounted display glasses that connect to the Internet.

Market Orientation[32]   Market orientation is defined as an organization's generation and dissemination of market information regarding customer needs and wants and the incorporation of that market analysis into strategic planning initiatives,[33] such as market launch. A firm that incorporates voice of the customer (VOC) studies, marketing studies, and competitive actions/reactions into the corporate strategic planning is seen as having a strong market orientation.[34,35] These types of activities have been found to positively impact execution of launch tactics and ultimately new product performance. Sara Blakely started Spanx, a women's undergarment business, in her home with $5,000 based on her own need and the needs of women worldwide for comfortable body shaping undergarments. Spanx are now sold through high-end retailers

Neiman Marcus, Nordstroms, Saks Fifth Avenue, and Bloomingdales as well as in stand-alone retail stores. Blakely's intimate connection with the target market allowed her to build a billion dollar business.

Industry Concerns

In addition to internal organizational issues, firms must be concerned with environmental factors that may impact launch. The intensity of the competition, bargaining power of suppliers and customers, need for logistics partners, threat of product substitution, and entry/exit barriers all affect firm performance.[36] Uncertainties in the market environment can arise from rapidly changing technology, changing market demands, and competitive reactions. These uncertainties, which are not controllable by the firm, must be considered when generating cult launch strategies. Questions the new product manager must ask include:

How quickly is the technological environment changing?
Is there a single powerful competitor or several smaller players?
Are there barriers to entry or exit for the firm or high switching cost for the consumer?
Is the market mature or evolving?

Maintaining an ability to be flexible as the market environment changes is critical for the success of the product launch. Dynamic industries present challenges that the team must be ready to address with speed and confidence. Examples of industry concerns are noted below.

Technical Environment[37] Turbulence creates a situation where considerable marketing and technological uncertainty and unpredictability confront NPD managers. In turbulent environments, NP managers must cope with uncertainty regarding their customers' needs, uncertainty as to which are the best long-term technologies and market paths to follow, and uncertainty as to levels of resources to commit to various endeavors. Competitive advantage lies in a firm's ability to quickly adapt to these changing environments. Technology is continually evolving. For example, evolving cloud computing services have opened up numerous market opportunities for firms. UberSense, a Boston-based start-up, capitalized on this technology by allowing its users to improve their biomechanics of various sports, such as tennis, track and field, gymnastics, and weightlifting, through video recordings. UberSense allows players and coaches to instantly connect via a cloud server enabling sharing and discussion of videos over mobile devices from anywhere in the world.

Competitors It is crucial to monitor competitors' new product launches as well as anticipate their moves. Being able to predict the strength and speed of reactions from competitors is important as their responses to your new product launch can seriously harm the performance of a new product.[38] When first unveiled in 2007, Google's Android smartphone operating system received a lukewarm welcome. Well-established competitors were skeptical. Nokia was quoted as saying, "We don't see this as a threat."[39] By the third quarter of 2012, Android had a 75% share of the global smartphone market.[40,41] Google's competitors no longer ignore new product or service introduction made by the company.

Partners/Suppliers   Sometimes partners and suppliers have more power than a firm would like to allow. In the opening vignette, we saw that Boeing's suppliers had a significant impact on the quality issues the aircraft manufacturer faced. To mediate the problems it was having with it suppliers, it even bought one of the companies to try to overcome delivery issues.

Customers   We have extensively explored in previous chapters how the voice of the customer can lead to innovative new products that meet the needs and wants of consumers/end users. MakerBot Industries, creators of a 3D printer, stays close to its customers by hosting an online community. In 2008, Thingiverse.com was launched as a place for anyone anywhere to share digital design files and instructions for "printing" things using MakerBot's 3D printers. Designs for hard-to-find automotive parts, cat toys, intricate artwork, etc., are posted by users of the 3D printers.

Logistics   Logistics involves the management of the flow of resources (parts, finished product, etc.) between the point of origin and the point of destination. Critical to firm success is making sure that the product is available for customers at launch so logistic functions should be integrated with marketing, manufacturing, and operations for seamless distribution of new products. However, even some of the best companies stumble in this area. In June 2010, it was reported that dealers and customers had not received Ford's new Fiesta cars designed for the North American market that they expected in May.[42] In July, Ford said initial shipments were delayed for up to two weeks by Hurricane Alex and, subsequently, by Tropical Storm Bonnie. By August 2010, Ford delayed further shipments because of a "quality problem."[43] These glitches in product launch are not all that uncommon and contingencies should be in place for such occurrences.

Tactical Launch Planning

Numerous research studies have shown that proficiency in launch tactics is positively associated with new product performance. Product launch should be planned with as much forethought as the new product development process. Steven Haines, new product manager guru, groups launch activities into three categories: arrange, activate, and announce.[44] **Arrange** takes into consideration all the concerns evaluated in the prelaunch strategic to establish a plan of action. **Activate** puts into place these

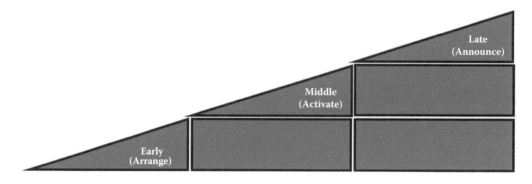

Early (Arrange)	Middle (Activate)	Late (Announce)
• Meet with executive sponsor to ensure approval of plans	• Arrange sales training dates	• Carry out sales training
• Confirm market window	• Get marketing material written and approved	• Carry out customer service agent training
• Synchronize with Production/Development	• Arrange any publicity/PR events	• Process pre-orders
• Update any changes to Launch Plan	• Evaluate product or market test information	• Meet with industry analysts for launch announcement
• Review Launch plan compared to Business Case and Marketing Plan	• Arrange meetings with industry analysts	• Ensure that no outstanding issues are left from product tests
• Finalize findings from product or market test	• Secure product codes and ensure product collateral is up to date	• Verify all regulatory approvals have been secured
• Finalize Pricing	• Provide status updates to sponsors and top management	• Load prices and product codes into appropriate systems
• Submit prototype to standards groups, regulatory bodies or product safety organization	• Prepare to activate customer service, repairs and returns processing, IT support, etc.	• Tune up all operational systems
• Decide on product name/brand association	• Review launch program risks and problems	• Load channels and ship products
• Align Marketing Plan with creative organizations responsible for copywriting, design, etc.	• Recheck value proposition, positioning, pricing, promotional program and channel readiness	• Validate post-launch metrics
		• Recommend Go/Kill decision
		• Announce New Product/Service to Market!

Figure 10.4 Launch activities. (From Endnote 45.)

plans of actions that will be implemented during launch, and **announce** is the point where a final go-kill decision to launch the new product into the marketplace is made. Figure 10.4 provides a tactical list of tasks that are to be completed. Each list builds upon the previous, which is emphasized in the upward slope of the triangles to indicate that the activities ramp up over time. Most of the tasks noted in this figure have been addressed in previous chapters or in principles of marketing books.[46] We highlight a few important issues for the new product manager: product naming, brand management, pricing, and launch timing.

Figure 10.5 The importance of product naming Bigbelly versus Seahorse Power. (From BigBelly Solar: The Smart Grid for Waste and Recycling™ (Endnote 47.)

Product Name and Branding Strategies

A product's name can lead to its success or contribute to its demise. For example, Seahorse Power, a Massachusetts-based company, started producing solar powered trash bins, which provide solar-powered self-compacting waste containers for public places. The containers hold five times the capacity of standard receptacles and will automatically notify the trash collection company using Wi-Fi when they are full. The first machine was installed in Vail, Colorado, in 2004, but was slow to take off. Seahorse Power didn't adequately describe the product so the company changed its name to BigBelly Solar, which was more distinguishable than Seahorse. By 2013, BigBelly, as shown in Figure 10.5, had bins in 30 different countries.

So, what is in a name and how are names given to new products? The major activities in naming a product include:[48]

1. Identifying and engaging the decision makers and other stakeholders. Is top management approval needed? Is the brand manager involved or just the NPD team? Is the consumer queried? Who is the *ultimate* decision maker?
2. Do the preparation: Set up the context within which the name is relevant. Who are buyers? Will the product be sold internationally? What makes the product unique? Who are competitors?
3. Develop the initial list: Using brainstorming and the creativity exercises introduced in previous chapters to develop a long list of potential names. Consider different types of product names, including personal (Ben & Jerry's), descriptive (Rollerblades), functional (Ty-D-Bol, toilet cleaner), emotional (Jaguar), invented (Spanx, women's shaping undergarments), abbreviations (TNT,

---

Checklist of Criteria for a Good Product Name

- Simplicity
- Memorable
- Ease of spelling
- Ease of pronouncing
- Ease of understanding
- Evokes positive image
- Does not have hidden meanings
- Translates well into other cultures
- Does not have negative meanings in other languages
- Can be graphically represented appropriately
- Fits with corporate mission and brand strategy
- Complements other product names
- Does not conflict other product names, nor dilute brand name

Check list of things to avoid

- Names with cult meanings
- Names based on what may be short-tern fads
- International faux pas
- Names offensive to religious or ethnic goups
- Names with two or more differing connotations

Figure 10.6 Naming short-list criteria. (From Endnote 49. With permission.)

Turner Network Television), or humorous (BigBelly trash compactors). "A consideration when generating and screening names is the common associations of letters and sounds. Words with consonants are considered more masculine and suggest hardness and sharp angles. Vowels are softer and more feminine. 'X' implies high-tech or extreme, but proceed with caution because the letter is currently overused."[50]

4. Select the short list: Figure 10.6 has a checklist of criteria for a good product name and a separate one for avoiding bad names. Select those names from the "long" list to create a "short" list of viable options.

5. Pick the name/register the trademark: Use a skilled attorney to register the trademark. At this time, it is also reasonable to conduct in-depth customer research to get reactions to the name. The graphic detail of the logo also may impact the decision, so looking at one or two finalists from a graphical perspective may be useful in the decision process.

6. Protect the trademark: Register the product name as a trademark and defend as needed. McDonald's has been very aggressive in protecting its corporate name.

Table 10.2 provides some pitfalls to avoid when naming a product. Some major concerns are making sure that the name does not infringe on others' trademarks and that the name is not offensive in other languages or cultures. A couple of famous unfortunate names were the Rolls Royce Silver Mist where "mist" refers to manure

Table 10.2   A Summary of Product Naming Pitfalls

Failing to anticipate the future uses of the name: Names may be unsuitable when the company expands and wants to use the equity in the product name for product extensions or when the company decides to sell internationally.

Not allowing sufficient lead-time for the process: Product naming can be a lengthy process, particularly if the product must be researched from a legal perspective in several countries.

Failure to allocate sufficient resources: A good naming process takes work and effort. Often, the person in charge of naming (e.g., the product manager) has many other responsibilities, and he/she fails to set aside enough time to perform the job properly.

Failure to identify the decision makers in advance: Product naming can be an emotional endeavor, everyone has an opinion, and more people than you might expect consider themselves to be stakeholders.

Deciding on a name that is comfortable: The best names may be provocative and controversial first. The iPad by Apple was ridiculed as being associated with feminine hygiene products.

Using too many decision makers: Too many decision makers bog you down, lengthen the process, and produce a decision that is safe, but not necessarily the best.

Picking an early favorite and running with it: If you become attached to a particular name, it can be tempting to focus the attention on that name and short circuit much of the testing.

Cutting corners on the trademark attorney: A qualified trademark attorney can navigate through the intricacies of the process and deal with the nuances that would not be obvious to a layperson.

Failure to identify negative meaning in other cultures: The formal translation of a word or phrase may appear good, yet there may be slight or odd meanings that are negative and/or inappropriate. When Mercedes-Benz was initially translated for the Chinese market, it was rendered Bensi, which means "rush to die." The brand became Benchi, or "run quickly as if flying," a much more appropriate name.

Failure to keep trademark registration current: Once you registered your trademark, you need to take steps to ensure you don't lose it. Observe usage rules, which vary by country, and keep your address current with the various registration offices.

Source: Endnote 51. With permission.

in German and the Ikea Fartfull workbench in the United States,[52] which refers to a cart full of flatulence. Not an entirely pleasant thought. Once a *good* name has been selected, its place with the brand architecture should be evaluated.

Checklist of Criteria for a Good Product Name

- Simplicity
- Memorable
- Ease of spelling
- Ease of pronouncing
- Ease of understanding
- Evokes positive image
- Does not have hidden meanings
- Translates well into other cultures
- Does not have negative meanings in other languages
- Can be graphically represented appropriately
- Fits with corporate mission and brand strategy
- Complements other product names
- Does not conflict with other product names, nor dilute brand name

Figure 10.7   Brand associations. Nestlé Toll House is a registered trademark of Nestlé.

Checklist of Things to Avoid

- Names with cult meanings
- Names based on what may be short-term fads
- International faux pas
- Names offensive to religious or ethnic groups
- Names with two or more differing connotations

Branding Decisions   An important branding decision is whether the new product should take on an existing brand name or receive its own (Figure 10.7). The American Marketing Association defines a brand as a: "Name, term, sign, symbol or design, or a combination of them intended to identify the goods and services of one seller or group of sellers to differentiate them from those of competition."[53] However, brand guru Marty Neumeier[54] defines a brand as a person's gut feeling about a product, service, or company. It's whatever the customers say it is. For example, the Nestlé brand is associated with numerous consumer products in Europe, but is generally associated with Nestlé's chocolate in the United States.

Because brand equity can be one of the most important assets that a company can hold, building brand equity with new product extensions or conversely diluting brand equity with failed new products is a major concern for firms. There are five ways of providing value from brand equity: high brand loyalty, high brand awareness, high perceived quality, more/better brand associations, and other brand assets. Figure 10.8 shows some of the world's top brands and their brand valuations. A brand with high marketplace valuation can help to support the launch of a new product and, likewise, it is expected that the new product should further build brand equity. Coke Zero is a product that has built brand equity for the company within a hard-to-reach segment of young males. In fact, Coke Zero, which was launched in the United States in 2005, has reportedly been Coca-Cola's biggest U.S. product launch in more than 20 years.[55] The success of this new product continues to build Coca-Cola's brand equity. Conversely, OK Soda, another Coca-Cola product, was launched in 1993 without the Coca-Cola brand support. It too was targeted to young buyers (the Generation X demographic in this case). The OK's slogan was: "Things are going to be OK." It performed poorly in

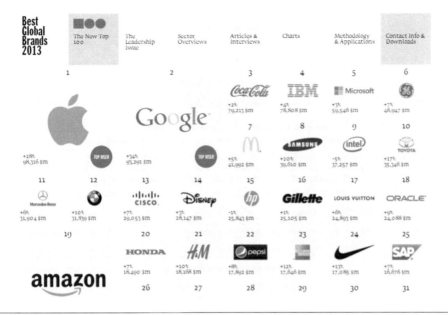

Figure 10. 8 Best global brands and their valuations in $millions. (From Endnote 56. With permission.)

select test markets and never reached national distribution before being completely discontinued in 1995. Because OK Soda was not associated with the Coca-Cola company, its brand equity was not negatively impacted with the failure of the new product. So, what decision goes into determining whether a new product should share the **family brand** name under the **brand architecture** or when should a new brand be established? Keep in mind building a new brand can cost millions.

Brand Extensions   Much care must be taken when determining if a new product should share an existing brand family or if a new brand should be created just for the new product/service. If the brand falls under the family brand, the manager must decide where within the brand hierarchy the brand should rest (Figure 10.9). The highest level in a brand hierarchy is the corporate brand, such as General Mills. A brand family would be Cheerios cereal. An individual brand is Frosted Cheerios, and the modifier would be Frosted Cheerios breakfast bars.

Corporate brands are rarely created for new products, although entrepreneurial companies may need to do so. Most new products are introduced as a new individual brand or a new brand modifier.

Brand extensions are used to reach multiple market segments with differing value propositions. For example, Toyota has three brands: Toyota, Lexus, and Scion. Scion is the lower-end Generation Y brand and Lexus is the high-end brand. This strategy allows them to use different price segments. Other reasons to use brand extensions are for different channels of distribution and different geographic boundaries. Tide laundry detergent is sold as Ace in Brazil and Alo in Turkey. There are five principles to consider in building a brand extension strategy:

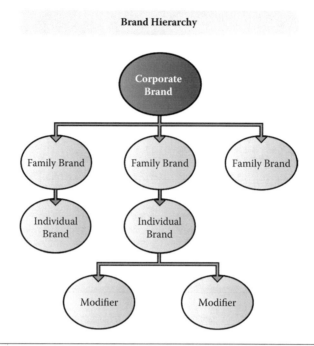

Figure 10.9   Brand hierarchy.

- *Principle of Simplicity*: Using as few hierarchical levels as possible so that brand extensions enhance the equity of the parent brand.
- *Principle of Relevance*: Using global associations that are relevant across as many individual products as possible.
- *Principle of Differentiation*: Creating distinctiveness between different brands so that they can be differentiated from one another with varying levels of awareness.
- *Principle of Prominence*: Building linkages between brands from different levels so that each is associated with the brand family, yet, can stand alone.
- *Principle of Commonality*: Sharing of common elements (brand names, logo, symbols, spokesperson, jingles, packaging, etc.), but maintaining differentiation.

When these principles are considered together, it provides a clear foundation to determine whether the brand should fall under the family brand or require an individual brand name. However, a major concern should be brand dilution, where the addition of a new product to a brand family actually decreases brand equity instead of increases it. Extension of existing brand names into inappropriate product categories can be detrimental for the entire brand family. A way to avoid brand dilution is by using a new brand name, such as Scion, which Toyota did for its lower-end product line.

The **brand architecture** also may influence the decision on whether the new product will fall under the family brand or take on its own individual brand. Brand architecture is the organizing structure that specifies the type, number, relationship, and purpose of brands within the brand portfolio.[57] The **breadth of the product line** is the

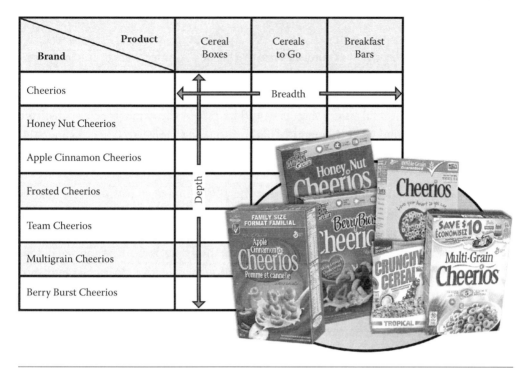

Product / Brand	Cereal Boxes	Cereals to Go	Breakfast Bars
Cheerios		Breadth	
Honey Nut Cheerios			
Apple Cinnamon Cheerios			
Frosted Cheerios	Depth		
Team Cheerios			
Multigrain Cheerios			
Berry Burst Cheerios			

Figure 10.10 Brand depth and breadth. (From Endnote 58. With permission.)

number and nature of different products with same brand name (number of different products or product mixes). The **brand depth** is number and nature of different brands in the same product class (number of different brands or brand mixes). Decisions on breadth and depth must be made so as not to dilute brand equity (Figure 10.10).

The major goal of a **brand extension** is to maximize brand coverage but minimize brand overlap. When the brand name begins to lose its brand equity value, brand overlap is too extensive and the new product manager may want to think about creating a new brand.

The right product name can contribute to the success of the company. Google, Xerox, Nike, Disneyland, Porsche, and Häagen Dazs have held up over time to identify great products and build brand equity.

Price

Price setting may be one of the most critical tasks in setting the marketing mix. It is the only "P" in which revenue is generating, all others are expenses. It is also one of the most difficult tasks a new product manager will have. Pricing too low can lose the company money; pricing too high also can result in lost sales, particularly if competition underprices you. It has been said that pricing is an art and a science as it requires both qualitative and quantitative inputs.

Value pricing is a common pricing strategy used by firms. It can be viewed as:

$$P_{new} = P_{substitution} + V_{new} \qquad\qquad (10.1)$$

where $P_{substitution}$ is the price of alternatives to the new product, such as competitor products or existing firm offerings. New is the value provided by the new product. Added together, the firm can set the price of the new product. If the new product or service adds no additional value, then it would be priced the same price as substitute products. For example, when General Mills introduced Chocolate Cheerios in 2011, it priced it the same as other Cheerio flavors as consumers viewed it as being equal to the other Cheerio offerings.

There is a wide spectrum of approaches the NPD manager can use to quantify the new product's incremental value, V_{new}. Managerial insights or intuition is one approach,[59] either based on the NPD manager's opinions or it may be come from a group discussion, for example, using the Delphi method approach. This approach does not take into consideration the customer's valuation perspective. We have explored previously how consumer insights can be collected regarding price through conjoint analysis, contingent valuation, and Vickery auctions. Price sensitivity meter questions can be asked as well in a concept test. These questions include:[60]

- At what price would you consider the product to be inexpensive where you would have doubts about its quality?
- At what price would you still feel the product was inexpensive yet have no doubts about its quality?
- At what price would you begin to feel this product is expensive, but still worth buying?
- At what price would you feel that the product is so expensive regardless of its quality that it is not worth buying?

As with any consumer research on pricing, there is likely to be response bias as consumers may state lower prices if they believe their opinions will be used to set the price or they may state higher prices, particularly for environmentally friendly products, so they are seen as socially responsible or having the luxury of not being concerned with prices. Qualitative methods include direct questioning through voice of the customer interviews, focus groups, and even ethnographic observations on buying choices.

There are three pricing strategies that are commonly used with new products: skimming, penetration, and freemium. Each of these pricing strategies has distinct goals.

Skimming   **Skimming pricing** is setting a premium price for a new product with the goal of extracting maximum return before competitors emerge. Its goal is to generate revenues to more quickly offset R&D expenses. It is typically used for luxury or status goods and for high technology products. A 50-in. flat screen TV cost more than $10,000 during its first year of introduction to the U.S. market. Today, you can buy them for a low as $550. A skimming strategy can be used when the product's quality and image can support its higher price. Competitors should not be able to enter the market easily and undercut the high price. For this reason, Internet pricing using a skimming strategy is not always beneficial. Due to shopping bots (price comparison services), online buyers can easily find competitors' products at lower prices.

Economies of scale is also a factor when utilizing a skimming strategy. Manufacturers can take advantage of learning, which comes with production experience to cut costs. As more units are produced, the marginal costs of producing the next unit will drop. Therefore, firms can reduce their prices as more consumers purchase the new product. As prices are lowered, more customers are likely to accelerate their purchase decision, which ultimately could add a greater ability to reduce manufacturing costs. The firm may, however, choose not to lower the price if there are no fierce competitors in the marketplace.

Penetration   Another tactic is to price low in anticipation of penetrating the market with a low-priced option in order to establish good word-of-mouth diffusion and accelerate sales. This strategy can help a firm gain a dominant share in the market place and drive economies of scale. A **penetration pricing strategy**[61] can be done when the market is highly price-sensitive, so a low price produces more market growth, and production and distribution costs fall as sales volume increases. The risk is that competitors will engage in a price war, so the low price position may only be temporary. As an Internet pricing strategy, penetration may not be as effective. Low prices on the Internet are often seen as being of lower quality. Price-to-value is important to convey in the online environment.

Freemium   **Freemium** has become a popular pricing strategy on the Internet. The strategy was popularized by venture capitalist Fred Wilson in 2006 via his blog posting:

> Give your service away for free, possibly ad[vertisement]supported, but maybe not, acquire a lot of customers very efficiently through word of mouth, referral networks, organic search marketing, etc., then offer premium priced value-added services or an enhanced version of your service to your customer base.[62]

The freemium strategy suggests giving an online service for "free" but charging a "premium" for advanced features, functionality, or virtual goods. It is closely related to tier pricing or *paywalls*, which is a pricing strategy used by online periodicals, such as *The New York Times*. In that model, Internet users are prevented from accessing Web page content without a paid subscription. A hard paywall allows no content to be displayed without a subscription and a soft paywall allows displaying of some content but not all, with full-featured content requiring the subscription. The freemium model works when the firm wants to build a consumer base and the marginal cost of producing extra units is low.[63]

Pricing for New-to-the-World Products   Pricing of **really new products**, where there may be no direct substitutes for the new product, makes pricing more challenging.

Web-based technologies have begun to ease the process of pricing new-to-the world products. As previously noted in our discussions on **information acceleration**, the Web provides a vehicle to educate a consumer on the benefits of the product that may be entirely new to them. Procter & Gamble has emerged as one of the leaders in utilizing online price testing.[64] Through the use of dedicated product Web sites, P&G has tested products and prices, and gauged response by directing corresponding with the customers. "The launch of Crest Whitestrips was a first foray into the use of virtual testing. The product, introduced in the year 2000, is a typical new-to-the-world product. It was due to be sold under the Crest brand umbrella; however, priced at over $40, it was at a significantly higher price point than any other Crest product. This high price point was a sticking point for Whitestrips managers, and it was realized that market testing was necessary."[65] P&G created whitestrips.com so the new product could be test marketed online over a period of nine months.[66] When using the Internet for testing price, response bias has shown to be lessened but sample bias does increase as technically savvy, high income consumers are more likely to respond to Internet-based surveys.[67]

In the appendix at the end of the chapter is a list of resources for the manager who wishes to dive deeper into the variety of pricing strategies that a firm may take.

Launch Timing

In launch planning, timing is an important issue for consideration. Should the firm rush a product to market or take time to make sure that the product is foolproof, production is ready, and the marketing mix is in alignment? Delaying launch may shorten the windows of opportunity if competitors are soon to launch their own product or if the industry is one of rapidly evolving technology, such as computers, mobile phones, or information communication systems. A company will want to speed to market with innovative products if short product life cycles are of concern. With the globalization of the marketplace, there are more opportunities for competitors to launch similar products building barriers to entry for followers who have delayed launch decisions. Blackberry, of Canada, has had a number of launch missteps. In 2013, it launched its new Z10 smartphone one day *after* the launch of Samsung's newest Galaxy S IV smartphone. CNET writer Brian Bennet further explains the misstep, "I suppose if BlackBerry had a wealth of clout in the form of customer demand and product loyalty, then it could intimidate all the big, bad American carriers into line. Sadly, that's not the case judging from the now staggered Z10 rollout, first AT&T, then T-Mobile and Verizon following at a later date. This cellular fragmentation is likely the first of many snafus in the eventual and unavoidable downfall of BlackBerry."[68]

Generally firms will take one of four launch decision strategies: general availability, limited availability, controlled introduction, and clustered launches.[69] **General availability** (GA) is when the product is available to the general public. Often, this will involve both physical retail outlets as well as an online presence. **Limited availability**

(LA) allows for a slightly wider audience to purchase the product, yet, limits who may have access to the product. For example, in order to manage its growth rate, Spotify, the commercial music streaming service, launched in October 2008 in Sweden with free accounts available by invitation only, but with paid subscriptions open to everyone. This approach allowed close monitoring of the service to quickly respond to any complications. Boeing, similarly, also used a limited availability launch by limiting initial sales to All Nippon Airways. An LA launch strategy permits marketing efforts and operational support systems to be easily fine-tuned.

Controlled introductions (CI) are more common in B2B marketing. It allows for a new product to be available to a very tightly controlled market area or customer segment. CIs are for products in a beta trial state with key clients or for test marketing in segments or geographical areas that are easily accessible for feedback. If the CI phase results in any undue risk, or the product does not work properly, successive launch phases can be canceled and the product can be redesigned with limited exposure to the organization. Figure 10.11 shows a hypothetical set of time and cost curves based on the proposed launch strategies. In controlled introductions, the risk is minimized, yet, so are revenues. In general, in availability launch, both the risk and rewards are greater. As costs and risks increase, revenues typically do not keep pace.

When deciding on launch timing, all risks should be considered. Sometimes getting the product "100% right" is essential, such as when there are low opportunity costs and high risks. Other times it makes sense to speed to market, such as when there are high opportunity costs and low risks. These tradeoffs are usually made judgmentally,

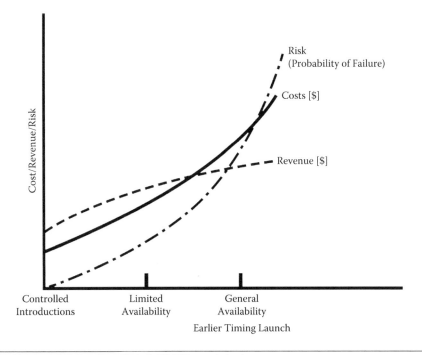

Figure 10.11 Launch timing. (From Endnote 70. With permission.)

but formal decision analysis could be applied.[71] Boeing must have felt the risks of failure were offset by potential revenues when it launched its 787 with known mechanical issues. In 2011, the company was already three years behind schedule, so pressure to conduct the first commercial flight was strong.

Another launch strategy, clustered launches, takes advantage of firm synergies. Some organizations have a standard launch timeframe (quarterly, semiannually, or annually) and coordinate across division or business units for a collective launch. These **clustered launches** are done so that a large company can make an impactful announcement to the market. They are frequently timed to industry trade shows, such as IFA Berlin (the world's leading trade show for Consumer Electronics and Home Appliances), budget windows, purchasing decision cycles, or a seasonal catalog. Clustered launches require product managers to race to meet deadlines and ensure that the organization will be fully prepared to sell and support the product. This is not an easy task if several divisions are simultaneously launching new products thereby straining resources.

In determining which launch timing strategy to undertake, some quantification of the gains to be achieved through early introduction should be made.[72] Among the questions to be considered are: How much harm can the competitor do if they enter before you or shortly after you with a better product? Are production capabilities in place for an early launch? Are suppliers and other partners on standby for your "go" decision? Is the marketing program in place? Is the sales team trained and committed to the new product? As Boeing witnessed, an overly aggressive launch could result in quality issues or production mishaps. Yet, an overly cautious approach can result in missed market windows of opportunity. The launch manager's task is to weigh the expected benefits against the risk to propose a launch window. How much risk? is a good question.

Launching of durable consumer and industrial goods does not require the marketing expenses that fast-moving consumer goods (FMCGs) require at launch, but does entail more production expenses to set up manufacturing compared to consumer goods. For example, it has been estimated that Boeing's new Dreamliner jet cost more than $32 billion to develop[73] and costs $200 million each to manufacture.[74] A general availability full-scale launch of industrial products is similar to launch of consumer goods. The marketing and production teams must be coordinated, timing relative to competitors is crucial, critical path analysis to achieve the objectives needs to be establish, and monitoring of the outcomes is essential for potential damage control. The primary difference will be the role of the customer in launch activities. They will be more actively involved in the entire process and their feedback is critical to keeping the launch on track. Boeing had an intimate partnership with Air Nippon Airways (ANA), which became even more important to Boeing as problems surfaced with the Dreamliner. Although product satisfaction is not directly reflected in repeat purchasing rates, positive word-of-mouth information from current buyers to prospective customer is critical to success. New prospective buyers will be talking to ANA about their experiences with Boeing, particularly on how well Boeing responded to the grounding of their planes. In durable and industrial products, service is of great importance. Pricing is another difference between frequently purchased and durable consumer and industrial products as the price of fast moving consumer goods is relatively stable. As we have seen above, the price of new conservative durables or industrial products can fall over time due to economies of scale, industrial learning, and competition. Price changes can be used as a strategic advantage more easily for durable and industrial goods compared to FMCGs.

Launch Management

After prelaunch strategy and tactical planning, the NPD manager must make a final go/kill decision. Should the product/service be moved into launch? If the answer is a "go," then all the plans that have been put in place are implemented; however, this does not relieve the manager of his/her duty. Monitoring launch activities is crucial to ensure the success of the new product/service.

Monitoring Launch

No matter how much planning the NPD team has undertaken, unexpected events will inevitably take place before and during the launch of the new product/service. Events outside the control of the firm are likely to occur. The NPD manager should be prepared for any type of catastrophic contingency that may occur. The marketing plan was designed to be adaptive in the sense that it should be continually updated by market response data. This type of updating makes a plan less susceptible to failures in execution. Prudent monitoring for potential problems can be quickly identified and rectified. Usually monthly reviews are exercised, but key metrics are tracked weekly during the first months of launch. A 4-step launch management system has been proposed with this in mind:[75]

1. **Monitor** launch activities in order to spot potential problems; this does not mean monitoring sales, but the root of what may be causing low sales or even higher than expected sales. Until the cause of any problem can be identified, the proper contingency plan cannot be implemented.
2. **Select** "mission critical" items to control, which are those activities that will cause launch to fail if left undone or those items that if occurring could cause failure; it is impossible to anticipate all problems, but it is possible to consider where major issues may occur and to be ready should they surface. Typical launches will have 6 to 8 "critical watch points."
3. **Develop** contingency plans for the control problems that can be put into place in a moment's notice; working under "surprise" panic conditions rarely leads

to the best results. Taking time before a potential problem occurs provides an opportunity to explore the best alternatives as opposed to being reactive.

4. **Design** a tracking system whereby launch and marketing metrics are measured to monitor whether goals are being met or not; thresholds must be set so that it becomes obvious that a problem has occurred, which can then trigger the activation of the contingency plan.

There are numerous events that can impact the new product launch. Economic changes may occur between test market and launch; consumer tastes may change, international markets may expand or collapse, or competitors may launch a counter attack. It is important to monitor the environment between the go decision and the national launch to be sure the environmental assumptions underlining the go decision have not been changed. What contingency plan do you have if competitors drop their price, conduct an aggressive marketing campaign, or speed-to-market a similar new product? For example, to counter a strong sampling campaign by Aim toothpaste, Procter & Gamble undertook an advertising campaign, which appealed to Crest users and discouraged them from trying the unfamiliar brand sample delivered through the mail.[76]

Postlaunch Analysis

After action reviews (AARs) are becoming more common in business applications.[77] AARs originated in the military to evaluate the success of military actions. The goal of the AAR is to identify how to correct deficiencies, sustain strengths, and focus on performance of specific mission essential tasks. AARs differ from a postmortem in that the NPD team jointly evaluates the role of key players involved in product develop and subsequent launch. It is meant to provide a review of each stage of the stage-gate process to try to understand how the product progressed through the steps. "The goal is to identify what went right (so it can be duplicated) and what went wrong (to identify weak areas in the firm's processes that need to be fixed). A good AAR includes statements of planned objectives and actual results, and attempt to rationalize the observed variances, a statement of what has been learned, and an outline for the next steps."[78] The AAR need not be formal and the review team may include a key customer, but it is not essential. If there is a large team or globally dispersed team, it may require running several AARs with relevant subgroups. Some firms delay the AAR for a year after product release in order to assess how well the new product fared in the marketplace and to determine if planned targets were achieved. This is another reason why monitoring metrics is important; the NPD manager cannot evaluate success or failure if measures are not taken.

Product Life Cycle Management

The product life cycle (PLC) refers to either the unit sales or revenue from a new product/service. It starts with the introduction stage, which refers to the product just after launch as sales progress through to the end-of-life of the product. This is sometimes referred to as **cradle-to-grave**. The goals of the firm will differ at the different stages of the PLC.

- *Introduction* (also sometimes referred to as Research & Development or the pioneer stage): Stage where the product or service upon completion of R&D and testing first enters the market. At this time, the goal of the firm is to create awareness and encourage trial of the new product. The primary expenditures will be on getting production running smoothly and in marketing advertising and PR.
- *Growth*: In the growth stage, sales start to escalate and new competitors enter the market with "me-too" versions of your product, or worse, with innovative new products that outshine your own. The primary focus in the growth stage is to build distribution channels whether they are domestic or international. Product quality needs to be maintained to fend off competition. This is also a time of focusing on building brand equity.
- *Maturity*: The maturity stage is about product standardization. Firms should be focused on building customer relationships to be used when launching the next version of the product or an entirely new product. The primary goal is strengthening market share. This also may be a time that the firm focuses on process innovations in order to drive down manufacturing costs and, subsequently, prices.
- *Decline*: As diffusion reaches it saturation point, revenues will start to slow done. With luck, it will be a cash cow for the firm generating profit with minimal efforts. Product extensions and low-cost manufacturing will be the focus as the product moves toward retirement and the end of the life cycle. In the chapter on Sustainability in Innovation, we will look at how a product can be designed for a *cradle-to-cradle PLC*, instead of the usual cradle-to-grave, (Figure 10.12) where products are disposed of at the end of their life.

Product Failure

Launch products sometimes do fail before going through all the steps in the PLC. Coca-Cola's OK Soda had a very short life. Despite positive market test results, Pepsi Crystal, which was a clear-colored cola drink, never gained market traction,[79] neither did aspirin made by Ben-Gay (the muscle ache cream), underwear by BIC (the pen company), or McDonald's Arch Deluxe premium burger. When a product appears to be failing, a firm must decide on the next steps. Should it pull the product from the market? Should it conduct a new marketing campaign? Or should it reformulate the

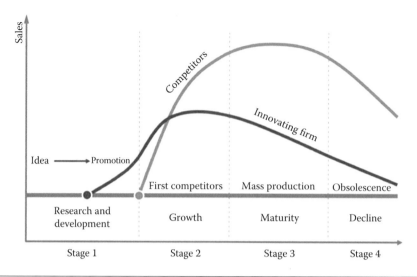

Figure 10.12   Product life cycle from cradle to grave. (From Endnote 80. With permission.)

product? All of these solutions are viable. Coca-Cola pulled its New Coke completely from the market within weeks of its original introduction due to public outcry to the flavor change. Procter & Gamble's Febreze was relaunched with a new marketing campaign after consumers couldn't understand its purpose, and WebTV metaphorsized into MSN TV. How can an NPD manager decide which is the best course of action for a failing product?

A four-step process to address this issue is suggested in Figure 10.13.[81] When establishing the launch management system, the team designed a tracking system

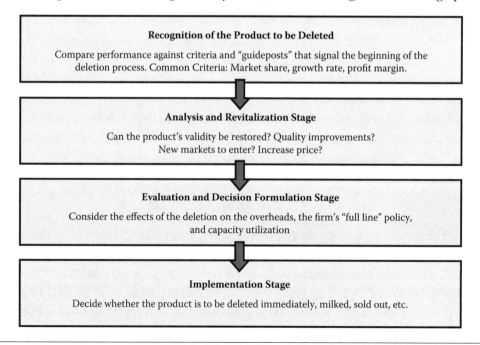

Figure 10.13   A stepwise product deletion process. (From Endnote 82. With permission.)

to monitor a set of key performance metrics, which would let the NPD manager know how well the project goals were being met. For example, these metrics may include weekly sales, monthly revenue, market share, growth rate, or profit margin. A threshold, such as weekly sales below a certain level, was set as a trigger point in which to activate a contingency plan. That contingency plan, also called the *analysis and revitalization stage*, starts first by identifying the source of the problem (such as was done in the AAR) and then attempts to implement a solution that addresses the issue. The contingency plan may include changing the price, changing the product, or substituting the product with an alternative offering until a solution can be found to fix the product that is floundering. Sometimes it may be necessary to temporarily pull the product from the market or stop promoting the product. This may happen if sales are higher than expected; hopefully, a good problem to have.

If the contingency plan fails to turn around the situation, then an evaluation must occur to determine if the product should be removed from the product line, foregoing the market opportunity. Of course, if large financial commitments were put in place, such as those required for durable and industrial goods, abandoning the market may not be that easy. A product deletion decision is a complex one that is likely to have ripple effects throughout the organization. No manager will want to be associated with a product failure and may either completely abandon the sinking ship or there may be a tendency to hold on and continue to channel resources to a possible losing proposition (called escalation of commitment[83]). New-to-the world-product projects have been shown to be harder to shut down. In these cases, managers tend to be more optimistic about the chances of the product finally gaining traction, be more emotionally tied to the project, and, thus, more likely to commit resources to the project. The Segway, the revolutionary human transporter, and the Apple Newton, one of the first personal digital assistants (PDAs), were both cases where serious problems arose early during product development, but were ignored.

A thorough evaluation should be conducted, preferably by a team not intimately connected to the project, to determine the repercussions of abandoning the product. Before making the deletion decision, the firm must systematically evaluate the full extent of the deletion on overheads, expenses, and capacity utilization and also determine whether the deletion leaves a major hole in the firm's product line.[84] Finally, if deletion is necessary or inevitable, speed must be determined: Should the product be deleted immediately, slowly drawn from the market milking any possible revenues, or, perhaps, reworked and released under a new name? Sometimes there may be an alternative to sell the product design to another party or perhaps even the whole business unit.

There have been a number of market failures that have been successfully turned around. The technology behind the Apple Newton became the foundation for the iPad. Procter & Gamble's Febreze, odor eliminator, was incorrectly accused of killing pets, but once endorsed by animal rights organizations, it became a household brand name. Clorox's Anywhere® cleaning product also had an unfortunate start, but has gone on to be one of Clorox's success stories. New Coke was a product that was quickly removed from the market, but the failed product was used to build brand loyalty with Coca-Cola customers. The key to these eventual turnarounds was following the plan by monitoring the product through launch, recognizing the possibility of the problem, evaluating the alternatives, and executing on a chosen plan with as much commitment as the original launch plan. Figure 10.13 summarizes this process.

Chapter Summary

Launch planning activities are made up of strategic planning and tactical planning. Strategic plans typically are devised prior to R&D and tactical plans after R&D. Some of the crucial tactical plans discussed in this chapter were naming the product and aligning the brand of the new product/service with the existing brand family. Pricing strategies were considered as well with skimming, penetration, and freemium prices as possible new product pricing alternatives. Launch timing was also discussed. Timing the launch of a new product/service to the window of opportunity is crucial for its success. After the product is launched, it should continue to be monitored by tracking key performance measures. This monitoring is the first step towards realigning a launch strategy that may have not gone according to plan. Market conditions are continually changing, and the product manager must be proactive instead of reactive in order to keep the product on track. However, each product/service, even if very successful at launch, does have an end-of-life. It is up the product manager must decide how to best pull a product from market.

Glossary

Activate: Putting into place product launch plans of actions.

Announce: Activities taken to introduce a new product to the marketplace.

Arrange: Product launch activities that take into consideration all the concerns evaluated in the prelaunch strategy to establish a plan of action.

Balanced matrix: An NPD team structure where a project manager is assigned to oversee the project and shares the responsibility and authority for completing the project with a functional manager. Project and functional managers jointly direct workflow segments and jointly approve decisions.

Brand architecture: The structure of brands within an organization and the ways in which they are related to each other.

Brand depth: The number and nature of different brands in the same product class (number of different brands or brand mixes).

Brand extension: Brands used to reach multiple market segments with differing value propositions, distribution channels, or geographic areas.

Breadth of the product line: The number and nature of different products with the same brand name (number of different products or product mixes).

Clustered launch: A launch strategy that takes advantage of a firm's synergies by coordinating collective product launches across division or business units, making a more impactful announcement to the market.

Controlled introduction (CI): A product introduction strategy that allows for a new product to be available to a very tightly controlled market area or customer segment. CIs are for products in a beta trial state with key clients or for test marketing in segments or geographical areas that are easily accessible for feedback.

Cradle-to-grave: Refers to a product or service's complete life cycle, starting from the introduction of the new product/service, through to the end-of-life of the product/service.

Degree of projectization: The extent to which participants of a project see themselves committed to or independent from the project.

Economies of scale: Advantages of learning and/or lower production costs that come with longer product lives or larger product volumes.

Family brand: A brand that encompasses many products or groups of products with possibly differing product images.

Freemium: An Internet pricing strategy that involves giving a service away for free, acquiring many customers, then offering premium priced value-added services or an enhanced version of the service to the customer base.

Functional: An NPD team structure with the least projectization, where a project is divided into segments and assigned to relevant functional areas (marketing, finance, sales, development, production, etc.) and/or groups within functional areas. Team members report to their own functional managers and the team to top management.

Functional matrix: An NPD team structure where a project manager with limited authority is designated to coordinate the project across different functional areas and/or groups. The functional managers retain responsibility and authority for their specific segments of the project. The team reports to the functional managers as well as top management.

General availability: When the product is available to the general public. Often this will involve both physical retail outlets as well as an online presence.

Industry concerns: Strategic planning concerns over environmental factors that may impact launch.

Information acceleration: The use of virtual reality to educate a consumer on the benefits of the product that may be entirely new to them.

Launch process: The development and implementation of a new product tactical and strategic launch program.

Limited availability: Allows for a slightly wider audience than *General Availability* to purchase the product, yet, limits who may have access to the product.

Market orientation: An organization's generation and dissemination of market information regarding customer needs and wants, and the incorporation of that market analysis into strategic planning initiatives.

Market planning: Tactical concerns in strategic launch planning.

Organizational concerns: Concerns in strategic launch planning over aligning the company's goals and overall strategy with the new product development process.

Paywalls: An Internet tier pricing strategy where users are prevented from accessing content without a paid subscription.

Penetration pricing strategy: A tactic involving pricing low in anticipation of penetrating the market with a low-priced option in order to establish good word-of-mouth diffusion and accelerate sales.

Principle of commonality: Sharing of common elements (brand names, logo, symbols, spokesperson, jingles, packaging, etc.) while maintaining differentiation.

Principle of differentiation: Creating distinctiveness between different brands so that they can be differentiated from one another with varying levels of awareness.

Principle of prominence: Building linkages between brands from different levels so that each is associated with the brand family, and can stand alone.

Principle of relevance: The use of global associations that are relevant across as many individual products as possible.

Principle of simplicity: The use of as few hierarchical levels as possible so that brand extensions enhance the equity of the parent brand.

Product concerns: Concerns in strategic launch planning that involve making sure that the product/service is physically ready to launch.

Project matrix: An NPD team structure where a project manager is assigned to oversee the project and has primary responsibility and authority for completing the project. Functional managers assign personnel as needed and provide technical expertise.

Project team: An NPD team structure where a project manager is put in charge of a project team composed of a core group of personnel from several functional areas and/or groups, assigned on a full-time basis. Neither functional managers nor top management typically have formal involvement.

Really new products: Products with no direct competitive substitutes, which are new to the market and industry.

Skimming pricing: Setting a premium price for a new product with the goal of extracting maximum return before competitors emerge.

Skunk works: An NPD team structure where a team is given a high degree of autonomy and is unhampered by bureaucracy, tasked with working on innovative, usually new-to-the-world projects. Project teams may be situated in buildings outside of main firm locations.

Review Questions

1. Why should a firm develop both prelaunch strategy and prelaunch tactics? Hasn't the strategy already been set early in the fuzzy front end of the NPD process?
2. Why should management spend the time and money to monitor new products during launch? What should they be checking for?
3. Who should be in charge of naming a product? What product naming pitfalls should an NPD manager avoid?
4. When should a new product be folded under the family brand? When should it be given its own brand name.
5. What special problems does a new product manager face when launching durable and industrial products? Services?
6. Production costs and prices drop rapidly as a product diffuses in markets for high-technology products, such as mobile phones and computers. Why? What implications do these decreases have in the product life cycle management of a new high-technology product?

Assignment Questions

1. You have spent two years developing a new pie mix, but six months prior to full launch, your major competitor finds out about your new product concept. Should you begin a crash program to get to market in three, rather than six, months? What information do you need to make a decision and how would you go about obtaining the necessary information?
2. Scandinavian Industries, a manufacturer of large appliances, has developed a new line of European-looking dining room furniture. Develop a hypothetical tactical plan to help the product manager, Mr. Rummel, launch a new line of furniture, which is outside their core industry of durable white goods.
3. You are the product manager in charge of launching a new line of Lego's building blocks for girls. Recent media have been negative, particularly from women. "As members of SPARK Movement to end the sexualization of girls, and partners of Powered By Girl, we are spreading the word that you, LEGO, are selling out girls."[85] What will you do to counter this negative publicity? Why?
4. You are the product manager for Zaarly.[86] Usage of the service was much lower than expected after launching in Boston, Chicago, and San Francisco. The company has changed its core benefit proposition from being consumer focused to being a storefront for small businesses. The transition has not been as smooth as you had hoped. What contingencies should have been in place from the beginning of launch for the likelihood that the service would not be well received after launch?

5. Your engineering buddies have just developed a 3D printer that allows imbedding of LEDs (light emitting diodes). This makes it unique to other desk-to 3D printers. They have asked you to come up with a product name for the printer, which is targeted to the hobbyist market. What issues will you take into consideration? Come up with a short list of names and explain why these are good product names.

6. General Mills missed the opportunity to catch the wave of consumer interest in Greek yogurt, a thicker, healthier yogurt. Chobani, FAGE, and Oikos came out as front runners because they were not associated with the traditional highly sweetened yogurt sold in the United States. General Mills must now decide (a) whether they should enter the market late with their own version of Greek yogurt and (b), if they do, should it fall under a family brand and, if so, which one, or (c) should they create a new family brand to differentiate it from their Yoplait line of yogurts. Provide an argument for what they should do.

Appendix: Pricing Resources

Chris Anderson, *Free: The Future of a Radical Price* (New York: Hyperion Books, 2009).

For software, see Neil Davidson, *Don't Just Roll the Dice: A Usefully Short Guide to Software Pricing* (Cambridge, U.K.: Redgate Books, 2009).

Jagmohan Raju and Z. John Zhang, *Smart Pricing: How Google, Priceline, and Leading Businesses Use Pricing Innovation for Profitability* (Upper Saddle River, NJ: Pearson Prentice Hall, 2010).

Joseph Zale, Thomas Nagle, and John Hogan, *Strategy and Tactics of Pricing*, 5th ed. (Upper Saddle River, NJ: Prentice Hall, 2010).

Endnotes

1. Online at: http://www.independent.co.uk/travel/news-and-advice/qa-what-is-the-impact-of-the-boeing-787-dreamliner-safety-concerns-8454919.html (accessed July 1, 2013).

2. Online at: http://www.newairplane.com/787/design_highlights/#/exceptional-value/dreamliner-advantages/dreamliner-advantages (accessed July 1, 2013).

3. Online at: http://www.independent.co.uk/travel/news-and-advice/qa-what-is-the-impact-of-the-boeing-787-dreamliner-safety-concerns-8454919.html (accessed July 1, 2013).

4. Online at: http://www.newairplane.com/787/design_highlights/#/exceptional-value/dreamliner-advantages/dreamliner-advantages (accessed July 1, 2013).

5. Tom Clark, "ANA says Denver Still in Hunt for Non-Stop to Tokyo," April 8, 2009. Online at: http://www.metrodenver.org/blog/ANA-Japan-Denver-international-flight.html (accessed July 1, 2013).

6. All_Nippon_Airways_Boeing_787-8_Dreamliner_JA801A_OKJ.jpg: Spaceaero2, Creative Commons Attribution-Share Alike 3.0. Online at: https://commons.wikimedia.org/wiki/File:All_Nippon_Airways_Boeing_787-8_Dreamliner_JA 801A_OKJ_in_flight.jpg

7. Dave Carpenter, "Boeing Delays First 787s by Six Months," Associated Press, October 10, 2007. Online at: http://web.archive.org/web/20071012021920/http:/biz.yahoo.com/ap/071010/boeing_787.html (accessed July 1, 2013); Nicola Clark, "Boeing Delays Deliveries of 787," *The New York Times,* October 10, 2007. Online at: http://www.ny times.com/2007/10/10/business/10cnd-boeing.html?_r=1& (accessed July 1, 2013); "Boeing

Reschedules Initial 787 Deliveries and First Flight," *Boeing*, October 10, 2007. Online at: http://www.boeing.com/news/releases/2007/q4/071010d_nr.html11 (accessed July 1, 2013).

8. "Boeing Buys Vought Share of Fuselage Builder," *Aviation Week*, March 28, 2008. (accessed June 14, 2011).

9. Online at: http://www.independent.co.uk/travel/news-and-advice/qa-what-is-the-impact-of-the-boeing-787-dreamliner-safety-concerns-8454919.html (accessed July 1, 2013).

10. Online at: http://www.independent.co.uk/news/business/news/boeings-dream liner-night-mare-gets-worse-as-japan-reveals-more-on-faults-8507247.html (accessed July 1, 2013).

11. Online at: http://money.cnn.com/2013/02/28/news/companies/boeing-dreamliner-apol-ogy/index.html?hpt=hp_t3 (accessed viewed February 28, 2013).

12. Charles Beard and Chris Easingwood, "New Product Launch: Marketing Action and Launch Tactics for High-Technology Products," *Industrial Marketing Management* 25(2) (1996): 87–103; Erik Jan Hultink and Susan Hart, "The World's Path to the Better Mousetrap: Myth or Reality? An Empirical Investigation into the Launch Strategies of High and Low Advantage New Products," *European Journal of Innovation Management* 1(3) (1998): 106–22; Erik Jan Hultink and Henry S. Robben, "Launch Strategy and New Product Performance: An Empirical Examination in the Netherlands," *Journal of Product Innovation Management* 19(6) (1999): 545–56; Erik Jan Hultink, Susan Hart, Henry S. J. Robben, and Abbie Griffin, "Launch Decisions and New Product Success: An Empirical Comparison of Consumer and Industrial Products," *Journal of Product Innovation Management* 17(1) (2000): 5–23.

13. Steven Haines, *The Product Manager's Desk Reference* (New York: McGraw-Hill, 2009).

14. Hultink and Hart, "The World's Path to the Better Mousetrap"; Hultink and Robben, "Launch Strategy and New Product Performance"; Hultink, et al. "Launch Decisions and New Product Success."

15. Marion Debruyn, Rudy Moenaert, Abbie Griffin, Susan Hart, Erik Jan Hultink, and Henry Robben, "The Impact of New Product Launch Strategies on Competitive Reaction in Industrial Markets," *Journal of Product Innovation Management* 19(2) (2002): 159–170.

16. Online at: http://web.archive.org/web/20060718020320/ (accessed February 23, 2013); http://www.thezeromovement.org/quotes_on_the_coke_zero_movement.html (accessed February 23, 2013).

17. Erik Jan Hultink, Abbie Griffin, Susan Hart, and Henry S. J. Robben, "Industrial New Product Launch Strategies and Product Development Performance," *Journal of Product Innovation Management* 14(4) (1997): 243–257; J. P. Guiltinan, "Launch Strategy, Launch Tactics, and Demand Outcomes," *Journal of Product Innovation Management* 16(6) (1999): 509–529.

18. Based on Yi-Chia Chiu, Benson Chen, Joesph Z. Shyu, and Gwo-Hshiung Tzeng, "An evaluation model of new product launch strategy." *Technovation* 26 (2006): 1244–1252.

19. Etsy is a Web site that supports microproducers of handcrafted goods and vintage items. Competitors include DaWanda, based in Germany; Bonanza in the United States, which focuses on clothing and fashion; Zibbet is based in Australia; iCraft is based in Canada.

20. Glen L. Urban and John R. Hauser, *Design and Marketing of New Products* (Upper Saddle River, NJ: Prentice Hall, 1980).

21. Danny Miller, "Configurations of Strategy and Structure: Towards a Synthesis," *Strategic Management Journal* 7(3) (1986): 233–249.

22. Vivek Kaul, "What Steve Jobs Did When He Was Fired from Apple," *DNA* (newspaper). Online at: http://www.dnaindia.com/money/1254757/report-what-steve-jobs-did-when-he-was-fired-from-apple (accessed July 1, 2013).

23. Online at: http://live.wsj.com/video/apple-product-launches-over-the-years/B33CF200-5EC8-45B9-92D2-3D4FDFCFCBC8.html#!B33CF200-5EC8-45B9-92D2-3D4FD FCFCBC8 (accessed February 26, 2013).

24. Katrin Talke and Erik Jan Hultink, "The Impact of the Corporate Mind-Set on New Product Launch Strategy and Market Performance," *Journal of Product Innovation Management* 27(2) (2010): 220–237.

25. Gene N. Landrum, *Profiles of Genius: Thirteen Creative Men Who Changed the World* (Buffalo, NY: Prometheus Books, 1993).

26. Gregory G. Dess and G. Tom Lumpkin, "The Role of Entrepreneurial Orientation in Stimulating Effective Corporate Entrepreneurship," *The Academy of Management Executive* 19(1) (2005): 147–156.

27. Adam Hartung, "Netflix—The Turnaround Story of 2012," *Forbes* (January 29, 2013). Online at: http://www.forbes.com/sites/adamhartung/2013/01/29/netflix-the-turnaround-story-of-2012/ (accessed June 17, 2013).

28. Roger J. Calantone and C. Anthony Di Benedetto, "Performance and time to market: accelerating cycle time with overlapping stages." *IEEE Transactions on Engineering Management* 47(2) (May 2000): 232–244.

29. Ibid.

30. Charles Merle Crawford and C. Anthony Di Benedetto, *New Products Management* (New York: McGraw-Hill Irwin, 2011).

31. Erik W. Larson and David H. Gobeli, "Organizing for Product Development Projects," *Journal of Product Innovation Management* 5(3) (1988): 180–190.

32. Erik Jan Hultink and Fred Langerak, "Launch Decisions and Competitive Reactions: An Exploratory Market Signaling Study," *Journal of Product Innovation Management* 19(3) (2003): 199–212.

33. Bernard J. Jaworski and Ajay K. Kohli, "Market orientation: antecedents and consequences." *Journal of Marketing* (1993): 53–70.

34. Fred Langerak, Erik Jan Hultink, and Henry S. J. Robben, "The Impact of Market Orientation, Product Advantage, and Launch Proficiency on New Product Performance and Organizational Performance," *Journal of Product Innovation Management* 21(2) (2004): 79–94.

35. Roger J. Calantone and C. Anthony Di Benedetto, "Clustering Product Launches by Price and Launch Strategy," *Journal of Business & Industrial Marketing* 22(1) (2007): 4–19.

36. Hultink and Langerak, "Launch Decisions and Competitive Reactions."

37. Roger Calantone, Rosanna Garcia, and Cornelia Dröge, "The Effects of Environmental Turbulence on New Product Development Strategy Planning," *Journal of Product Innovation Management* 20(2) (2003): 90–103.

38. Hultink and Langerak, "Launch Decisions and Competitive Reactions."

39. "Technology Q&A: Google's Android," *BBC News*, June 11, 2007. Online at: news.bbc.co.uk/2/h i/7080758.stm (accessed July 1, 2013).

40. Jon Brodkin, "On Its 5th Birthday, 5 Things We Love about Android," *Ars Technica*, May 11, 2012. Online at: http://arstech nica.com/gadgets/2012/11/on-androids-5th-birthday-5-things-we-love-about-android/ (accessed July 1, 2013).

41. "Android Marks Fourth Anniversary Since Launch with 75.0% Market Share in Third Quarter, According to IDC." Online at: http://www.idc.com/getdoc.jsp?containerId=prUS23771812 (accessed July 1, 2013).

42. Jamie Lareau, "Ford Fiesta Not Yet in Dealerships; Dealers and Customers Get Antsy," Autonews via *Autoweek*, June 18, 2010.

43. Chris Shunk, "Ford Fiesta Shipments Flowing Again after Faulty Part Fixed," *Autoblog*, August 8, 2010. (http://en.wikipedia.org/wiki/Ford_Fiesta-cite_note-68).

44. Haines, *The Product Manager's Desk Reference.*

45. Based on Haines, *The Product Manager's Desk Reference.*

46. Philip Kotler and Kevins Keller, *A Framework for Marketing Management* (London: Pearson, 2009).

47. http://www.bigbellsolar.com with permission.
48. Based on Leland D. Shaeffer and James S. Twerdahl, "Giving Your Product the Right Name: Do It Yourself While Avoiding the Pitfalls," in *The PDMA Toolbook 3 for New Product Development*, eds. Abbie Griffin, Stephen Somermeyer, and Paul Belliveau. (Hoboken NJ: John Wiley & Sons, 2007).
49. Based on Shaeffer and Twerdahl, "Giving Your Product the Right Name."
50. Ibid. p. 244.
51. Ibid.
52. Online at: http://www.i18nguy.c om/translations.html (accessed June 26, 2013).
53. Online at: http://www.marketingpower.com/_layouts/dictionary.aspx?dletter=b (accessed June 17, 2013).
54. Marty Neumeier, *The Brand Gap, 2nd ed.* (San Francisco: Peachpit Press, 2005).
55. Martin Hickman, "Introducing 'Bloke Coke'—Is This Now the Real Thing?" *The Independent* (London) July 4, 2006. Online at: http://www.independent.co.uk/news/media/introducing-bloke-coke-is-this-now-the-real-thing-406556.html (accessed July 1, 2013).
56. Interbrand's Best Global Brand 2013 is a look at financial performance of the brand, role of the brand in the purchase decision process, and the brand strength. http://www.bestglobal-brands.com for more information.
57. Online at: http://merriamassociates.com/2009/09/what-is-brand-architecture/ (accessed June 27, 2013).
58. Cheerios and all its brands are the trademark of General Mills. This diagram is for illustrative purposes only.
59. Kent B. Monroe, *Pricing: Making Profitable Decisions* (New York: McGraw-Hill, 1990); Jeffrey G. Covin, John E. Prescott, and Dennis P. Slevin, "The Effects of Technological Sophistication on Strategic Profiles, Structure and Firm Performance," *Journal of Management Studies* 27(5) (1990): 485–510; Eunsup Shim and Ephraim F. Sudit, "How Manufacturers Price Products," *Management Accounting–New York* 76 (1995): 37.
60. Kent B. Monroe, "Measuring Price Thresholds by Psychophysics and Latitudes of Acceptance," *Journal of Marketing Research* (1971): 460–464.
61. Kotler and Keller, *A Framework for Marketing Management*.
62. Online at: http://www.avc.com/a_vc/2006/03/my_favorite_bus.html (accessed June 27, 2013).
63. Chris Anderson, *Free: The Future of a Radical Price* (New York: Hyperion Books, 2009).
64. C. T. Heun, "Procter & Gamble Readies Online Market Research Push," *Information Week*, October 15, 2001: 26.
65. Heather Bergstein and Hooman Estelami, "A Survey of Emerging Technologies for Pricing New-to-the-World Products," *Journal of Product & Brand Management* 11(5) (2002): 303–319.
66. J. Gaffney, "How Do You Feel about a $44 Tooth-Bleaching Kit?" *Business 2.0* 2(7) (2001): 126127.
67. Heather and Estelami, "A Survey of Emerging Technologies for Pricing New-to-the-World Products."
68. Online at: http://www.thetelecom blog.com/2013/03/07/blackberry-release-misstep-ruins-u-s-market-hype/ (accessed June 27, 2013).
69. Haines, *The Product Manager's Desk Reference*.
70. Based on Urban and Hauser, *Design and Marketing of New Products*.
71. Ralph L. Keeney and Howard Raiffa, *Decision Analysis with Multiple Conflicting Objectives* (New York: John Wiley & Sons, 1976).
72. Based on Urban and Hauser, *Design and Marketing of New Products*.
73. Online at: http://seattletimes.com/html/businesstechnology/2016310102_boeing25.html (accessed June 28, 2013).

74. Online at: http://business.time.com/2013/01/17/is-the-dreamliner-becoming-a-financial-nightmare-for-boeing/ (accessed June 28, 2013).

75. Crawford and Di Benedetto. *New Products Management*.

76. Urban and Hauser, *Design and Marketing of New Products*.

77. Crawford and Di Benedetto, *New Products Management*.

78. Ibid.

79. Kate Bonamici Flaim, "Winging It," *Fast Company* (magazine), December 19, 2007. Online at: http://www.fastcompany.com/magazine/119/winging-it.html (accessed June 29, 2013).

80. Jean-Paul Rodrigue, Dept. of Global Studies & Geography, Hofstra University, New York, © 1998–2013.

81. Crawford and Di Benedetto, *New Products Management*.

82. Ibid., p. 472.

83. Jeffrey B. Schmidt and Roger J. Calantone, "Escalation of Commitment during New Product Development," *Journal of the Academy of Marketing Science* 30(2) (2002): 103–118.

84. Crawford and Di Benedetto, *New Products Management*.

85. Online at: http://www.change.org/petitions/tell-lego-to-stop-selling-out-girls-liberatelego (accessed June 29, 2013).

86. Online at: http://readwrite.com/2011/03/09/zaarly_is_this_the_future_of_mobile_money_and_mark#awesm=~oablxFMo7P4mZz (accessed June 29, 2013).

11

GLOBAL NEW PRODUCT DEVELOPMENT

GLORIA BARCZAK* AND ROSANNA GARCIA

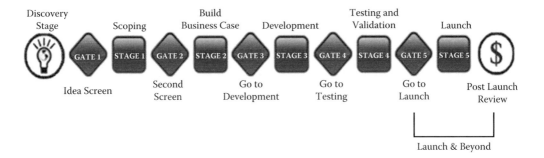

Learning Objectives

In this chapter, we will address the following questions:

1. What are the differences and benefits of global versus co-located teams? How do you effectively manage global NPD teams?
2. What is open innovation? What are examples of open innovation approaches?
3. What is the difference between reverse innovation and bottom-of-the-pyramid?
4. What is a global brand? How do you develop a global brand?
5. What do you need to consider when launching a new product globally? What different launch strategies can be used?

Global Innovation at GE Healthcare

Firms must be globally oriented to be competitive. GE Healthcare's clinical systems unit is keenly aware of the need to design products that can be used anywhere around the world. In 2008, "GE Healthcare engineer Davy Hwang's marching orders were straightforward. Take a 15-pound electrocardiograph machine that cost $5.4 million and that took three and a half years to develop. Then, squeeze

---

* Gloria Barczak, PhD, is professor of marketing and editor of The Journal of Production Innovation Management. She has published more than 30 articles, one book, and several book chapters on innovation.

the same technology into a portable device that weighs less than 3 pounds and can be held with one hand. Oh, and develop it in 18 months for just 60% of its wholesale cost. He thought I was crazy," says Hwang's boss, Omar Ishrak, CEO of GE Healthcare's clinical systems unit, based in Wauwatosa, Wisconsin.[1] Hwang's team successfully pulled off this crazy request by combining two-parts technical know-how mixed with one-part creativity to meet the deadline. The reason for the low cost was the new MAC 400 portable ECG (electrocardiogram) machine was headed to India to meet the needs of the fast-growing market of healthcare for rural communities.

Global sourcing of parts and globally located teams contributed to the success of the redesign. For the machine's printer, the team, which consisted of engineers both in the United States and in India, adapted a printer used in bus terminal kiosks in India because they were designed to withstand dusty outdoor environments. Further drawing on an open innovation strategy, the team also bought a commercially available processing chip at a quarter of the price it would have cost them to design one in-house. The team also drew upon other products developed by GE. "From the team responsible for GE's portable ultrasound machine, they learned about a low-cost source for technology, which can cut plastic mold prototypes far earlier in the process than usual. That let them get feedback from doctors before changes got costly." To deal with power outages in many parts of India and an acute shortage of healthcare professionals, the MAC 400 was powered by common batteries and was designed to be easy to use.

One of the first buyers was Dr. Girish Khurana, an internist in Bahadurgarh, near Delhi, who ordered one to take to his rural cardiac clinics. "I was surprised how light it was," he says. Another buyer, Dr. S. S. Ramesh, ordered one for his mobile cardiology lab he takes weekly to rural areas.[2] The $2,100 unit allows him to reach more people at half the price that he used to charge.

Dr. Girish Khurana, an internist in Bahadurgarh near Delhi, holds the $2,100 MAC 400. (From Endnote 3. With permission.)

However, the benefits of the durable, portable, and low-cost ECG are not limited to just emerging countries.[4] Because the unit is manufactured in India at a much lower cost, it can be sold into other countries at a lower price than competing products. After the success of the product in India, GE also decided to sell the unit in Germany and is eyeing other developed markets for the economical portable machine.

Organizing for New Product Development

In developing new products, cross-functional teams of individuals from functional areas such as marketing, R&D, engineering, and manufacturing are involved in what is called the *core team*. Other functions may be part of the core team as well depending on the industry. For example, in the telecommunications industry, the legal department is always part of the core team because of all of the state and national regulations that govern that industry. The core team works together from the beginning of the project, ideation, through the launch of the product into the marketplace. Additional team members may be brought in throughout the development process depending on the needs of the project. However, these members come and go and are not involved throughout the life of the project. The core team plus these ad hoc members form the **extended team**. In a benchmarking study of mid- and large-size business units in a variety of industries, 72% of survey respondents indicated that they used cross-functional teams for new product development.[5]

According to Wheelwright and Clark,[6] there are two main types of teams for new product development (NPD): *lightweight* and *heavyweight* teams. Lightweight teams are those in which functional members are assigned by their functional leaders and act as representatives of their functional area on a particular project. Members of lightweight teams are loyal to their functional area first and the project second. As a result, lightweight team members may undertake actions that benefit their functional group rather than the project. Such behavior can cause challenges in getting the project completed effectively and efficiently.

Heavyweight teams, by contrast, are those in which team members report directly to the project leader and are dedicated to the project for its duration. In these teams, participants are more likely to own the goals of the project and feel responsible for ensuring its effective completion than in lightweight teams. Thus, members of heavyweight teams show allegiance first to the project and second to their functional group.

Physical Proximity of NPD Teams

Many years ago, most NPD team members worked in the same building or at least in the same country. In today's world, it is not uncommon to find NPD team members working together even though they are located in different parts of the globe. Thus, when creating and managing NPD teams, it is important to consider their physical proximity. Physical proximity can be viewed as a continuum[7] where **co-located** teams are at one end and **global** teams at the other end. Co-located teams are composed of individuals who are physically located together. For example, team members might be on the same floor or in the same building. Global teams work across functional, geographic, national, and cultural boundaries.[8] As a result, global team members are likely to speak different languages, have different cultural values, work in different time zones, and use technology-mediated communication more than face-to-face communication.

Between these two extremes is another category of teams called *virtual* or *distributed* teams. These teams are physically dispersed and occasionally meet face-to-face. An example would be a team in which members are in different geographic regions of the same country. Table 11.1 lists the benefits of co-located and global teams.

Global teams are increasingly being used for new product development.[9] Three major factors account for this trend. First, many companies are developing products that share common materials, components, and subsystems, thus requiring little customization. However, at the same time, they are working to tailor those products to appeal to local market requirements. Such a "glocal" strategy attempts to balance global and local concerns by generating efficiencies from standardization yet simultaneously respond to and expand in particular markets by modifying the product to fit local tastes and conventions.[10] By using a global team in which team members come from diverse countries and have varied cultural perspectives, firms can use their perceptions and know-how to determine how to localize products for specific country markets.[11]

Second, there has been a rise in technological competencies in diverse countries around the world[12] leading to "centers of excellence" in specific areas. For example, Bangalore, India, is known for its expertise in information technology and business process outsourcing. China is renowned for low-cost manufacturing. The creation of these centers of excellence implies that for many product development efforts, these dispersed resources need to be brought together without increasing costs. Advanced information and communication technologies permit such geographically disparate team members to communicate and collaborate 24/7.

Table 11.1 Benefits of Co-Located and Global Teams

CO-LOCATED TEAMS	GLOBAL TEAMS
• Can solve problems and resolve issues as they arise	• Allows for diverse and specialized information and knowledge to be shared
• Easier to share information	• Brings multiple perspectives into decision making
• Opportunities for informal information sharing and gathering are more abundant	• Members from different countries may have more knowledge of local market requirements
• Helps build trust faster because of face-to-face time	• More creative in generating ideas and solving problems

Source: Endnote 13. With permission.

Finally, the trend toward **open innovation** necessitates using global teams.[14] As with centers of excellence, firms can share knowledge and utilize the expertise of various partners through technology-mediated communication. In addition, firms are able to obtain first-hand knowledge of local behaviors, regulations, product requirements, etc., throughout the project.

However, several characteristics of global teams make them more complex and more difficult to manage.[15] First, global teams are usually responsible for development projects that are global in scope and important to the organization.[16] Such projects are multifaceted and can involve many people throughout the world, adding to the complexity. Second, global team members are very diverse. Not only do they come from different functional areas, they may represent dispersed organizations and have varied languages and cultural values that can impact the collaboration, communication, and dynamics of the team. Creating effective global teams is possible; it just is more difficult[17] and takes a longer time to achieve. Third, the geographical dispersion of team members means that communication needs to be coordinated around time zones. Although this is possible today with tools such as Skype and WebEx, such communication is tougher because it often lacks visual cues. Firms, such as Hewlett-Packard and Deloitte & Touche LLP,[18] have their own global electronic networks that they use to connect employees and to provide access to other online tools that can track and synchronize tasks.

To deal with the increased complexity of global teams, it is important that global project leaders take specific actions to ensure team effectiveness. Table 11.2 highlights some of the actions global project leaders should take in each of four areas to aid the team.

Table 11.2   Global Team Leader Behaviors and Actions

• Assist team communication efforts	• Create communication norms and rules
	• Develop common terminology
	• Send critical documents prior to meetings
	• Include nontask communication in meetings
• Ensure appreciation for and understanding of team diversity	• Develop team expertise directory
	• Provide a skills matrix for the team
	• Undertake cultural diversity training with all team members
	• Pair up members from varied backgrounds
• Manage project work including meetings	• Have regular, mandatory virtual meetings with specified agenda and time duration
	• Track team member participation in meetings and activities
	• Distribute minutes of meetings
	• Rotate team meetings among different time zones and geographic regions
	• Ensure that decision-making rules are established and followed
	• Create virtual workspace for team
• Supervise team and individual progress and behaviors	• Frequent and regular meetings throughout the project
	• Acknowledge team milestones and individual accomplishments
	• Monitor behavior through virtual workspace and through communication tools

Source: Endnote 19. With permission.

Open Innovation and Global Markets

It used to be that many firms practiced the strategy of closed innovation or what many referred to as "not invented here." Closed innovation meant that, if the firm itself did not create and develop an innovation, then the innovation was not of value. Today, however, due to our fast-paced, globally competitive environment, few if any firms have the time, money, or capabilities to develop new products all on their own. The advent of the Internet and other digital technologies makes partnering with customers, suppliers, distributors, and even competitors in developing new products much easier to do. As a result, many firms employ **open innovation**. Open innovation implies that firms combine external ideas and knowledge with internal ideas and knowledge to create new products. In fact, in 2009, *Business Week* declared that "it's a new world of collaboration across corporate and national boundaries."[20] Table 11.3 delineates the differences between a closed and an open perspective towards innovation.

Open innovation is a two-way street. Firms can receive ideas and technological knowledge from external entities as well as distribute their knowledge and expertise through partnerships, licensing, and joint ventures. Procter & Gamble is one company that has been a strong proponent of open innovation claiming that it wants 50% of its new products to come from outside the company.[21]

Outsourcing is often the first method of open innovation that companies use. Outsourcing is the strategic decision to reject the internalization of an activity.[22] For years, many companies, such as Hasbro and Apple, have outsourced manufacturing of their products. However, within the context of new product development, we are referring to the outsourcing of development and possibly the design of components of the new product or the product itself. For example, companies such as Dell, Motorola, and Phillips are buying complete designs of lower priced devices from Asian developers. Some question whether outsourcing one's R&D is a good thing in the long run. Recent research suggests that firms that focus on innovation activities and outsource

Table 11.3 Closed versus Open Innovation Principles

CLOSED INNOVATION PRINCIPLES	OPEN INNOVATION PRINCIPLES
The smart people in the field work for us.	Not all the smart people in the field work for us. We need to work with smart people inside and outside the company.
To profit from R&D, we must discover it, develop it, and ship it ourselves.	External R&D can create significant value; internal R&D is needed to claim some portion of that value.
If we discover it ourselves, we will get it to the market first.	We don't have to originate the research to profit from it.
The company that gets an innovation to market first will win.	Building a better business model is better than getting to the market first.
If we create the most and the best ideas in the industry, we will win.	If we make the best use of internal and external ideas, we will win.
We should control our intellectual property (IP) so that our competitors don't profit from our ideas.	We should profit from others' use of our intellectual property (IP) and we should buy others' IP whenever it advances our business model.

Source: Endnote 23. With permission.

peripheral tasks are more innovative.[24] Likewise, when there is little opportunity to develop a competitive advantage, firms will favor outsourcing their R&D.[25] These findings suggest that outsourcing R&D to external entities should occur only for activities that are not strategic for the firm and for which they cannot differentiate themselves from competitors.

Too much outsourcing of the design and development of parts and components can cause problems, such as loss of strategic competence. Take the case of the Boeing 787 Dreamliner. Thirty percent of the plane has foreign-made components coming from countries such as South Korea, China, Australia, and Italy. Rather than follow its standard practice of providing suppliers with detailed blueprints, Boeing gave suppliers less-detailed specifications and allowed them to design, as well as manufacture, components. Unfortunately, some of these suppliers were not qualified to design their components nor did many of the components and subsystems fit together well. Some parts were not ready when they needed to be. The high degree of outsourcing (about 50 suppliers) prevented Boeing from being able to closely supervise the design and manufacture of many of the parts and components, leading to the 787 being billions of dollars over budget and three years late to market. Boeing executives now admit that the level of outsourcing created more management issues than expected.[26]

Crowdsourcing is another approach to open innovation. Crowdsourcing means getting ideas from an undefined, generally large group of people in the form of an open call. In recent years, a number of companies have been using crowds to generate new product ideas. In 2008, Google launched its Project 10 to the 100 to mark its 10th anniversary of existence. They asked people around the globe to come up with ideas that would help society. Google received over 150,000 ideas from 170 countries and then narrowed those down to 16 idea themes. The public was then asked to vote for the top five ideas and the winners were announced in October 2011. Google gave a total of $10 million to the winners to help support their efforts.[27]

Staples instituted its EcoEasy Challenge for undergraduate and graduate students. Staples is interested in "inventions for a business or home office product that allows users to work more easily and in a manner that is more environmentally responsible than existing similar products. This concept could be a new design to an existing product that represents greater environmental responsibility or sustainability; a product or product line that uses an eco-innovative material; or a completely new type of product that promotes sustainability (either through the construction or use of that item)."[28] The first prize winner receives $25,000 toward developing his or her product and the two runnersup get $5,000 each. If Staples decides to produce a product and sell it in their stores, they provide a 6% royalty to the team.

While some companies and experts advocate the use of crowdsourcing to generate ideas and solutions to customer needs, others are more skeptical of the value of such crowds. In a recent study that compared users versus professionals (R&D, marketing, and design employees of a particular firm), user ideas scored higher in terms of novelty

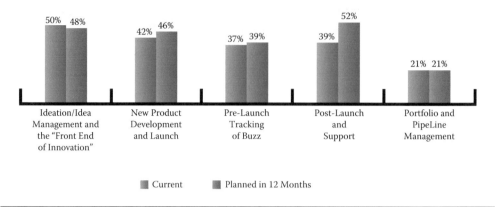

Figure 11.1 Crowdsourcing. (From Endnote 29. With permission.)

and customer benefit, but somewhat lower on feasibility than the professionals' ideas. Overall, though, the best ideas were judged to come from users.[30]

Companies also have emerged that aim to connect firms with problems to people who might have solutions to those problems. Innocentive (www.innocentive.com) and Nine Sigma (www.ninesigma.com) are two prominent companies that promote this type of crowdsourcing. Additionally, social media is just beginning to be used for crowdsourcing. In a recent study of 90 companies by Kalypso (www.kalypso.com), it found that most firms are using social media for idea generation and launch activities (Figure 11.1). However, most companies expect their usage of social media for all aspects of the new product development to increase over the next few years.

When firms partner with their customers to create new products and services, it is referred to as *co-creation*. According to Prahalad and Ramaswamy, who coined the term,[31] customers have become sophisticated enough that they are not simply satisfied with what companies provide to them. Rather, customers want to be involved with the firm in co-creating value (i.e., in helping design and develop new products). The My Starbucks Idea Web site lets customers submit ideas and vote on the ideas of others. More importantly, Starbucks lets you know whether an idea is being reviewed, what's been approved, and what ideas are being implemented. As a result of many comments on the Web site, on July, 1 2010, Starbucks started providing free and simpler Wi-Fi service within its stores in the United States. Previously, customers could only use the Wi-Fi free for two hours and they had to provide a username and password.[32]

Local Motors Inc. provides an even more explicit example of co-creation. Local Motors' mission is: "To lead the next generation of crowd-powered automotive manufacturing, design, and technology in order to enable the creation of game changing vehicles."[33] Through their Web site, you can upload car designs, tinker with the designs of others, vote on designs, and purchase a car designed by the crowd. The Rally Fighter is the first production car from Local Motors and it is currently being built in a small microfactory in Phoenix, Arizona. If you purchase one of these cars, you must spend six days at the factory helping to build your car.[34]

Innovation in Emerging Markets

With the growth of emerging markets, such as the BRIC countries (Brazil, Russia, India, and China), many companies have entered or are looking to enter these and other growing markets. Sometimes a company can take its existing products and with no, or minor, adaptations, launch that same product in a foreign country. General Motors, for example, even though it was late entering the Chinese automobile market, has had success with its Buick brand.[35] Coca-Cola has had success in India since relaunching its brand there in 1992, with many of its brands being the top brand in their segment.[36]

Often, however, companies run into legal, cultural, and/or economic barriers that force them to alter their existing products significantly or create entirely new products. As an example, in December 2009, China instituted a restriction that all new cosmetic ingredients need to be approved by the government. However, for two years now, Beijing has not approved any new ingredients including skin whiteners and moisture retainers because of previous problems with foreign-made cosmetics. As a result, Shiseido made the decision to use ingredients already approved by the government and, thus, the cosmetic products it sells in China have different ingredients from its products sold in other countries. By contrast, P&G has been unable to sell some of its SK-II products in China because they refuse to alter the components.[37]

Reverse Innovation

Reverse innovation is built on the belief that you cannot simply take an innovation from one country, such as the United States, and simply adapt it for an emerging market. Rather, firms need to understand the customers in a particular country, and design products and services expressly for them. Thus, a reverse innovation is first adopted in a developing market. However, it then must be diffused to other countries to make it a global product.[38] Table 11.4 highlights the major principles behind reverse innovation.

An example of reverse innovation is the Tata Nano launched in India in 2009 by Tata Motors, the largest automobile company in India. The Nano was designed to appeal to the mass market of Indian consumers who used two- rather than four-wheel

Table 11.4 Reverse Innovation Principles

- Reverse innovation requires a decentralized, local-market focus.
- Most, if not all, of the people and resources dedicated to reverse innovation efforts must be based and managed in the local market.
- Local Growth Teams (LGTs) must have P&L responsibility and the decision-making authority to choose which products to develop, how to make, sell, and service them.
- LGTs must have the right (and support) to draw from the company's global resources.
- Products developed using reverse innovation must be taken global.

Source: Endnote 39. With permission.

Figure 11.2　Tata Nano. (From Endnote 40. With permission.)

vehicles. It is a small, basic car and the cheapest model costs about US$2,900[41] as shown in Figure 11.2.

GE Healthcare also has developed products, such as a handheld ECG (electrocardiogram) machine for India, that costs about $1,000 and a portable, PC-based ultrasound (as shown in Figure 11.3) designed for China that sells for about $15,000 (versus a traditional ultrasound that can cost 10 to 20 times more[42]). Both of these innovations have made their way into other country markets around the world as well.

Bottom-of-the-Pyramid

Bottom-of-the-pyramid, or BOP, refers to the 4 billion people (2/3 of the population) in the world living on $1,500 or less per capita income per year (Figure 11.4). The sheer size of the BOP has attracted the attention of a number of companies, such as Hindustan Lever, a subsidiary of Unilever PLC.

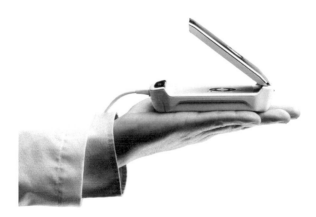

Figure 11.3　Vscan-pocket-sized ultrasound from GE Healthcare. (From Endnote 43. With permission.)

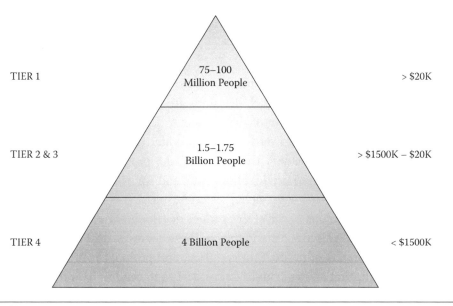

Figure 11.4   Base of the pyramid.

After seeing the success of a local Indian firm, Nirma, in the laundry detergent category, Hindustan Lever developed a new detergent that had a lower oil to water ratio, was low priced, and available in small, local outlets where most rural consumers shop. Today, both companies have about equal market share in this product category in India.

As another example, First Energy Pvt. Ltd (originally a subsidiary of British Petroleum) launched the Oorjastove, which uses smokeless fuel pellets for cooking. The product, used by households, restaurants, and other commercial establishments, is easier to use, more cost-effective, and healthier than traditional alternatives, such as using smoky wood or coal.[44]

There is some disagreement about whether or not targeting the BOP is advantageous to firms. Some argue that creating technologies and products for these poor people will help lift them out of poverty and bring them into the formal market economy. At the same time, such innovations will yield growth and long-run profits for companies if they focus on volume and manufacturing efficiency.[45] However, others contend that pursuing the BOP without improving the living conditions of the locals is irresponsible and that offering more product variety can lead these poorer consumers to make bad purchase decisions (i.e., make impulse purchases).[46] BOP2.0 takes a more holistic approach to designing products for developing countries by including the locals to be a part of the solution, by involving them as distribution partners, by educating them, and by channeling some of the profits back to the local community.

Launching Global New Products

Global New Product Launches

In developing a new product, firms need to consider two types of launch decisions: strategic and tactical. Strategic launch decisions are those that need to be made prior to the actual physical development of the product.[47] They include decisions regarding the firm's overall orientation toward its new product development efforts (first-to-market, fast follower, cost leader), the nature of the product to be developed (e.g., new-to-the-world, addition to product line, etc.), and the nature of the market where the product will be launched (e.g., niche, selective, mass).[48]

Tactical launch decisions occur after product development and answer "how" the product will be launched into the marketplace. These decisions deal with the traditional marketing mix elements of product, price, place, and promotion.[49] It is important to remember, though, that strategic and tactical decisions need to be integrated. A particular strategy may dictate the implementation of specific marketing mix strategies.

A study of launch decisions of consumer products firms[50] found that the combination of various strategic and tactical decisions led to different performance outcomes. Table 11.5 shows the four clusters of strategic and tactical decisions. Cluster 2 had the highest performance with 88% of the new products being successful. This cluster also had the highest level of customer acceptance and financial performance. Cluster 3 was the second most successful group with 68% of the new products being successful. Cluster 4 had a 58% success rate and Cluster 1 had the lowest success rate at 34%. Similar results were found in a study of industrial firms.[51] Thus, the most successful launch strategy, whether for a consumer product or a B2B product, appears to be one in which a firm develops an innovative product driven by both technology and market needs that is launched into a niche market with few competitors. At the tactical level, success is governed by high distribution expenditures and a skimming pricing strategy.

Global Brands

A global brand is one in which its positioning, communication strategy, packaging, and brand identity are nearly identical from country to country. Looking at Table 11.6 of the Top 10 Global Brands for 2012, these are all brands you probably can easily recognize.

Branding Strategies

During launch, the NPD manager must consider brand strategies when naming the product. When the product is to be launched internationally additional branding issues must be considered. There are a number of global branding strategies from which marketers can choose. *Umbrella* branding implies that a firm uses a particular brand name for all its products in a given product line worldwide. Kellogg's uses its

Table 11.5   Clusters of Strategic and Tactical Decisions

	CLUSTER 1	CLUSTER 2	CLUSTER 3	CLUSTER 4
STRATEGIC LAUNCH DECISIONS				
PRODUCT				
Relative innovativeness	More innovative	More innovative	??	Equal
NPD cycle time	>3 years	1–3 years	??	<1 year
Product advantage	High	High	Low	Low
MARKET				
Targeting strategy	??	Niche	Selective	Mass market
Number of competitors	More than 4	None	1–3	1–3
FIRM				
Driver of NPD	Technology	Mix	??	Market
TACTICAL LAUNCH DECISIONS				
PRODUCT				
Branding strategy	New brand	??	??	Company name
PLACE				
Distribution intensity	Selective	Intensive	??	Intensive
PROMOTION				
Promotion expenditures	Lower	Higher	??	Equal
PRICE				
Price strategy	Other	Skimming	Penetration	Penetration

Note: Innovativeness, distribution expenditures, sales force intensity, breadth of product assortment, promotion
         expenditures, and price level were measured in comparison with competing products on the market.
Source: Endnote 52. With permission.

corporate name on all of its cereal brands. Kraft uses names such as Planters, Di Giorno, and Maxwell House for particular product categories.[53] Procter & Gamble, by contrast, tends to use *individual* branding for its products in different countries: Tide is called Ace in Latin America and Alo in Turkey. **Co-branding** is another strategy to match one brand not well known with a more popular brand in another

Table 11.6   Brand Equity of Top 10 Global Brands

1	Coca-Cola	77,839 ($million)
2	Apple	76,568 ($million)
3	IBM	75,532 ($million)
4	Google	69,726 ($million)
5	Microsoft	57,853 ($million)
6	GE	43,682 ($million)
7	McDonald's	40,062($million)
8	Intel	39,385 ($million)
9	Samsung	32,893($million)
10	Toyota	30,280 ($million)

Source: Endnote 54. With permission.

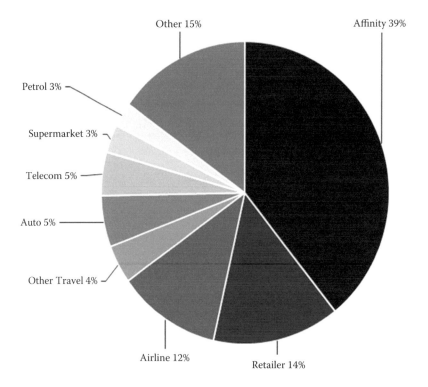

Note: Study Based on 723 credit card programs in 31 countries.
Other travel includes hotel, auto and all other travel besides airline. Other
includes telecomunications, media, insurance , and all others not shown here.
An affinity group is a university, social movement, or other group with which a card
member wishes to align themselves

Figure 11.5 Co-brand credit card landscape in Europe. (From Endnote 55. With permission.)

country. Co-branding represents an arrangement between two or more companies that agree to associate their respective logos, brand names, or other brand identifiers on a single product or service. In 1995, the BMW Z3 Roadster was featured in the James Bond movie, *Golden Eye*, and showcased in the Nieman Marcus Holiday Catalog, followed by inclusion in an Apple advertisement shown during NFL's Super Bowl.[56] British Airlines has partnered with Visa, and American Express with Delta Airlines. Figure 11.5 shows the type of co-branding situations that arise for credit cards in Europe. Mobile phones also typically are co-branded with the name of the phone manufacturer and the name of the service provider, such as T-Mobile Nokia Lumia 710 or the Vodaphone Samsung Galaxy Nexus. In fact, many mobile phone manufacturers actually use multiple co-branding where they partner with a number of service providers. Apple, as an example, provides the iPhone to both AT&T and Verizon in the United States and Vodaphone partners with T-Mobile and Orange in Europe.

Standardization or Adaptation

When going global, managers need to decide whether to standardize elements of the brand, such as the name, package, and positioning, or adapt one or more of these elements to the local market. Standardization can lead to economies of scale, cost efficiencies, and less brand confusion, but adaptation can lead to higher customer satisfaction because the product is tailored to local market needs. "P&G's Gillette division uses the same brand name and positioning worldwide ("The Best a Man Can Get"), and typically launches a new blade in 100-plus countries worldwide, to achieve significant scale economies in marketing and manufacturing."[57]

McDonald's, on the other hand, localizes its menu and its interior as well. In Berlin, there is a regular McDonald's on the first floor with a McCafe, which serves coffee, on the second floor. In India, McDonald's does not sell Big Macs, but does sell a variety of vegetarian options as well as a lamb burger.[58] Firms must be careful when creating a brand name for a new product as words can mean different things in diverse countries. For example, would you be willing to buy a drink product called Pocari Sweat? Probably not, but it is the name of a sports drink sold in Japan. Or, what about a product named Barf? Barf is the name of a line of cleaning products in Iran where barf means "snow."[59]

Packaging

A standardized package can help create brand awareness and generate brand recognition. For example, the contour shape of the Coca-Cola bottle, which officially became the standard in 1920, is recognized worldwide.[60] The same is true for the green glass Perrier bottle, shown in Figure 11.6.

Besides the look of the package, marketers need to think about the text that goes on the package. Today, on some products, you will find packaging copy in multiple

Figure 11.6   Classic bottle shapes. (From Endnote 61. With permission.)

languages. Home Depot has for years had suppliers provide both English and Spanish text on products sold in their stores because of the large Hispanic population in the United States. Dr. Oetker's frozen pizza, sold in Europe, has 12 different translations of the directions for cooking the pizza including Dutch, French, German, Spanish, and English.

Consumer Perceptions of Global Brands

The country-of-origin of a brand can have a strong impact on consumers' evaluative judgment regarding a particular brand. Country-of-origin refers to the country where the headquarters of the company marketing the product is located; usually, this is the home country for the company.[62] However, in recent years, because product design, parts manufacture, and final product assembly can occur in different countries, country-of-origin has been broken into country-of-design, country-of-parts, and/or country-of-assembly.[63]

Country–of-origin can be an important cue to buyers when they are unfamiliar with a product[64] or when their knowledge about the product is limited.[65] Country-of-origin can positively impact consumer choices, such as car brands from Germany, or it can negatively impact choices, such as the Yugo car built by the Yugoslav/Serbian Zastava Corporation. Global brands tend to be associated with higher product quality and prestige resulting in greater purchase likelihood.[66] However, consumers value product quality as most important, suggesting that marketers should emphasize creating and communicating product quality rather than status or prestige.[67] It is important to recognize, though, that not all consumers perceive global brands in the same way. Some consumers are less open to foreign cultures and less cosmopolitan and may see purchasing foreign products as hurting the domestic economy. In fact, these consumers will often forego quality to purchase a product made in their home country. Such consumers are low on what is called *consumer ethnocentrism*. Consumer ethnocentricism is defined as "the beliefs held by consumers about the appropriateness, indeed morality, of purchasing foreign-made products."[68] As an example, a report in the autumn of 2011 by *ABC News*, called "Made In America," showed that very few products in our homes are made in the United States. This led to a movement for consumers to buy at least one holiday present that was made in the United States, appealing to U.S. consumers who are potentially low on ethnocentrism.

Consumers' perceptions of global brands also are influenced by the stereotypes they hold of particular countries.[69] These stereotypes come into play in evaluating products even though the individual may have had no intention to use such information.[70] Sometimes companies will try to imply a country-of-origin for a product through the brand name. For example, Haagen Dazs sounds foreign, but it was originally started by the Pillsbury company. Similarly, Yoplait yogurt is owned by General Mills. Marketers will use such tactics to develop a particular image for a product that is related to the perceived origin of the product.[71]

Protecting Your Global Brand

A major challenge facing global brands is *counterfeiting*, which means illegally imitating something. We all have seen (and maybe purchased) fake copies of well-known movies, software, handbags, and clothing. However, today, many other types of products are being copied including pharmaceuticals, automobile parts, cigarettes, and baby food.[72] With the growth of global brands, the Internet, and digital technology, it is easy for others to copy, imitate, or otherwise infringe upon one's brand. In 2011, at least 22 fake Apple stores were found in Kunming, China. The stores and products were so authentic that even employees thought that they were working for Apple. In March 2012, 26 people were arrested for trying to import "knockoff" goods that imitated Ugg boots, Timberland boots, Nike shoes, and Burberry scarves from China to the United States.[73] It is believed that the majority of counterfeited goods come from countries that include China, North Korea, Taiwan, India, and Russia.[74]

The benefits of counterfeiting are clear. It allows the counterfeiters to seemingly satisfy customer needs for certain globally branded products and generate large revenue and profits by illegally manufacturing "knockoff" products that look similar to the actual brand, but which they sell at discounted prices. The Counterfeiting Intelligence Bureau of the International Chamber of Commerce® claims that counterfeiting is the fastest growing crime in the twenty-first century. It is estimated that counterfeiting accounts for 5 to 7% of world trade; an estimated $600 billion a year.[75] Table 11.7 lists market values for different types of products. Have you ever bought a counterfeit good? You might have without even knowing.

So, what can a firm do to protect its global brand? Brands like Kate Spade and Louis Vuitton are using private investigators to find where knockoffs of their brands

Table 11.7 Global Counterfeit Market Values

Electronics	$100 billion
Prescription Drugs	$72.5 billion
Food	$49 billion
Automotive Parts	$45 billion
Toys	$34 billion
Clothes	$12 billion
Shoes	$12 billion
Sporting Goods	$6.5 billion
Tobacco	$4 billion
Cosmetics	$3 billion
Weapons	$1.8 billion
Alcohol	$1 billion
Watches	$1 billion
Pesticides	$735 million
Purses	$70 million
Lighters	$42 million
Batteries	$23 million

Source: Endnote 76. With permission.

are being sold.[77] Inkjet technology is being used for pharmaceutical, tobacco, and other products to authenticate products.[78] German fashion retailer brand Gerry Weber is deploying radio frequency identity (RFID) chips across its entire inventory to keep track of suppliers that were purposely producing excess stock for unauthorized sale.[79] Most experts agree that companies need to attack the problem from multiple perspectives and levels as no one approach will come close to combating the problem of counterfeiting.

Chapter Summary

In this chapter, issues surrounding global new product development are discussed. The increasing use of global new product teams indicates the need for managers of such teams to understand the added complexity of teams that cross national, cultural, and geographic boundaries. Open innovation is playing a bigger role as firms strive to remain competitive. Similarly, reverse innovation and innovations aimed at the bottom-of-the-pyramid are gaining in importance as companies try to take advantage of emerging markets. Finally, when launching global brands, firms need to make a number of important strategic and tactical decisions that can influence the success or failure of their new product.

Glossary

Bottom-of-the-pyramid (BOP): Refers to the 4 billion people in the world living on $1,500 or less per capita income per year.

Closed innovation: Using only internal ideas and knowledge to develop an innovation. Also referred to as the "not-invented-here" (NIH) syndrome.

Co-branding: Arrangement between two or more companies that agree to associate their respective logos, brand names, or other brand identifiers on a single product or service.

Co-creation: Firms partnering with their customers to create new products and services.

Co-located team: Composed of individuals who are physically located together.

Core team: A cross-functional team that works together from the beginning of the project, ideation, through the launch of the product into the marketplace.

Counterfeiting[80]: The process of fraudulently manufacturing, altering, or distributing a product that is of lesser value than the genuine product.

Crowdsourcing: Getting ideas from an undefined, generally large group of people in the form of an open call.

Distributed (virtual) team: Members are physically dispersed and occasionally meet face-to-face.

Extended team: Core team plus team members from various functions brought in as needed during the new product development process.

Global team: Team members come from different functional, geographic, national, and cultural backgrounds.

Heavyweight team: Team members report directly to the project leader and are dedicated to the project for its duration.

Individual branding: Creating a distinct brand name for a product with no attachment to the firm/unit producing the product.

Lightweight team: Functional members are assigned by their functional leaders and act as representatives of their functional area on a particular project.

Multiple branding: Two or more brands offered by the same company that compete in a product category.

Open innovation: Combining external ideas and knowledge with internal ideas and knowledge to create new products and services.

Outsourcing: Decision not to internally perform a particular activity but rather have another firm undertake that activity.

Reverse innovation: Developing products specifically for an emerging country market and then launching that product in developed countries.

Umbrella branding: An over-arching brand across related product categories.

Review Questions

1. What are the characteristics of lightweight versus heavyweight teams?

2. Why are global teams being used more often for new product development?

3. What are some actions that global team leaders need to take to facilitate the global team?

4. What is open innovation? What are its advantages and disadvantages?

5. What is the difference between reverse innovation and bottom-of-the-pyramid?

6. What is a global brand?

7. What are the advantages and disadvantages of standardization versus adaptation of various marketing mix elements?

Assignment Questions

1. Describe the efforts that a specific company is using to engage in open innovation. Some companies to consider include: Procter & Gamble, IBM, Hewlett-Packard, Dow Chemical, DuPont, AT&T, Beiersdorf (Nivea brand), Nokia, Starbucks, Microsoft, Unilever, Tata, Threadless, Dell, Adidas, BMW.

2. Find examples of companies focusing on the bottom-of-the-pyramid markets. What kinds of products are these companies marketing to these customers? Some companies to consider include: Nokia, Unilever, Nestle, GE Healthcare.

3. Do some research to find out what kinds of things companies are doing to prevent counterfeiting of their products and/or to find counterfeiters.

Endnotes

1. Jena McGregor, "GE: Reinventing Tech for the Emerging World," April 16, 2008. Online at: http://www.businessweek.com/stories/2008-04-16/ge-reinventing-tech-for-the-emerging-world (accessed July 2, 2013).
2. Online at: http://knowledge.wharton.upenn.edu/india/article.cfm?articleid=4476 (accessed July 2, 2013).
3. Image from Jena McGregor, "GE: Reinventing Tech for the Emerging World."
4. Online at: http://images.businessweek.com/ss/09/04/0401_pg_trickleup/5.htm (accessed July 2, 2013).
5. Robert G. Cooper, Scott J. Edgett, and Elko J. Kleinschmidt, "Improving New Product Development Performance and Practices," Benchmarking Study (Houston, TX: American Productivity and Quality Center, 2002).
6. Stephen Wheelwright and Kim Clark, *Revolutionizing Product Development: Quantum Leaps in Speed, Efficiency and Quality* (New York: Free Press, 1992).
7. G. T. Hertel, S. Geister, and U. Konradt, "Managing Virtual Teams: A Review of Current Empirical Research," *Human Resource Management Review*, 15 (2005): 69–95.
8. M. Maznevski and N. Athanassiou, (Guest eds.) "Introduction to the Focused Issue: A New Direction for Global Teams Research," *Management International Review* 46(6) (2006): 631–646.
9. Edward McDonough, Kenneth Kahn, and Gloria Barczak, "An Investigation of the Use of Global, Virtual and Collocated New Product Development Teams," *Journal of Product Innovation Management* 18 (2001): 110–120; K. Sivakumar and Cheryl Nakata, "Designing Global New Product Teams: Optimizing the Effects of National Culture on New Product Development," *International Marketing Review* 20(4) (2003): 397–445.
10. For example, Thomas Begley and David Boyd, "The Need for a Corporate Global Mind-Set," *Sloan Management Review*, Winter 2003; Jeffrey Immelt, Vijay Govindarajan, and Chris Trimble, "How GE Is Disrupting Itself," *Harvard Business Review*, October 2009; Goran Svensson, "'Glocalization' of Business Activities: A 'Glocal Strategy' Approach," *Management Decision* 39(1) (2001): 6–18.
11. McDonough, Kahn, and Barczak, "An Investigation of the Use of Global, Virtual and Collocated New Product Development Teams."
12. Sivakumar and Nakata, "Designing Global New Product Teams."
13. D. Ancona and D. Caldwell, "Demography and Design: Predictors of New Product Team Performance," *Organization Science* 3(3) (1992): 321–341.

14. Sivakumar and Nakata, "Designing Global New Product Teams."
15. Gloria Barczak and Edward McDonough, "Leading Global Product Development Teams," *Research Technology Management* 46(6) (2003): 14–18; Maznevski and Athanassiou, "Introduction to the Focused Issue."
16. Maznevski and Athanassiou, "Introduction to the Focused Issue."
17. McDonough, Kahn, and Barczak, "An Investigation of the Use of Global, Virtual and Collocated New Product Development Teams."
18. Jeremy Lurey and Mahesh Raisinghani, "An Empirical Study of Best Practices in Virtual Teams," *Information & Management* 38(8) (2001):1–22.
19. Gloria Barczak, "Designing and Forming Global Teams," in *Handbook of Technology Management*, ed. Hossein Bidgoli (Hoboken, NJ: John Wiley & Sons, 2010).
20. Steve Hamm, "A Rethink of R&D," *Bloomberg Businessweek*, August 26, 2009) ("http://www.businessweek.com/stories/2009-08-26/a-radical-rethink-of-r-and-d" www.businessweek.com/stories/2009-08-26/a-radical-rethink-of-r-and-d).
21. Larry Huston and Nabil Sakkab, "Connect and Develop: Inside Procter and Gamble's New Model for Innovation," *Harvard Business Review*, March 2006.
22. K. Matthew Gilley and Abdul Rasheed, "Making More by Doing Less: An Analysis of Outsourcing and Its Effects on Firm Performance," *Journal of Management* 26(4) (2000): 763–790.
23. Online at: http://www.openinnovation.eu/open-innovation/ (accessed July 1, 2013).
24. K. Matthew Gilley and Abdul Rasheed, "Making More by Doing Less: An Analysis of Outsourcing and Its Effects on Firm Performance," *Journal of Management* 26(4) (2000): 763–790.
25. Michael Stanko and Roger Calantone, "Controversy in Innovation Outsourcing Research: Review, Synthesis and Future Directions," *R&D Management* 41(1) (2010): 8–20.
26. Michael Hiltzik, "787 Dreamliner teaches Boeing costly lesson on outsourcing," *LA Times*, February 15, 2011. Online at: http://articles.latimes.com/2011/feb/15/business/la-fi-hiltzik-20110215_ (accessed, 2013).
27. To see the winners, go to http://www.google.com/onceuponatime/project10tothe100/ (accessed July 1, 2013).
28. Online at: https://ecoeasychallenge.com/pages/Rules.html (accessed June 1, 2012).
29. Amy Kenly and Bill Poston, "Social Media and Product Innovation," White paper, Kalypso: Beachwood, OH (2010).
30. Marion Poetz and Martin Schreier, "The Value of Crowdsourcing: Can Users Really Compete with Professionals in Generating New Product Ideas," *Journal of Product Innovation Management* 29(2) (2012): 245–256.
31. C. K. Prahalad and Venkatram Ramaswamy, "Co-opting Customer Competence," *Harvard Business Review*, January 2000.
32. Online at: http://survey.cvent.com/blog/market-research-and-survey-basics/starbucks-listens-to-customer-feedback-a-case-study (accessed July 1, 2013).
33. Online at: http://www.local-motors.com/about/mission/ (accessed June 1, 2012).
34. Online at: http://www.rallyfighter.com/buy-a-rally-fighter (accessed July 1, 2013).
35. Online at: http://knowledge.insead.edu/strategy-china-edward-tse-100416.cfm (accessed June 1, 2012).
36. Online at: http://www.coca-colaindia.com/ourcompany/company_history.html (accessed July 1, 2013).
37. Online at: http://ajw.asahi.com/article/economy/business/AJ201203080107 (accessed July 2, 2013).
38. Vijay Govindarajan, Blog, 2009. Online at: http://www.tuck.dartmouth.edu/people/vg/blog-archive/2009/10/what_is_reverse_innovation.htm (accessed July 1, 2013).
39. "Tata's Nano, the Car That Few Want to Buy," nytimes.com, December 9, 2010.

40. Original photo by Anugrah Adams; this image is licensed under the Creative Commons Attribution-Share Alike 2.0 Generic.

41. "Tata's Nano, the Car That Few Want to Buy," nytimes.com, December 9, 2010.

42. Online at: http://www.gereports.com/reverse-innovation-how-ge-is-disrupting-itself/ (accessed July 1, 2013).

43. Image from: http://news.bbc.co.uk/2/hi/health/8521067.stm (accessed July 2, 2013).

44. Online at: http://www.firstenergy.in/about-us/innovation (accessed July 1, 2013).

45. Prahalad and Hart, "The Fortune at the Bottom of the Pyramid."

46. Aneel Karnani, "Romanticizing the Poor," *Stanford Social Innovation Review* (2009). Online at: http://www.ssireview.org/articles/entry/romanticizing_the_poor (accessed July 1 2013).

47. Erik Jan Hultink, Abbie Griffin, Susan Hart, and Henry Robben, "Industrial New Product Launch Strategies and Product Development Performance," *Journal of Product Innovation Management* 14(4) (1997): 243–257.

48. Ibid.

49. Erik Jan Hultink, Abbie Griffin, Susan Hart, and Henry Robben, "Industrial New Product Launch Strategies and Product Development Performance," *Journal of Product Innovation Management* 14(4) (1997): 243–257.

50. Based on Erik Jan Hultink, Susan Hart, Henry Robben, and Abbie Griffin, "New Consumer Product Launch: Strategies and Performance," *Journal of Strategic Marketing* 7 (1999):153–174.

51. Hultink, Griffin, Hart, and Robben, "Industrial New Product Launch Strategies and Product Development Performance."

52. Based on Erik Jan Hultink, Susan Hart, Henry Robben, and Abbie Griffin, "New Consumer Product Launch: Strategies and Performance," *Journal of Strategic Marketing* 7 (1999):153–174.

53. C. Anthony Di Benedetto, "Global Branding Strategies," White Paper (2012). Online at: http://www.temple.sg/executive-education/white-papers/global-branding-strategies-white-paper (accessed July 1 2013).

54. Online at: "http://www.interbrand.com/en/best-global-brands/2012/Best-Global-Brands-2012.aspx" http://www.interbrand.com/en/best-global-brands/2012/Best-Global-Brands-2012.aspx. To find out more about how Interbrand determines its rankings, go to http://www.interbrand.com

55. 2011 First Annapolis Consulting Study Driving Growth in Europe via co-brand partnerships, Stephen Mendelsohn and Erik Howell, Cards International, September 15, 2011. (accessed at http://www.firstannapolis.com/driving-growth-europe-co-brand-partnerships (accessed July 2, 2013).

56. Online at: http://www.inc.com/michelle-greenwald/innovative-co-branding-and-creative-partnerships.html (accessed July 2, 2013).

57. C. Anthony Di Benedetto, "Global Branding Strategies," White Paper (2012). Online at: http://www.temple.sg/executive-education/white-papers/global-branding-strategies-white-paper (accessed July 1 2013).

58. Online at: http://businesstoday.intoday.in/story/mcdonalds-paneer-burger-mcspicy-paneer-mccurry-pan/1/15778.html (accessed July 1, 2013) .

59. Online at: http://www.deseretnews.com/article/700078810/Some-brand-names-dont-translate-well.html (accessed July 1, 2013).

60. "1916 … Birth of the Contour Bottle," The Coca-Cola Company. Archived from the original on January 18, 2010. Online at: http://www.thecoca-colacompany.com/ourcompany/historybottling.html (accessed July 1, 2013).

61. Coca-Cola, Inc. owns copyright on the design of its bottles. Perrier owns copyright on its bottle design as well.

62. J. K. Johansson, S. P. Douglas, and I. Nonaka, "Assessing the Impact of Country of Origin on Product Evaluations: A New Methodological Perspective," *Journal of Marketing Research* 22 (November 1985): 388–396.

63. Paul Chao, The Moderating Effects of Country of Assembly, Country of Parts, and Country of Design on Hybrid Product Evaluations," *Journal of Advertising* 30(4) (2001): 67–81; Gary S. Insch and J. Brad McBride, "The Impact of Country-of-Origin Cues on Consumer Perceptions of Product Quality: A Binational Test of the Decomposed Country-of-Origin Construct," *Journal of Business Research* 57(2) (2004): 256–265.

64. C. M Han and V. Terpstra, "Country-of-Origin Effects for Uni-National and Bi-National Products," *Journal of International Business Studies* (Spring 1988): 535–555.

65. D. Maheswaran, "Country-of-Origin as a Stereotype: Effect of Consumer Expertise and Attribute Strength of Product Evaluations," *Journal of Consumer Research* 21 (1994): 354–365.

66. J. E M. Steenkamp, R. Batra, and D. L. Alden, "How Perceived Brand Globalness Creates Brand Value." *Journal of International Business Studies* 34(1) (2003): 53–65.

67. Ibid.

68. T. A. Shimp and S. Sharma, "Consumer Ethnocentrism: Construction and Validation of the CETSCALE." *Journal of Marketing Research* 24 (1987): 280–289.

69. Scott Liu and Keith Johnson, "The Automatic Country-of-Origin Effects on Brand Judgements," *Journal of Advertising* 34(1) (2005): 87–97.

70. Ibid.

71. Mrugank Thakor and Anne Lavack, "Effect of Perceived Brand Origin Associations on Consumer Perceptions of Quality," *Journal of Product & Brand Management* 12(6) (2003): 394–407.

72. APCO, Global Counterfeiting Background Document, January 27 2003. Online at "http://www.apcoworldwide.com" http://www.apcoworldwide.com (accessed July 10, 2012).

73. Online at: http://www.cnbc.com/id/46602177/Top_Luxury_Brands_Targeted_in_Counterfeiting_Scheme. (accessed July 1, 2013).

74. Online at: http://ezinearticles.com/?With-Counterfeiting-on-the-Rise,-Brand-Security-Plays-a-Vital-Role-to-Both-Large-and-Small-Brands&id=6907545 (accessed July 2013).

75. Online at: http://www.icc-ccs.org/index.php?option=com_content&view=article&id=29&Itemid=39 (accessed June 1, 2012).

76. International Chamber of Commerce; 2011 Mark Monitor® Brand Jacking Index; Online Piracy and Counterfeiting report, January 2011. Online at "http://www.markmonitor.com/resources/brandjacking_index.php" http://www.markmonitor.com/resources/brandjacking_index.php (accessed June 1, 2012).

77. Online at: http://www.brandchannel.com/features_effect.asp?pf_id=187. (accessed July 1, 2013).

78. Online at: http://www.dnatechnologies.com/news/ (accessed July 1, 2013).

79. Online at: http://www.marketingweek.co.uk/analysis/cover-stories/the-new-copycats/3030320.article (accessed July 2, 2013).

80. Online at: http://legal-dictionary.thefreedictionary.com/Counterfeiting.

SUSTAINABILITY IN INNOVATION

MARIUS CLAUDY[*] AND ROSANNA GARCIA

Learning Objectives

In this chapter, we will address the following questions:

1. What drives companies to develop sustainable products and services?
2. What makes a product or service sustainable?
3. What are the levels of eco-efficiency in NP design?
4. What are sustainable product systems, including the cradle-to-cradle strategy?
5. What different product–service systems strategies can a firm take?

---

[*] Marius Claudy is associate professor of marketing at University College in Dublin, Ireland. His PhD is from the Dublin Institute of Technology. Dr. Claudy's research is in marketing of innovations and sustainability in the innovation process.

Sustainability Efforts Decompose at Frito-Lay[1]

In 1999, manufacturer of the popular SunChips snack food, Frito-Lay (a wholly owned subsidiary of PepsiCo), set a corporate-wide goal to become more sustainable at its manufacturing facilities by reducing water usage by 50%, natural gas by 30%, and electricity by 25%. To tout this green initiative, the company ran an ad campaign announcing its facility was utilizing solar power to make SunChips. In a billboard campaign, the letters spelling out the SunChips brand name were placed as a cutout on top of the signs upside down and backward. When the sun came out, the SunChips name would appear as a shadow across the top of the signs.

Frito-Lay didn't want its sustainability efforts limited to manufacturing. In April of 2009, the company announced that by Earth Day 2010, it would begin selling its snacks in fully compostable packaging. In a press release, Frito Lay reported, "This month, the SunChips brand is taking the first step toward this transformational packaging. The outer layer of packaging on 10½ oz.-size SunChips snacks bags will be made with a compostable, plant-based renewable material, polylactic acid (PLA). By Earth Day 2010, PepsiCo's Frito-Lay North America division plans to roll out a package for its SunChips snacks where all layers are made from PLA material so the package is 100% compostable [so] it will fully decompose in about 14 weeks when placed in a hot, active compost pile or bin."[2] Frito-Lay utilized an open innovation strategy by partnering with NatureWorks, the innovator of Ingeo™, a biopolymer made from corn-based sources. Green-product enthusiasts eagerly awaited the introduction of this sustainable packaging.

Yet, the marketplace acceptance of the new packaging was less than enthusiastic. Shortly after launch of the fully decomposable packaging, year-on-year sales of SunChips started to plummet. According to the SymphonyIRI group, a Chicago market research firm that tracks sales at retailers, SunChips sales declined more than 11% over a 52-week period. The reason for the drastic decrease was that consumers found the new bag to be too "noisy." The noise of the bag—due to the molecular structure of Ingeo, which made the bag more rigid—registered at more than 100 decibels, where a lawn mower registers 90 decibels and a motorcycle 95 decibels. Apparently, consumers liked to be secretive about their snacking habits and the noise drew attention to their indulgences. Public opinion was fervent as there was even an active Facebook group named "Sorry But I Can't Hear You Over This SunChips Bag" with more than 44,000 friends. Within 18 months after launching its biodegradable bag, Frito-Lay discontinued the use of the Ingeo material from the packages of five of six SunChips flavors, returning to its former nonrecyclable, quieter bags. This experience showed that while consumers say they want companies to be more environmentally conscious, they still want products to be convenient, predictable, and consumer-friendly, and will shun products that don't meet these characteristics.

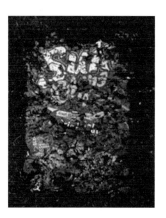

Introduction

In 1954, the management guru, Peter Drucker, famously stated that the business enterprise has two—and only two—basic functions: marketing and innovation. Innovation, as an organizational function is the provision of better and more economic goods and services. In today's marketplace, simply providing *more economic* (i.e., faster, cheaper, bigger) products and services is no longer enough. A steadily growing number of companies incorporate *social* and *environmental* objectives into the new product development process. Global corporations like General Electrics (GE), Walmart, Nike, Unilever, and Tiffany & Co., to name just a few, have launched extensive sustainability initiatives. The aim is to enhance their value proposition by (a) achieving economic goals (e.g., sales, profit, or market share), (b) helping to restore ecosystems, and (c) decreasing global poverty. These three key elements of sustainability, *people, planet,* and *profit,* are commonly referred to as the **triple-bottom line,**[3] and are illustrated in Figure 12.1.

New product development is directly linked to sustainability; both are oriented toward change and the future. Sustainability is concerned with the well-being of the future. Product innovation is concerned with creating new products and services that

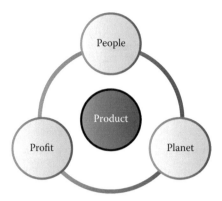

Figure 12.1 The triple bottom line and the new product development process.

generate value only if they fit in this future. The *Harvard Business Review* recently referred to sustainability as an emerging megatrend[4] and the mother lode of organizational and technological innovation.[5] Market research data show that across industries, environmental sustainability is now a key driving force of product innovation. For example, launches of green products in the United States doubled between 2007 and 2008, and continue to grow exponentially.[6] Further, a global survey found that about 50% of executives are taking climate change issues into consideration when developing new products.[7]

This sustainability focus has been recognized as a sixth wave of innovation following the fifth wave of digital networks, as shown in Figure 12.2. Behind the sixth wave of innovation lie two interconnected sources of pressure: constraints of finite resources and increasing expectations from stakeholders. These interconnected trends

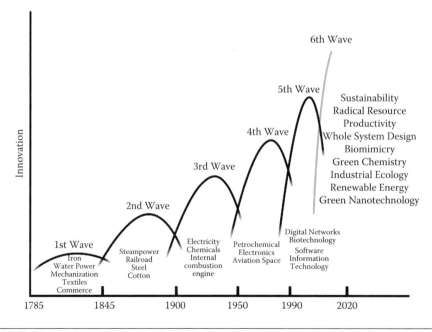

Figure 12.2 Waves of innovation. (From Endnote 8. With permission.)

will increasingly drive the way businesses create value and ultimately establish long-term competitive advantage. It is these trends and their impact on sustainable new product development that we will examine in this chapter.

Finite Resources

Planet Earth holds finite supplies of natural resources. Increasing prices for oil and minerals, scarcity of water, and elevated levels of carbon pollution not only threaten society as a whole, but also endanger companies' daily business operations and global competitiveness. Interrelated societal changes like growing poverty, aging populations, or stressed healthcare systems compound these issues. *Harvard Business Review*[9] stated that "most executives know that how they respond to the challenges of sustainability will profoundly affect the competitiveness—and perhaps even the survival—of their organizations." Environmental and social problems will not simply disappear, and companies that respond to the challenges most effectively will ultimately gain competitive advantage.

Stakeholder Pressure and Growing Transparency

Companies face increasing pressure from key stakeholders. Worldwide, governments continue to introduce a growing number of environmental regulations. Consumers are more conscious and demand greener and fairer products. Business partners are requiring suppliers to follow sustainability practices. Employees want a sense of purpose and don't want to work for companies who pollute the environment or mistreat their workers. In the past decade, Coca-Cola, for example, faced protests from consumers in India over its excessive water consumption, was criticized for using substances that were harmful for the ozone layer, and was penalized by the European Union for failing quality tests for bottled water.

Transparency is becoming a competitive advantage. Consumers and other stakeholders want to know where the "stuff" they buy comes from, what it contains, and where it's going at the end of the product life cycle. New information technologies and communication tools, such as the Internet, enable people to obtain accurate information about a product's life cycle, i.e., resources extraction, production, distribution, usage, and disposal. Using Sourcemap®, the upstream (with suppliers) and downstream (with customers) flow of product development can be mapped, as shown in Figure 12.3, using the laptop computer.

The Business Case for Sustainability

The relationship between corporate economic success and sustainability is not always straightforward and there is no best strategy or approach that companies can adopt.

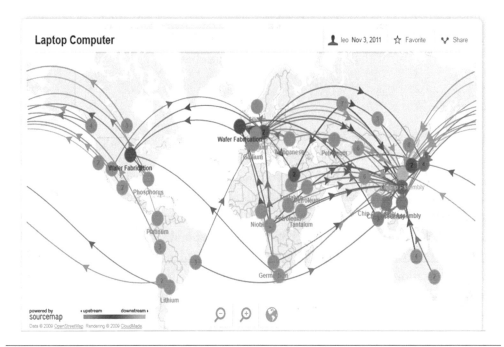

Figure 12.3 Sources of inputs for a laptop computer. (From Endnote 10. With permission.)

However, there are some general drivers of a business case for sustainability that have a direct influence on companies' economic performance.

Costs Reduction

Sustainability offers an opportunity to reduce costs by focusing on using fewer resources, including raw materials, toxic substances, and/or water. By designing and producing less resource-intensive products and services, companies not only save materials, but also can minimize waste and pollution throughout the product life cycle. Walmart, for example, reengineered the design of milk containers to more rectangular, stackable jugs that were easier to pack and transport. Reportedly, this halved labor, cut water use by up to 70%, and curbed the number of delivery trips, resulting in savings of fuel and CO_2 emissions. Walmart was able to reduce prices by $0.40 a gallon, improving its competitive position in the market.

Complying with Regulation

A key motivation for companies to develop greener products is to comply with new environmental regulations. In recent decades, numerous new regulations and environmental legislations have been introduced on the national level and increasingly on regional (e.g., European Union (EU)) or international levels (e.g., Kyoto Protocol). By September 2010, the EU had implemented 681 acts in relation to the environment alone, with many of them having direct or indirect implications for companies across different industries. Hewlett Packard (HP) anticipated the ban on lead solders, and by the time the EU had introduced new regulations in 2006, HP already had already developed innovative solutions, giving it an advantage over its competitors.

Reputation and Brand Value

Another reason for businesses to engage in sustainability is to increase its brand value. This is particularly true in a digital age, where companies' negative actions can easily become the target of bloggers, or social media campaigns on Twitter or Facebook. On the other hand, companies who are authentic and genuine in their sustainability efforts can boost their reputation and increase brand value. For example, the Swiss government recently awarded the rail technology leader, Bombardier, the "Tell Award" for leading the way in economically sustainable and environmentally responsible transport solutions. The award sent a strong signal to this and other industries to focus on sustainability efforts.[11]

Differentiation

Sustainability as a differentiation strategy can help to increase customer loyalty, allowing businesses to charge premium prices. Marks and Spencer (M&S), for example, has embarked on a journey to become the world's most sustainable major retailer. In its so-called Plan A ("Because there is no Plan B"),[12] M&S dedicates itself to achieve 180 sustainability goals and become the most sustainable retailer by 2015. However, the U.K.-based retailer not only responds to changing consumer preferences, but tries to engage customers in behavioral change. For example, by partnering with major energy providers, M&S created incentive schemes that offer shopping vouchers for all new customers who reduce their energy usage by 10% in the first year.

Attract and Retain Employees

Companies' employees are also a key source of competitive advantage. Businesses that are actively working to become more sustainable often seem to focus more on creating healthier, more stimulating, and more productive work environments. As a result, they might find it easier to attract and retain talented employees.

Attract Capital Investment

According to the *Harvard Business Review*,[13] "venture investing in clean tech reached a nearly $9 billion annual run rate in 2008 and shows signs of growing again after a slowdown in 2009. The flow of private-sector investment into the tech marketplace has been estimated at more than $200 billion a year—with fast growth not just in the United States and Europe, but in China, India, and the developing world. And G20 governments (G20 are the 20 countries with the highest gross domestic product) have earmarked some $400 billion (to their $2.6 trillion) in stimulus funds for clean tech and sustainability programs." There also is a growth in socially responsible and ecological funds as well as new industry standards like the Dow Jones Sustainability Index, FTSE4Good, or the Equator Principles.

Developing Sustainability Strategies

Principles Underlying Sustainability Product Design

The UN's Brundtland Report of 1987 defined *sustainable* as "development that meets the needs of the present without compromising the ability of future generations to meet their own needs." Karl-Henrik Roberts, initiator of *The Natural Step*, defines sustainability as human societies' ability to continue indefinitely. Or, to put it differently, sustainability is the ability of a system to maintain or renew itself perpetually. Sustainability can take on an ecological perspective, in regards to perpetuity of natural resources; a social perspective, in regards to the perpetuity of societies; or a business perspective, in regards to perpetuity of the organization. Yet, environmental, social, and economic dimensions are entwined and interdependent, or to put it in the words of the biologist Barry Commoner: "Everything is connected to everything else."[14] Thus, sustainability strategies must take a systemic perspective. This means that all aspects of the environment, society, and the organization must be considered in planning for sustainable innovation.[15]

Grasping the full complexity of a production system and its interaction with the social and natural environment can be an overwhelmingly complex task. Managers thus need heuristics or principles, which guide decisions and ultimately allow for the development of sustainable products and services. These principles should guide a new product manager in the development of new products and services. We outline these principles in Table 12.1 and discuss them below.

1st Principle: Reducing Use of Finite Materials   The first principle focuses on reducing the use of finite resources, such as fossil fuels or heavy metals energy and water that is used

Table 12.1   Strategies for Sustainable Product Design

1ST PRINCIPLE: REDUCE USE OF FINITE MATERIALS	2ND PRINCIPLE: ELIMINATE USE OF TOXIC MATERIALS	3RD PRINCIPLE: MINIMIZE PHYSICAL DESTRUCTION	4TH PRINCIPLE: LOOK TO SOLVE SOCIAL PROBLEMS
1. Responsible manufacturing	5. Reduce toxicity	8. Design in safety for disposal	12. Fair trade
2. Resource use reduction	6. Organic farming/ production	9. Compostable	13. Think global, act local
3. Energy efficiency	7. Sustainable harvesting and mining	10. Recyclable	14. Extend product life
4. Reusable/Refillable		11. Recycled content	15. Source reduction

Source: Endnote 16.

Pepsi Reveals World's First 100% Plant-Based Bottle

"Today, PepsiCo unveiled a breakthrough in bottle technology with the announcement of the world's first petroleum-free plastic bottle in an effort to further reduce its carbon footprint by piloting production of the new bottle in 2012. The snack and soft drinks company will use raw materials such as switch grass, pine bark and corn husks, and has identified methods to create a molecular structure that is identical to polyethylene terephthalate, or PET, which is the plastic material used to produce a majority of bottles. The race to create a petroleum-free bottle was triggered in 2009 when Pepsi's rival Coca-Cola unveiled a bottle that was 30 percent plant-based. Environmentalists have been critical of petroleum-based plastic because of their high carbon emissions and for being slow to biodegrade.

Pepsi said the new bottle will eventually allow for by-products from its food businesses, including potato peels from chips and orange peels from juice to be recycled into bottles. PepsiCo Chairman and CEO, Indra Nooyi adds the company will ultimately have "a sustainable business model that we believe brings to life the essence of Performance with Purpose."

Figure 12.4   Pepsi's eco-bottle. *Source*: Endnote 17.

in the production of products and services. What does this mean for developing new products and services? Further, it implies using alternative recyclable or biodegradable nonfinite materials. Frito-Lay, with the installation of solar panels on its SunChip manufacturing plants, was reducing the use of fossil fuel. PepsiCo recently announced the launch of the first petroleum-free plastic bottle, thus reducing noxious emissions and dependence on oil (Figure 12.4). However, scientists have already warned that plant-based plastics might take more energy to produce and that environmental effects from pesticides and chemicals used in agricultural production need to be taken into consideration before calling it a "sustainable alternative." This leads to the second principle: reducing the use of toxic materials.

2nd Principle: Eliminate Use of Toxic Materials   The systematic increase of manmade substances is poisoning the natural system. Studies show that about 750,000 synthetic chemicals are on the market; many that cannot be broken down by nature, and, thus, accumulate in the Earth's ecosphere and remain there for hundreds of years. There are solutions, and companies have begun to develop sustainable product alternatives that contain no or significantly less toxic chemicals. A great example for sustainable

Professional Wet Cleaning Explained

"The first clearly identifiable green technology to emerge in the garment care industry was a water-based alternative known as professional wet cleaning. Professional wet cleaning industrialized the practice of hand-laundering delicates by using computer-controlled washers and dryers, specially formulated detergents, and specialized finishing equipment to create a cost-effective alternative to dry cleaning. Professional wet cleaning has shown to be an energy efficient, non-toxic technology that eliminates hazardous air emissions, hazardous waste production, and the potential for soil and groundwater contamination. From a performance standpoint, dry cleaners who have switched to professional wet cleaning have been able to maintain their level of service and customer base while increasing profits."

Figure 12.5   Professional wet cleaning to minimize toxic materials. *Source:* Endnote 18.

product innovation is so-called *wet clean washers*, which were developed in response to stricter regulations and bans on carcinogenic chemicals used in traditional dry-cleaning services (Figure 12.5).

Another industry that has been active in eliminating the use of toxic materials is agriculture, with the growing trend toward organic and sustainable farming. The U.S. Department of Agriculture defines organic production as "a production system that … by integrating cultural, biological, and mechanical practices that foster cycling of resources, promote ecological balance, and conserve biodiversity."[19] According to Organic Monitor, "global organic sales reached $54.9 billion in 2009, up from $50.9 billion in 2008. The countries with the largest markets are the United States, Germany, and France."[20] Denmark, Switzerland, and Austria have the highest per capita consumption of organic foods.

The reduction of toxic material in production has not been limited to the cleaning and agricultural industries. An assessment of the Toxic Use Reduction Act found that from 1990 to 2001 more than 600 companies worldwide took reduction measures resulting in:[21] 45% reduction in chemical use, 69% reduction in toxic chemical byproducts, 92% reduction in releases of toxics, while resulting in an overall increase in production of 17%.[22]

3rd Principle: Minimize Physical Destruction   The third principle refers to systematic physical destruction (harvesting and manipulation) of the ecosystem. Frito-Lay, with its SunChip compostable bag, made an effort at minimizing the physical destruction of our planet by encouraging disposal of bags in compost bins instead of dumps. Human economic activity has led to an accelerating decline of productive surfaces and biodiversity, often with unforeseeable effects. It is estimated that since the 1950s about half of the Earth's mature tropical forests have been cleared, and some estimates show that, unless drastic measures are taken, by 2030 only 10% of healthy forests could remain.[23] Innovation is vital to dematerialize products and services, and to effectively minimize waste. A great example of a service innovation is RecycleBank, a company that incentivizes households to recycle by awarding them point credits they can use in over a thousand stores. RecycleBank is not only a profitable enterprise, but it provides value to customers, while helping to reduce wastage of natural resource. (See the Appendix at the end of this chapter for a link to hear an interview with Ron Gonan, CEO of RecycleBank.)

4th Principle: Look to Solve Social Problems   The fourth principle "recognizes that humans, when struggling to meet their basic needs, will understandably ignore the needs of nature rather than see their families suffer or perish."[24] The human population is growing exponentially and with it the demand for goods and services. The world population is expected to reach a staggering 9 billion in 2050.[25] Therefore, in order to protect the natural environment, we need to make sure that people's basic needs are met through innovative products. Unilever's Shakti entrepreneurs, who are women in rural India who sell UniLever products to local villages, are one example of a company using both business model innovations along with product innovation (single-use packets) to look to solve social problems.[26] This approach is often referred to as social innovations. Social innovations bring in the end user as a partner in designing products that have minimal impact on the environment.

In summary, managers often find it difficult to translate the notion of sustainability into the new product development process. In order to cope with the complexity, managers can use the four principles of sustainability as guidelines in the new product development process. That is, to develop new products and services that:

- reduce the use of nonrenewable materials, such as petroleum, water, energy, metals, minerals, etc.;
- eliminate the use of toxic materials;
- minimize the degradation of biodiversity and ecosystems; and
- seek solutions to solve social issues, such as poverty, pollution, deforestation, and other social and environmental issues.

Four Paradigms for Sustainable New Product Development

Designing products for eco-efficiency increases the longevity of products, utilizes recyclable and nontoxic materials, or reduces the amount of pollutants and waste emitted during production, distribution, usage, and disposal.[27] Eco-efficiency is about creating more value with less impact. The ecological impact of eco-efficiency improvements is often measured as a factor increase in resource productivity. For example, a Factor 4 gives a fourfold increase in resource productivity,[28] thereby giving the company some competitive advantage (via cost leadership or differentiation) over rival businesses. Figure 12.6 demonstrates four paradigms that sustainable new products can embody: *product improvement*, *product redesign*, *function innovation*, and *system innovation*. It is important to note that eco-efficiency can occur at all stages of a product's physical life cycle (i.e., at the extraction, manufacturing, distribution, usage, and disposal stages).

Product Improvement and Redesign

Products at the low end of the factor scale, less than factor 4, are considered to be products of low eco-efficiency. In this range, sustainable products are generally seen as incremental innovations and can be broadly categorized into two strategies:[29]

1. **Product improvement:** Partial changes and upgrades of certain aspects of a product. The product itself and production techniques remain the same. For example, changes in the type of coolant or substances used in a television set to improve its environmental performance do not change the product architecture.

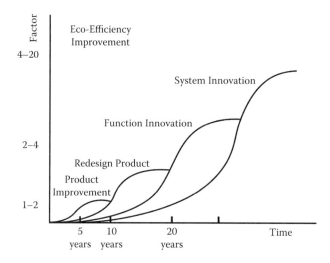

Figure 12.6 Four levels of innovation in the design path towards sustainability. (From Endnote 30. With permission.)

2. **Product redesign:** Typology of the product remains unchanged, but components are improved or replaced in order to facilitate disassembly, recycling, re-use procedures, or energy use reduction. Examples include energy-saving systems in a television set, the introduction of recyclable plastic parts, and the redesign of a chip bag to be compostable, such as the Frito-Lay SunChip example.

Eco-efficiency has sometimes been criticized for lacking **eco-effectiveness**. For example, innovative solutions can make products more energy efficient. Yet, energy savings might be neutralized by consumers' reckless usage behavior or increased material usage. A good example is an energy-saving refrigerator. Innovation has led to significant improvements in the energy efficiency of durable goods. Today's refrigerators are (on average) bigger and cost less resulting in exponential growth in the global market for refrigerators. As a result, the energy required to produce, distribute, and dispose of new refrigerators is likely to significantly outweigh the benefits resulting from improved energy-saving technologies. This is known as the *rebound-effect* (Figure 12.7).

Thus, low-level eco-efficiency might be unsustainable as it merely helps to prolong some of the potentially catastrophic consequences (e.g., climate change, loss of biodiversity, over-fishing) of our current, unsustainable actions. Many advocates of sustainability argue that Factor 4 strategies are not far-reaching enough and that it takes more radical changes in order to address true sustainability efforts. One solution to overcome the "incremental redesign" perspective of low eco-efficiency is to take a more holistic view and consider issues of consumer behavior, recyclability, or embodied-energy (i.e., energy that it takes to produce the materials and components of a new product) early in the design process. This often requires more radical approaches, such as function innovation and system innovation, which can bring about a 4 to 20 factor improvement in eco-efficiency.

Is a Top Rated Energy Saving Refrigerator a Sustainable Product?

Figure 12.7　Rebound effect.

Functional and System Innovation

Two strategies that go beyond low-level eco-efficiency and relate to the **design for sustainability** paradigm include:

1. **Function innovation:** Change is no longer confined to the existing product concept. The old television tube, for example, is replaced with other image carriers (LCD monitor instead of CRT-based monitor or replacement of the television with a light projector). Through these innovations, environmental performance in fulfilling the function can be improved by as much as a factor of 10.

2. **System innovation:** The entire sociotechnical system (i.e., the product, the production chain, associated infrastructures, and institutions) is replaced by a new system. One example of this is cloud-computing replacing mainframe data storage, potentially achieving improvements by a factor of 20. This new service can replace labor demands, physical transport, service maintenance, and end-of-life disposal of expensive equipment that can quickly become outdated.

System innovation is a more radical approach to new product development and includes innovation along the entire sociotechnical system (i.e., the product, the production chain, associated infrastructures, and institutions). New product development (NPD) managers thus become facilitators who stimulate exchange and innovation processes within a complex network of different stakeholders. To understand how a system innovation can result in a factor 20 eco-efficiency improvement, it is useful to examine the product life cycle (PLC) of an innovation through to the end-of-life (EOL). The PLC starts with the extraction, processing, and supply of the raw materials and energy needed for the product. It then covers the production of the product, its distribution, use (and, possibly, but not likely, reuse and recycling), and its ultimate disposal, typically through incineration or landfill dumping as demonstrated in Figure 12.8a. This is referred to as the **cradle-to-grave life cycle**.

A system innovation approach looks at how the PLC can become sustainable. Michael Braungart and William McDonough proposed the **cradle-to-cradle** (C2C) approach. In C2C product design, products are designed so that their end-of-life disposal is the input for a new product, as shown in Figure 12.8b. C2C aims to meet humans' needs while minimizing the negative consequences of production and consumption activities. The basic idea of C2C is that all materials are either fed back into the "natural" cycle (i.e., biological nutrients, such as biodegradable products) or into the "technical" cycle (i.e., technical nutrients, such as metals or polymers). A number of companies have begun to embrace this concept of making products that are completely reusable, recyclable, or compostable. Puma CEO, Franz Koch, announced in 2011 that the company was looking at closing the loop on its products by following the cradle-to-cradle product design philosophy:[31]

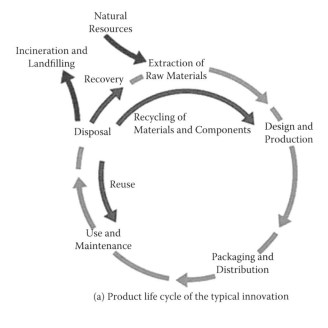

(a) Product life cycle of the typical innovation

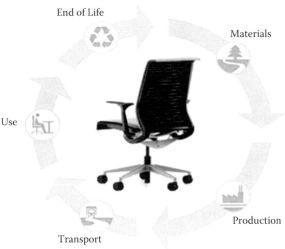

(b) Cradle-to-cradle product life cycle

Figure 12.8 Product life cycles. (a) Product life cycle of the typical innovation; (b) cradle-to-cradle product life cycle.

We are confident that in the near future we will be able to bring the first shoes, T-shirts, and bags, that are either compostable or recyclable, to the market," Koch told the German business magazine *Wirtschaftswoche*. On the principle of the "cradle-to-cradle" design, he explained, "It follows two circuits, the technical and the biological: I can use old shoes to make new ones or something completely different, such as car tyres," said Koch. "In the biological cycle, I can make shoes and shirts that are compostable so I can shred them and bury them in the back garden. We are working on products that meet these two criteria."

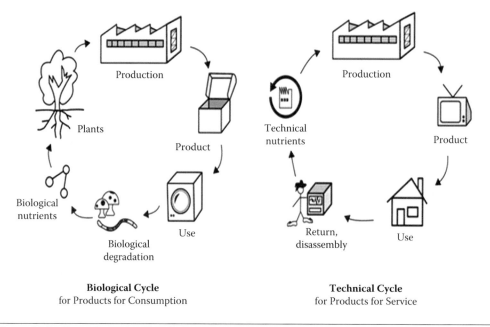

Biological Cycle
for Products for Consumption

Technical Cycle
for Products for Service

Figure 12.9 The nutrient cycles. (From Endnote 32.)

As noted by CEO Koch, all materials used in C2C product/service production fall into one of two categories: technical or biological nutrients. As shown in Figure 12.9, **technical nutrients** are inorganic or synthetic materials, such as plastics and metals, which can be used many times over without any loss in quality, theoretically staying in a continuous cycle of reuse. The goal is to reuse materials over and over again instead of "downcycling"[33] them into lesser products, which ultimately becomes landfill waste. **Biological nutrients** are organic materials that can decompose into the natural environment as soil, water, or other natural elements providing nourishment for bacteria and microbiological life. The ability of biological nutrients to not have a negative impact on the environment is dependent on the ecology of the region; for example, organic material from one country or landmass may be harmful to the ecology of another country or landmass.

A company that has been adhering to the idea of CDC design is the company, Preserve, with its Gimme 5 program. The program collects #5 plastics, typically used for yogurt and other food products, for recycling, as demonstrated in Figure 12.10. Even though many cities collect #5 plastics, this type of plastics is often put into landfills because it is difficult to make into high-quality plastic again. Preserve uses the collected plastics to create toothbrushes and other kitchen containers, which can be recycled back through Preserve at the end of its life, thus closing the loop on the PLC using C2C techniques.

C2C product design allows for only three types of products:

Figure 12.10 Preserve's Gimme 5 C2C approach to recycling #5 plastics.

Biodegradable products: The first category comprises products that can be literally consumed or are made of materials that are 100% biodegradable and can thus be fed back as "nutrients" into the natural cycle.

Recyclable products: The second group is durables, made from technical nutrients, which, after they provide a service to the user, get recycled and fed back into the technical cycle. These products always remain the responsibility of the maker and, therefore, can only be rented or licensed to consumers. However, it is important that the recyclability of products is considered at the very early design stage, so that individual components and materials can later be effectively separated, without losing quality.

Unmarketable products: The third category consists of unmarketable products that are made from toxic materials that should not be sold at all. Unmarketable products cannot be consumed in a sustainable way and, thus, need to be replaced completely.

However, there has been much criticism of the approach because of the practicability of the concept. C2C products must be designed for disassembly so that all components (product chunks) reenter the cycle either as a biological nutrient or a technical nutrient. But, not all materials can be used indefinitely. Glass and metals (technical nutrients) can be recycled over and over, theoretically staying out of the landfill. Paper (biological nutrient) also can be recycled many times over. Most plastic, however, degrades significantly in the recycling process and cannot be recycled in such a "closed-loop" manner at the same quality as the initial product. For C2C to become a reality, the plastics industry must become a partner in C2C design by innovating recaptured plastic that can be recycling back into original quality packaging. This need for cooperation between partners is what makes C2C design a systems innovation. All entities of the supply chain from suppliers to manufacturer to the end user must be involved in the reuse/recycling effort. Should any one of the partners fail to do their part, C2C cannot be successful. It has been reported that only 28% of U.S. consumers recycles on a regular basis, with some EU countries recycling only 10%, and only a handful of countries, such as Austria, have recycling efforts of approximately 60%. If consumers are not willing to reuse and recycle, C2C closed-loop product design is meaningless. Companies must make it easy for consumers to be a part of the C2C process by making it easy to recycle products, such as Preserve is doing with #5 plastic.

System Innovation and the Role of Services

Incorporating system innovation and, particularly, service innovation into the NPD, therefore, calls for a different approach. It requires NPD managers to:

- take a systemic approach to innovation by including all stakeholders in the NPD process;
- integrate their supply chains in the NPD process;
- see consumers as value co-creators in the provision of services;
- link profits to sustainable services instead of product revenues;
- increase awareness of sustainability issues in order to build value in new service offerings; and
- focus sustainability on back-stage, daily operations, as well as front-stage, customer touchpoints.

System innovation also places greater emphasis on services. Services already form an increasing proportion of the world economy and continue to change the way businesses interact with consumers and stakeholders in the marketplace. In marketing, this trend has been formalized in a new Service-Dominant (S-D) Logic,[34] which argues that wealth and value creation are increasingly dependent on *operant* (i.e., technologies, knowledge, and skills) as opposed to *operand* resources (i.e., raw materials).[35] Recently firms have recognized the importance of services as a sustainability strategy through three different types of strategies: **product-orientated services, use services, and results-focused services**.

Product–Oriented Services

Products are increasingly converging with services as part of a total system that includes how the product-service is specified, delivered, explained, installed, repaired, replaced, and recycled.[36] In the product-oriented services model, the end user retains ownership of the product while the firm takes over stewardship. Product services can extend the life of a product through maintenance services, repair, warranties/guarantees, educational programs, upgrading (i.e., modernizing by changing parts and components), and product trade-ins/take-backs (i.e., enabling users to return and recycle products or components). Product-oriented services are well suited to products that are complex, need technical expertise, require regular maintenance, or require supporting infrastructures. Automatic maintenance services for industrial equipment and even durable white-good home appliances (i.e., washing machines, refrigerators) can increase the longevity of these products. Despite the loss of sales of the product, monthly services fee can generate a new source of revenue.

User-Oriented Services

Sometimes consumers have no desire to own a product. User-oriented services focus on providing a service to consumers instead of selling a product. Examples include product renting, product leasing (long-term rental), or product pooling (simultaneous usage of a product). By renting or leasing the product, the consumer receives the benefit of use without the obligations associated with ownership, such as maintenance and disposal. Bicycle-sharing and car-sharing programs have become ubiquitous worldwide. Zipcar, a U.S. and U.K. car-sharing program, rents cars billable by the hour or day. These types of sharing programs result in diminished demand for the production of durable goods. Eco-efficiency is reached through the product's high-use intensity and through the hourly payment system, which can lead end users to be more conscientious of usage.

Results–Focused Services

With results-focused services, the product is owned and operated by the service provider. In this model, providers work toward satisfying customer needs (the end result), while reducing the need for material products altogether. Public transportation replaces the need for a car, online banking and automated bill pay replace brick-and-mortar bank institutions, and cloud computing replaces the need for IT equipment. Pay-per-services, such as billing copiers by prints made instead of selling the copier, is another example of results-focused services. The product-services itself takes on a minor role, and instead the focus is put on satisfying the consumers' needs.[37] Revenues are based on efficiently offering a service with high satisfaction so that it replaces the need for a material product; this is referred to as *dematerialization*. The offering of a *result* instead of a prespecified product makes it possible for sustainable solutions (i.e., low material and energy consumption) to be included from the beginning.[38]

Marketing Sustainable Products[39]

Now that we have an understanding of how sustainable design should be considered in product development, we should look at the important role of the marketing team when promoting new products. The consumer must be an integral part of the product life cycle in order to move products from cradle-to-grave toward C2C. Without the consumer doing their part at the end of life of a product, the discarded product cannot become a biological or technical nutrient.

A model of a green marketing-mix should, of course, contain all four Ps (product, price, place, promotion). In their book, *Sustainability Marketing*, Belz and Peattie go one step further than the four Ps in terms of not just marketing, but operating "green" by transforming the four Ps into the four Cs: customer solutions, customer cost, communication, and convenience for the customer:

- *Customer solutions*: Present solutions to customer problems. Results-oriented product design means knowing your customers and their needs to offer products and services that satisfy these needs, but that take into account social, as well as environmental aspects.
- *Customer Cost*: Consider psychological, social, and environmental costs of obtaining, using, and disposing of a product into the cost for the consumer. Lowering these costs will encourage more sustainability efforts on the consumer-partner.
- *Communication*: Build an interactive dialogue with customers to foster trust and credibility. With the continued focus of firms on customer relationship management, these conversations create better relationships.
- *Convenience*: Design products and services that meet consumers' needs and are easy and convenient to access and use.[40]

An important challenge facing marketers is to identify which consumers are willing to pay more for environmentally friendly products. It is apparent that an enhanced knowledge of the profile of this segment of consumers would be extremely useful. Research[40] results portray "green consumers" as highly socially conscious females, premiddle aged, with a high level of education, and above average socioeconomic status.[41] Lifestyles of Health and Sustainability (LOHAS) has been used to label an integrated, rapidly growing market for goods and services that appeal to consumers whose sense of environmental and social responsibility influences their purchase decisions. The Natural Marketing Institute's (NMI) estimates the LOHAS consumer market of products and services to be US$209 billion.[42] NMI defines five types of "green" consumer segments:

1. *LOHAS*: Active environmental stewards dedicated to personal and planetary health. These are the heaviest purchasers of green and socially responsible products and the early adapters who influence others heavily.

2. **Naturalites**: Motivated primarily by personal health considerations. They tend to purchase more LOHAS consumable products versus durable items.

3. **Drifters**: Drifters follow "green" trends when it is easy and affordable. They are currently quite engaged in green purchasing behaviors when it is convenient.

4. **Conventionals**: Pragmatists who embrace LOHAS behavior when they believe they can make a difference, but are primarily focused on being very careful with their resources and doing the "right" thing because it will save them money.

5. **Unconcerned**: Either unaware or unconcerned about the environment and societal issues mainly because they do not have the time or the means; these consumers are largely focused on "getting by."

The level of green marketing a company takes is heavily dependent upon its sustainable product strategy. Grundey and Zaharia[43] lay out three approaches a firm can take towards green marketing: tactical, quasistrategic, and strategic greening, as shown in Table 12.2. Companies that operate at the tactical level are often accused of "greenwashing." The primary objective of greenwashing is to provide consumers with the feeling that the organization is taking the necessary steps to responsibly manage its ecological footprint. In reality, the company may be doing very little that is environmentally beneficial.[44] Since Greenwashing has become a central feature of debates about marketing communications and sustainability with numerous campaigns and laws developed in an attempt to reduce or curb it.[45] Terrachoice, an environmental and consulting firm, has established the seven sins of greenwashing, as shown in Table 12.3. For a firm to be a believable partner in the sustainability system, innovative companies must avoid these sins of greenwashing.

Eco Labels

There is an abundance of labels that a firm can put on its products to indicate its eco-friendliness. In recent years, there has been an explosion in the numbers of different eco-labeling programs worldwide and across business sectors, with many labels covering social, ethical, and safety issues as well as ecological. Of as writing, there were 437 eco-labels in 197 countries and in 25 business sectors.[46] It's no wonder that brand awareness of most eco-labels remains low. How many can you name? Eco-labels are meant to add credibility to messages on the environmental benefits of brands. Certification by a third party allows a brand to carry a particular eco-label.[47]

In the United States, the two most commonly recognized labels are the Energy Star and the "chasing arrows," as shown in Figure 12.11a. What is not known by many consumers is that the chasing arrows recycling label is not meant to imply the item is recyclable, but as an indicator of what type of plastics were used in the packaging. As previously noted, #5 plastics are difficult to recycle. In Europe, the most

Table 12.2   Green Marketing Activities

	TACTICAL GREENING	QUASISTRATEGIC GREENING	STRATEGIC GREENING
Targeting	Ad mentioning green features are run in green-focused media.	A firm develops a green brand in addition to its other brands.	A firm launches a new strategic business unit (SBU) aimed at the green market.
Green design	A firm switches from one raw material supplier to another with more eco-friendly processes.	Life-cycle analysis is incorporated into the eco design process to minimize eco-harm.	For example, Fuji Xerox develops its Green Wrap paper to be more eco friendly from the ground up.
Green positioning	For example, a mining company runs a public relations campaign to highlight its green aspects and practices.	For example, British Petroleum (BP) Amoco redesigns its logo to a sun-based emblem to reflect its view to a hydrogen/solar-based future of the energy industry.	For example, the BODY SHOP pursues environmental and social change improvements and encourages its consumers to do so as well.
Green pricing	Cost savings due to existing energy-efficiency features are highlighted for a product.	For example, a water company switches its pricing policy from a flat monthly rate to a per-unit-of-water-used basis.	A company rents its products rather than selling; consumers now pay only for use of the product.
Green logistics	A firm changes to a more concentrated detergent, which lowers manufacturing costs.	Packaging minimization is incorporated as a part of a firm's manufacturing review process.	A reverse logistics system is put into place by Fuji Xerox to reprocess and remanufacture copiers.
Marketing waste	A firm improves the efficiency of its manufacturing process, which lowers its waste output.	For example, TELSTRA (a phone company) has internal processes so that old telephone directories (waste) are collected and turned into cat litter products by other companies.	For example, a Queensland sugarcane facility is rebuilt to be co-generation based, using sugarcane waste to power the operation.
Green promotion	An oil company runs a PR campaign to highlight its green practices in order to counter an oil spill getting bad press coverage.	A company sets a policy that realistic product eco-benefits should always be mentioned in promotional materials.	As a part of its philosophy, the BODY SHOP co-promotes one or more social-eco campaigns each year with in-shop and promotional materials.
Green alliance	A company funds a competition (one-off basis) run by an environmental group to heighten community awareness on storm water quality issues.	For example, SOUTHCORP (a wine producer) forms a long-term alliance with the Australian Conservation Foundation to help combat land-salinity issues.	A company invites a representative of an environmental group to join its board of directors.

Source: Endnote 48. With permission.

Table 12.3 Seven Sins of Greenwashing[49]

Sin of the Hidden Trade-off: A claim suggesting that a product is "green" based on a narrow set of attributes without attention to other important environmental issues. Paper, for example, is not necessarily environmentally-preferable just because it comes from a sustainably-harvested forest. Other important environmental issues in the paper-making process, such as greenhouse gas emissions or chlorine use in bleaching, may be equally important.

Sin of No Proof: An environmental claim that cannot be substantiated by easily accessible supporting information or by a reliable third-party certification. Common examples are facial tissues or toilet tissue products that claim various percentages of post-consumer recycled content without providing evidence.

Sin of Vagueness: A claim that is so poorly defined or broad that its real meaning is likely to be misunderstood by the consumer. "All-natural" is an example. Arsenic, uranium, mercury, and formaldehyde are all naturally occurring, and poisonous. 'All natural' isn't necessarily 'green'.

Sin of Worshipping False Labels: A product that, through either words or images, gives the impression of third-party endorsement where no such endorsement exists; fake labels, in other words.

Table 12.3 (Continued)

commonly recognized label is the EU Flower, as shown in Figure 12.11b. There are more than 300 different types of eco-labels worldwide. Many consumers are baffled by the different types of labels and, in the coming years, companies and governmental agencies will be forced to put into place standards regarding usage of these labels and certification.

Table 12.3 (Continued)

Sin of Irrelevance: An environmental claim that may be truthful but is unimportant or unhelpful for consumers seeking environmentally preferable products. "CFC-free" is a common example, since it is a frequent claim despite the fact that CFCs are banned by law.

Sin of Lesser of Two Evils: A calim that may be true within the product category, but that risks distracting the consumer from the greater environmental impacts of the category as a whole. Organic cigarettes could be an example of this Sin, as might the fuel-efficient sport-utility vehilce.

Sin of Fibbing: Environmental claims that are simply false. The most common examples were products falsely claiming to be Energy Star certified or registered.

Source: Endnote 50. With permission.

Chapter Summary

In this chapter, we introduced the types of sustainability strategies a firm can take. Designing environmentally friendly products is not any easy endeavor and firms must be dedicated to the green strategy to be successful at it. Four principles of sustainable product design were introduced: reducing use of finite materials, elimination of toxic materials, minimization of physical destruction, and looking to solve social problems. Most firms currently develop products for cradle-to-grave; however, advocates of sustainability encourage a cradle-to-cradle approach where all components of a product become biological or technical nutrients that become the materials for other new products. Longevity of the environment and, subsequently of the firm, requires embracing sustainability as an important strategy within the company. New product development strategies must begin to include sustainability issues in product design.

(a) US Energy star label and 'chasing arrows' recycling label

(b) EU Flower label

Figure 12.11   Popular labels. (a) U.S. Energy Star label and "Chasing arrows" recycling label; (b) EU flower label.

Glossary

1st principle of sustainability—Reducing use of finite materials: Reducing the use of finite resources, such as fossil fuels or heavy metals energy and water, that are used in the production of products and services.

2nd principle—Eliminate use of toxic materials: The systematic decrease of man-made substances used in the production of goods and services.

3rd principle—Minimize physical destruction: The systematic decrease in physical destruction (harvesting and manipulation) of our ecosystem.

4th principle—Look to solve social problems: Designing sustainable products that seek solutions to solve social issues, such as poverty, pollution, deforestation, etc. Sometimes referred to as social innovations.

Base of the pyramid: World markets, often referred to as survival markets, where individual annual income is less than $3,260.

Biodegradable products: Products that can be literally consumed or are made of materials that are 100% biodegradable and, thus, can be fed back as nutrients into the natural cycle.

Biological nutrients: Organic materials that can decompose into the natural environment as soil, water, or other natural elements providing nourishment for bacteria and microbiological life.

Conventionals: Category of consumers who embrace LOHAS behavior when they believe they can make a difference, but are primarily focused on being very careful with their resources and doing the "right" thing because it will save them money.

Cradle-to-cradle: Designing for sustainability where the waste (of one product) is the food (materials for another product); see also eco-effectiveness.

Cradle-to-grave life cycle: Life cycle of a product or service from its production, its distribution, use (and possibly, but not likely, reuse and recycling), and its ultimate disposal, typically through incineration or landfill dumping.

Eco-effectiveness: Looking at the life cycle of a product/service to determine its total environmental impact from its creation-use-disposal. The central design principle of eco-effectiveness is waste equals food.

Eco-efficiency: Management strategy of doing more with less through increasing product or service value; optimizing use of resources, reducing environmental impact at least to a level in line with the Earth's carrying capacity.

Dematerialization: Offering a service with high satisfaction so that it replaces the need for a material product.

Drifters: Category of consumers that follow "green" trends when it is easy and affordable. They are currently quite engaged in green purchasing behaviors when it is convenient.

Finite Resource: There are limited supplies of natural resources.

Function innovation: A change made within a product/service in making it sustainable so that it still has the same result, but the way in which it comes to that result is changed, for example, car sharing still gives individuals car ownership, it is just shared among a group of people.

LOHAS (lifestyle of health and sustainability): Category of consumers known as active environmental stewards dedicated to personal and planetary health. These are the heaviest purchasers of green and socially responsible products and the early adapters who influence others heavily.

Naturalites: Category of consumers motivated primarily by personal health considerations. They tend to purchase more LOHAS consumable products versus durable items.

Product improvement: Partial changes and upgrades of certain aspects of a product. The product itself and production techniques remain the same. For example, changes in the type of coolant or substances used in a television set to improve its environmental performance doesn't change the product architecture.

Product-oriented service: The end user retains ownership of the product while the firm takes over stewardship; i.e., product trade-ins.

Product redesign: Typology of the product remains unchanged but components are improved or replaced in order to facilitate disassembly, recycling, reuse procedures, or energy use reduction.

Rebound effect: The extent of the energy saving produced by an efficiency investment that is taken back by consumers in the form of higher consumption, either in the form of more hours of use or a higher quality of energy service.

Recyclable products: Durables, made from technical nutrients, which, after they provide a service to the user, get recycled and fed back into the technical cycle of cradle-to-cradle life cycle.

Results-focused services: Satisfying customer needs (the end-result) while reducing the need for material products altogether, i.e., online banking and automated bill pay replace brick-and-mortar bank institutions.

Sustainable [market]: "Development that meets the needs of the present without compromising the ability of future generations to meet their own needs," as cited by the UN's Brundtland Report of 1987.

System innovation: The replacement of an entire socio-technical system (the product, the production chain, associated infrastructures, and institutions) by a new system.

Technical nutrients: Inorganic or synthetic materials, such as plastics and metals, which can be used many times over without any loss in quality, theoretically staying in a continuous cycle of reuse.

Triple-bottom line: Three key elements of sustainability: *people*, *planet*, and *profit*.

Unconcerned: Category of consumers either unaware or unconcerned about the environment and societal issues mainly because they do not have the time or the means; these consumers are largely focused on "getting by."

Unmarketable products: Products made from toxic materials that should not be sold at all from a cradle-to-cradle perspective.

Use service: A service to consumers instead of selling a product, i.e., product leasing (long-term rental) or product pooling (simultaneous usage of a product).

Review Questions

1. How is the triple bottom line associated with sustainability?
2. Which factors are important in the business case of sustainability?
3. What are the four underlying principles of sustainable product design?
4. What are the four paradigms for sustainable new product development?
5. How does cradle-to-cradle differ from cradle-to-grave design?
6. What is the role of service innovation in sustainable product design?
7. What are the different types of "green" consumers?

Assignment Questions

1. Pick a company for which you have worked. What would be the economic benefits of helping to restore the natural environment and/or decreasing global poverty? How could the company's sustainability efforts be turned into long-term competitive advantage?

2. What do you think? Is Unilever's model ethically and morally justifiable? Are there any other ways you can think of how to improve people's lives in impoverished areas? What role does product and service innovation have to play?

3. Think of a product you recently bought. What do you think it is made of and where do the materials come from? Which of them are toxic? What proportion of the product gets recycled? Think about how could it be made more sustainable? Would that add to its value for you? Whom else would it benefit?

4. On average, what percentage of products gets recycled? What happens with the recycled material? If you wanted to introduce a 100% recyclable computer, what would you have to change in the sociotechnical system?

5. Take a look at the latest report of the Dow Jones Sustainability Index: www. sustainability-index.com. What companies are the most sustainable according to this index? Is this what you expected?

6. Pick a company that you would consider unsustainable. Check if this company has a sustainability campaign or if it makes claims to be green or socially responsible? Is this company on the greenwashing index (www.greenwashingindex.com)? Could you accuse it of greenwashing? And, if so, what would be the implications for that company?

7. Pick a company that you believe enjoys a competitive advantage in its market. Find out if the company engages in any sustainable marketing activities. What eco-labels does the company use? Look at a competitor's Web site. What opportunities exist for improving competitive advantage? How can this company use sustainability to create greater competitive advantage?

Appendix

Online Video Presentations

Interview with Ron Gonan, CEO RecycleBank. Online at: http://www.busine ssinsider.com/recyclebank-innovation-2010-2#ixzz1bu6hwyhz (accessed July 2, 2013).

Find out what it would mean to apply cradle-to-cradle principles to your orga- nization. Online at: http://www.mbdc.com/detail.aspx?linkid=2&su blink. An interview with William McDonough, the inventor of cradle-to-cradle. Online at: http://www.youtube.com/watch?v=ufZix3J3S1g&feature=related

or

William McDonough: The wisdom of designing Cradle to Cradle. Online at: http://www.youtube.com/watch?v=IoRjz8iTVoo

and if you time watch the full talk: http://www.youtube.com/watch? v=vyciEjLtiCM

H P Bulmers Ltd. Case Study

In 2001, U.K.-based cider producer, Bulmers, embarked on a journey toward mak- ing sustainability a core source of competitive advantage for the company. Bulmers is the world's leading cider producer with approximately $750 million in annual sales. Bulmers has served as an example for many major corporations as to what transform- ing an organization into a sustainable enterprise does entail. Bulmers' sustainability efforts came as a surprise to many because the company operates in a highly competi- tive market, which is generally not known for its social conscience. Yet, Bulmers never saw sustainability as an added-on cost, but as an opportunity to generate more value, lower costs, and increase efficiencies. According to the former head of sustainabil- ity Charlie Bower, sustainability strategies implemented during his time at Bulmers alone are generating about £7 million a year for the company. Bower managed to suc- cessfully transform Bulmers into a sustainable enterprise and was described by Rocky Mountain Institute CEO Amory Lovins as "the most effective change agent I have ever met."[51] However, Bulmers' journey toward sustainability is ongoing and has not always been easy.

In 2001, Bulmers' sustainability efforts were kick-started with a three-day work- shop with leading sustainability experts from all over the world. The workshop was facilitated by Amory Lovins and the aim was to identify a series of initiatives to make the company more sustainable. The first initiatives were predominantly cost-cutting and risk management exercises, which received "benign endorsement" from executives. In the first year, sustainability efforts generated $450,000 in savings. These savings were primarily generated from eliminating inefficiencies in manufacturing (reduction in energy, waste, water), transportation (switching from road to rail), and supplier/ community initiatives. The latter also resulted in intangible benefits, and, in 2002, Bulmers was ranked within the UK-based *Sunday Times* "Top 50 Best Companies to Work For." One initiative that appeared particularly promising was to transform the annual 1,500 tons of apple prunings from waste into substrate for growing agricultural crops, which was valued at a potential of $10 million annually—more than the value of the apples themselves.

However, extending these initiatives and implementing sustainability within core business strategies proved far more difficult. Despite early cost savings, Bower struggled to guarantee executives' support for a more far-reaching sustainability program. The lack of top-level support partly resulted from a steep slowdown in market for cider and other "long alcoholic drinks," which led to 200 jobs being cut and a $1.8 million loss in the fiscal year 2003. During these times, anything that was not seen as contributing to the bottom line was naturally viewed unfavorably. However, Bower did not give up and began to focus Bulmers' sustainability strategy on top-line growth opportunities: brand differentiation, new products outside the core business, new businesses, and new financial strategies.

For example, the company began to investigate possibilities for sustainable orcharding, which later resulted in Bulmers launching its first organic cider. Bulmers also introduced a sustainable (i.e., triple-bottom line) accounting framework, in order to identify potential hidden (externalized) costs the company may potentially impose on the social and natural environment. This new accounting approach proved helpful in attracting new support and investors, like ethical funds, who would have normally shied away from companies within the alcoholic drinks industry.

However, establishing Bulmers as a truly sustainable brand has proved far more challenging and is still ongoing. A key challenge Bulmers faces results from the fact that its cider products are sold under various brands (Strongbow, Scrumpy Jack, Woodchuck, and so on) and that marketing Bulmers as a sustainable company is unlikely to resonate with its customers. Further, the drinks industry in general is not seen as ethical, given that societies face a variety of drink-related problems. Another problem resulted from Bulmers' customers. Market research seemed to clearly suggest that cider consumers are not interested in sustainability. For Bower, the real issue was "how to communicate a sustainability message to people who may be drinking to forget it."

Endnotes

1. Based on reports at Frito-Lay Sustainability Efforts Save $55 Million On Water, Energy, October 8, 2008. Environmental Leader. Online at: http://www.environmentalleader.com/2008/10/08/frito-lay-sustainability-efforts-save-55-million-on-water-energy/ (accessed June 2, 2012); SunChips to Introduce Completely Compostable Bags. Discovery Communications, LLC. Online at: http://www.treehugger.com/files/2009/04/sun-chips-compostable-bags.php (accessed June 2, 2011); source: S. Elliott, "Trumpeting a Move to Put the Sun in SunChips," *The New York Times,* March 27, 2008. Online at: http://www.nytimes.com/2008/03/27/business/media/27adco.html?ref=business (accessed June 2, 2012); source of bag: "SunChips Solar Coming-Out Party," April 3, 2008. Environmental Leader LLC. Online at: http://www.environmentalleader.com/2008/04/03/sunchips-solar-coming-out-party/ (accessed June 2, 2012).

2. "SunChips Pleases Litterbugs and Worms with Compostable Bags," *Melodies in Marketing*, April 23, 2009. Online at: http://www.melodiesinmarketing.com/2009/04/23/sunchips -compostable-bag-pla-pepsico-frito-lay (accessed June 2, 2012).

3. For more on the triple bottom line, see J. Elkington, "Partnerships from Cannibals with Forks: The Triple Bottom Line of 21st-Century Business," *Environmental Quality Management* 8(1) (1998): 37–51. Online at: http://online library.wiley.com/doi/10.1002/ tqem.3310080 106/abstract; J. Elkington, "Enter the Triple Bottom Line," August 17, 2004, Chap. 1. Online at: http://www.johnelkington.com/TBL-elkington-chapter.pdf (accessed June 2, 2012).

4. D. A. Lubin and D. C. Esty, "The Sustainability Imperative," *Harvard Business Review* 88(5) (2010): 42–50.

5. R. Nidumolu, C. K. Prahalad, and M. R. Rangaswami, "Why Sustainability Is Now the Key Driver of Innovation, *Harvard Business Review* (September 2009): 57–64.

6. Greener Design Staff, "Green Product Trends: More Launches, More Sales," *GreenBiz Group*, April 23, 2009. Online at: www.greenbiz.com/news/2009/04/23/green-product-trends-more-launches-more-sales#ixzz12QGyzDGh (accessed April 2011).

7. McKinsey & Company, "How Companies Think About Climate Change," McKinsey Global Survey, 2007. Online at: www.mckinsery.com/clientservice/ccsi/pdf/climate_ change_survey.pdf (accessed April 2011).

8. The Natural Edge Project, 2004. Online at: http://www.naturaledg eproject.net/Keynote. aspx

9. Lubin and Esty, "The Sustainability Imperative."

10. Typical laptop computer created by Leonardo Bommani. Online at: https://sourcemap. com/view/744 (accessed April 3, 2013).

11. "Bombardier Honoured by Switzerland for Investment in Product Development," *Thomas Reuters*, October 27, 2011. Online at: http://www.reuters.com/article/2011/10/27/idUS 132132+27-Oct-2011+HUG20111027 (accessed June 2, 2012).

12. "Marks & Spencer Launches 5-Year Eco-Plan," *Environmental Leader*, January 16, 2007. Online at: http://www.environmentalleader.com/2007/01/16/mark-spencer-launches-5-year-eco-plan/ (accessed March 27, 2012).

13. Lubin and Esty, "The Sustainability Imperative."

14. B. Commoner, *The Closing Circle* (New York: Bantam Books, 1972).

15. "Sustainability Cartoon: RealEyesVideo," April 9, 2012. Sustainability explained through animation.YouTube. Online at: http://www.youtube.com/watch?v=B5NiTN0 chj0&feature=related (accessed June 2, 2012).

16. Based on Jacquelyn Ottman, *The New Rules of Green Marketing: Strategies, Tools, and Inspiration for Sustainable Branding*. San Francisco: Berrett-Koehler Publishers (2011).

17. Online at http://psfk.com/2011/03/pepsico-releases-worlds-first-100-percent-plant-based-bottle.html.

18. Online at "http://www.environment.ucla.edu/reportcard/article.asp?parentid=5726" http:// www.environment.ucla.edu/reportcard/article.asp?parentid=5726.

19. "Sustainable Table Issue," September 2009. *Sustainable Table*. Online at: http://www.sustainabletable.org/issues/organic/ (accessed June 2, 2012).

20. Organic Trade Association, "Industry Statistics and Projected Growth," *Organic Trade Association*, June 2011. Online at: http://www.ota.com/organic/mt/business.html (accessed June 2, 2012).

21. "TURI Results to Date," *Tura Data*, 2011. Online at: http://turadata.turi.org/Success/ ResultsToDate.html (accessed June 2, 2012).

22. B. Thorpe, "How Companies Can Eliminate Their Use of Toxic Chemicals," *Clean Production Company*, June 2009. Online at: http://www.cleanpr oduction.org/library/ Factsheet3_ToxicsUseReduction.pdf (accessed June 2, 2012).

23. Online at (2007): http://www.csiro.au/Portals/Publications.aspx (accessed July 2, 2013)

24. Diane Martin and John Schouten, *Sustainable Marketing* (London: Pearson, 2011).

25. Thoraya Ahmed Obaid, "State of World Population 2007: Unleashing the Potential of Urban Growth," United Nations Population Fund, 2007.

26. "Sustainable Living," Unilever, 2012. Online at: http://www.unilever.com/sustainability/casestudies/economic-development/creating-rural-entrepreneurs.aspx (accessed March 27, 2012).

27. World Business Council for Sustainable Development, *Eco-Efficiency: Creating More Value with Less Impact* (Geneva, Switzerland: WBCSD, 2000).

28. E. U. Von Weizsäcker, A. B. Lovins, and L. H. Lovins, *Factor Four: Doubling Wealth—Halving Resource Use: The New Report to the Club of Rome* (London: Earthscan, 1998).

29. J. C. Brezet, "Dynamics in Eco-Design Practices," *UNEP Industry and Environment* 20(1–2) (1997): 21–24.

30. Simona Rocchi, "Enhancing Sustainable Innovation by Design: An Approach to the Co-Creation of Economic, Social and Environmental Value" (PhD diss., Erasmus University, Rotterdam, 2005).

31. M. Wheeland, "Puma's New Tack on Sustainable Consumption: Compostable Clothes," November 14, 2011. Online at: http://www.green biz.com/news/2011/11 /14/pumas-new-tack-sustainable-consumption-compostable-clothes (accessed June 2, 2012).

32. EPEA (Environmental Protection Encouragement Agency), 2010. Online at: http://epea-hamburg.org/index.php?id=199&L=0 (accessed viewed July 2, 2013).

33. Downcycling is the process of converting waste materials or useless products into new materials or products of lesser quality and reduced functionality.

34. S. L. Vargo and R. F. Lusch, "Evolving to a New Dominant Logic for Marketing," *Journal of Marketing* 68(1) (2004): 1–17.

35. J. Rifkin, *The Age of Access* (New York: Putnam, 2000).

36. Sandra Rothenberg, "Sustainability through Servicizing," *MIT Sloan Management Review* 48(2) (2007): 83–91.

37. A. Tukker, "Eight Types of Product-Service Systems: Eight Ways to Sustain Sustainability? Experiences from Suspronet," *Business Strategy and the Environment* 13 (2004): 246–260.

38. F. van der Zwan and T. Bhamra, "Alternative Function Fulfillment: Incorporating Environmental Considerations into Increased Design Space," *Journal of Cleaner Production* 11 (2001): 897–903.

39. This section is based on: "Green Marketing." Online at: http://en.wikipedia.org/wiki/Green_marketing (accessed June 2, 2012).

40. F. Belz and K. Peattie, *Sustainability Marketing: A Global Perspective* (Hoboken, NJ: John Wiley & Sons, 2009).

41. L. Berkowitz and K. G. Lutterman, "The Traditional Socially Responsible Personality," *Public Opinion Quarterly* 32 (1968): 169–185; W. T. Anderson and W. H. Cunningham, "The Socially Conscious Consumer," *Journal of Marketing* 36 (July 1972): 23–31.

42. M. Laroch, J. Bergeron, and G. Barbaro-Forleo, "Targeting Consumers Who Are Willing to Pay More for Environmentally Friendly Products," *Journal of Consumer Marketing* 18(6) (2001): 503–520.

43. D. Grundey and R. M. Zaharia, "Sustainable Incentives in Marketing and Strategic Greening: The Cases of Lithuania and Romania," *Baltic Journal on Sustainability* 14(2) (2008): 130–143.

44. E. Orange, "From Eco-Friendly to Eco-Intelligent," *THE FUTURIST* (September–October 2010): 28–32.

45. UL Environment, Sins of Greenwashing, "http://sinsofgreenwashing.org/findings/the-seven-sins/" http://sinsofgreenwashing.org/findings/the-seven-sins/ (accessed April 4, 2013).

46. Online at: http://www.ecolabelindex.com/ (accessed April 4, 2013).

47. F. J. Montoro-Rios, T. Luque-Martinez, and M.-A. Rodriguez-Molina, "How Green Should You Be: Can Environmental Associations Enhance Brand Performance?" *Journal of Advertising Research* (December 2008): 547–563.

48. Belz and Peattie, *Sustainability Marketing*.

49. Terra Choice merged with UL Labs in 2013. UL Environment, Sins of Greenwashing, "http://sinsofgreenwashing.org/findings/the-seven-sins/" http://sinsofgreenwashing.org/findings/the-seven-sins/ (accessed April 4, 2013).

50. UL Environment, Sins of Greenwashing, "http://sinsofgreenwashing.org/findings/the-seven-sins/" http://sinsofgreenwashing.org/findings/the-seven-sins/ (accessed April 4, 2013).

51. A. B. Lovins and L. H. Lovins, "RMI Solutions," *Rocky Mountain Institute Newsletter* 17(3) (2011). Online at: http://www.rmi.org/Content/Files/RMI_SolutionsJournal_FallWint01.pdf (accessed July 2, 2013).

Index

A

AAR, *see* After action reviews (AARs)
abbreviated names, 301–302
ABC News, 342
acceptance, 266, 281
Ace laundry detergent, 305, 339
activate, 299–300, 318
adaptation, 341
additions to existing product lines, 17, 22
Adidas, 34
advantages, perceptual maps, 172, 173
advertising, 265–268
Advil pain reliever, 59, 170
affiliate marketing, 191, 205
after action reviews (AARs), 314
agent-based models, 186, 205
Aim toothpaste, 314
air conditioning needs, 157–158
ALCOA, 35, 60–61
Alexander, Rhoda, 180–181
alignment, opportunities with strategy, 71–72
alliances, 35–36
All Nippon Airways (ANA), 290, 311,
    312–313
Alo laundry detergent, 305, 339
alpha testing, 260–261
AltaVista, 39
alternative fuel vehicles, 77, *see also specific*
    type

Amazon online retailer
    affiliate marketing, 191
    growth potential, 74
    internal ideation sources, 59
    iPad comparison, 180
American Express, 113, 340
ANA, *see* All Nippon Airways (ANA)
Anacin pain reliever, 170
analogy, forecasting by, 200–202
analysis and revitalization stage, 317
Android smartphone
    attribute rating perceptual maps, 160
    competitive attractiveness, 76
    competitors, 299
Anheuser Busch, 40
announce, 300, 318
Apple, *see also specific products*
    co-branding, 340
    competitive attractiveness, 76
    corporate culture, 29
    failures, 18
    focus, 33, 55
    industrial setting, 37
    innovations, 2
    innovation strategy, 33, 42
    market/customers/product categories, 34
    mission statement, 31–32
    outsourcing, 332
    perfection before launch, 30, 31
    portfolio mix, 43

proactive processes, 41
protection of brand, 343
psychosocial cues, 264
reactive processes, 40
reactive *vs.* proactive strategies, 39
resources, 35
risk, 3, 36, 258
technology, 34
architecture, 230–231
archival analysis, 64
Ariel Cool Clean, 37
Ariel Excel Gel, 37
arrange, 299, 318
Arrid deodorant, 191
artist, 70, 81
Asahi, 40
Ask Jeeves, 39
aspartame, 167
aspirin, *see also* Pain relievers; *specific brand*
    product failures, 315
    repositioning, 17, 24
    value maps, 155
ASSESSOR, 273
assignment questions, *see end of each chapter*
Association of National Advertisers, 4
assumption-based modeling, 184, 206
ATAR model
    with cannibalization, 194–195
    defined, 206
    managerial use, 204–205
    overview, 193
ATM, *see* Automatic teller machines (ATMs)
AT&T
    attribute rating perceptual maps, 160
    co-branding, 340
    speed to market, 296
attention, 266, 281
attitude-change models, 272, 281
attribute dependency, 68
attribute listing, 68, 81
attribute rating perceptual maps
    defined, 173
    overview, 159–161
    producing, 164–165
attributes, 126, 144
AT&T Universal credit card, 17, 24
auctions, 270, 284, 308
automated bill pay, 369

automatic teller machines (ATMs), 133
autoregressive integrated moving average (ARIMA), 185, 206
autoregressive moving average (ARMA), 185, 206
availability, *see also* ATAR model
    consumer purchase decisions, 191
    general, 319
    limited, 319
awareness, 190, *see also* ATAR model

B

balanced matrix, 297, 318
"ballpark" estimates, 188
Ban deodorant, 191
Bank of America, 154
Barf, 341
Barnes and Noble's Nook, 180
Barton, Jim, 217
base of the pyramid, 375
BASES, 273
Bass diffusion model, 185, 197–198, 206, 210–212
Bayer, 155
behaviors and actions, team leader, 331
believability, 266–268, 281
benefit chains, 105–107, 116
benefits
    co-located/global teams, 330
    defined, 173
    perceptual maps, 154–155
Ben-Gay, 315
Benioff, Marc, 179
Ben & Jerry's, 301
Bennet, Brian, 310
best opportunities selection, 80–81
beta testing, 261–262
BIC pen company, 315
bicycle pumps, 130
bicycle-sharing programs, 369
BigBelly Solar, 301
BigBelly trash compactors, 302
biodegradable products, 367, 375
biological nutrients, 366–368, 375
Blackberry
    competitive attractiveness, 76
    launch timing, 310

Pre positioning, 151
    psychosocial cues, 264
Black & Decker, 34, 228–229
Blak beverage, 278–279
Blakely, Sara, 297–298
blind comparisons, 262, 281
blind tests, 263
Bloomberg BusinessWeek, 180
Bloomingdales, 298
BMW M3 coupe, 133
Boeing, 41, 311, *see also* Dreamliner
Bold cellphone, 151
Bombardier, 357
Booz Allen Hamilton, Inc., 3, 16
BOP, *see* Bottom-of-the-pyramid (BOP)
Borghetti, William, 254
bottom-of-the-pyramid (BOP), 336–337, 344
bottom-up segmentation, 279, 282
bounding boxes, 14, 23
Bower, Charlie, 379
Box-Jenkins models, 206
brainstorming
    defined, 81
    group creativity, 71
    opportunity identification and idea
        generation, 67–68
branding
    architecture, 306, 318
    co-branding, 339–340, 344
    decisions, 304–305
    depth, 307, 318
    extensions, 305–307, 318
    global, 338–341
    strategies, 338–340
    value, sustainability, 357
Braungart, Michael, 364
Braun shaver, 7–8
breadth of the product line, 306–307, 319
break-even analysis, 45
Brim, 80
Brin, Sergey, 41
British Airlines, 340
British Airways, 154
British Broadcasting Corporation (BBC),
        5–6
Britt Design Award, 124
Brookstone, 280
Brown, Tim, 236

Brundtland Report, 358, 377
Brut deodorant, 191
bubble diagrams, 43–44, *see also* Portfolio
        maps
Bud Dry, 40
Bufferin pain reliever, 170, 203
Buick vehicle, 335
Building the Business Case
    defined, 23, 116
    overview, 8, 111
    product definition, 112, 113–114
    project justification, 112, 114
    project plan, 112, 114–115
    scoping, 96
    situational analysis, 111–113
    stage-gate process, 10–12
    sustainability, 355–358
    voice of the customer analysis, 100
Bulmers, *see* HP Bulmers Ltd. case study
Burberry scarves, 343
Burger Chef, 36
Burger King, 36
business/financial assessment, preliminary,
        117
business rationale, 92, 116
business strategy, 44, 294
Business Week, 154, 332

C

Campbell's Soups, 73
Canny, Richard, 123
Canon LaserJet, 16, 24
capability, *see* Service/capability planning
capital investment, sustainability, 357–358
Capitalism, Socialism and Democracy, 50
car door, 224–225
Cargill, 189
car-sharing programs
    determinant gap maps, 158
    user-oriented services, 369
case study, HP Bulmers Ltd., 379–380
casual/regression modeling, 185
category management, 274, 282
Caterpillar, 129
causal/regression modeling, 184, 206
CBP, *see* Core benefit proposition (CBP)
cellphones, 4, *see also* Smartphones

centers of excellence, 330
CES, *see* Consumer Electronics Show (CES)
champagne, 153
channels
    of distribution, 61
    innovation strategy, 34–35
    segmentation, rollouts, 280
"chasing arrows" label, 371–372
Cheerios products, 228, 305
Cherry Coke, 41
Chevrolet Volt vehicle, 98, 124
chewing gum, 135
choice models, 102, 116
Chrome browser, 60
Chrysler, 296
chunks, 230–231, 244, 367
CI, *see* Controlled introduction (CI)
cider beverage, 380
Ciserani, Gianni, 37
classification, 16
Clorox products
    aligning opportunities, 71–72
    believability, 135, 142–143
    competitive risk, 78
    failure turned around, 317–318
closed innovation
    defined, 344
    global markets, 332
    global new product development, 332
cloud computing, 364
clustered launch, 311–312, 319
co-branding, 339–340, 344, *see also* Branding
Coca-Cola
    Blak beverage, 278–279
    brand loyalty, 318
    Coke Zero, 34, 41, 292, 304
    customer ideation, 41
    failures, 18, 255, 315–316, 318
    focus, 33
    "fridge" packs, 191
    geographical segmentation, 279
    innovation, emerging markets, 335
    intention translation model, 189
    packaging, 341
    technology, 34
    test market, 278–279
    transparency, 355
co-creation, 334, 344

coefficients, diffusion of innovation, 197–202, 210–212
coffee, 80–81, 107
Coke Black, 41
co-located/global teams, 330, 344
Columbia business school, 169
"comfort," 155–156, 158
commonality
    brand extensions, 306
    defined, 244, 320
    product architecture, 230–231
Commoner, Barry, 358
communication, 331, 370
company structure, 295–296
Compaq, 40
competition and competitors
    avoiding failures, 18
    external ideation sources, 60
    industry concerns, launch, 298–299
    innovation strategy, 34
    situational analysis, 112
    value maps, 171
competitive attractiveness, 76
components
    control, 68
    product platform, 228
comprehension, 266, 282
concept description, 144
concept development and refinement, 57, 81
concept diagnosis, 127–128
concept statement preparation
    conducting concept tests, 130–134
    information acceleration, 133–134
    presentation impact, 134
    virtual reality, 132–133
    visuals only, 131
    words and visuals, 131–132
    words only, 130–131
concept test
    concept diagnosis, 127–128
    concept statement preparation, 130–134
    concerns, 142–143
    conducting, 127–141
    customer/market research, 186
    defined, 144, 206, 282
    determining goal, 127–128
    diagnostics, 136–138
    forecasting, 128

frequency questions, 135
glossary, 144–145
information acceleration, 133–134
learning objectives, 123
likelihood of purchase ratings, 140–141
overview, 124–127, 256
positioning, 128, 141–142
presentation impact, 134
pricing, 137
product diagnosis, overall, 135–136
profiling variables, 137
purchase intention, 134–135, 138–139, 147
questionnaire development, 134–137
results, interpreting and reporting,
    137–138
sales forecasts, 139–141
sample size determination, 129
sampling guidelines, 128–129
specific attribute questions, 136–137
summary, 143–144
surveys, 128–130, 134–137
TH!NK vehicle, 123–124, 146–147
virtual reality, 132–133
visuals only, 131
words and visuals, 131–132
words only, 130–131
conducting concept tests
    concept diagnosis, 127–128
    concept statement preparation, 130–134
    conducting the survey, 134–137
    determining goal, 127–128
    diagnostics, 136–138
    forecasting, 128
    frequency questions, 135
    information acceleration, 133–134
    likelihood of purchase ratings, 140–141
    overview, 127
    positioning, 128
    presentation impact, 134
    pricing, 137
    product diagnosis, overall, 135–136
    profiling variables, 137
    purchase intent basis, 139
    purchase intention questions, 134–135,
        138–139
    questionnaire development, 134–137
    results, interpreting and reporting,
        137–138

sales forecasts, 139–141
sample size determination, 129
sampling guidelines, 128–129
specific attribute questions, 136–137
surveys, 128–130
virtual reality, 132–133
visuals only, 131
words and visuals, 131–132
words only, 130–131
conducting product use tests
    blind tests, 263
    experimental variations, 263
    issues, 263–264
    single product evaluation, 263
    summary, 264
conducting surveys, 134–137
conjoint analysis and studies, *see also* Stated
        preference
    defined, 144, 282
    price testing, 268
    purchase intention questions, 138
Conner, Raymond, 291
consideration set, 190, 206
Consumer Electronics Show (CES), 151, 295
consumer ethnocentrism, 342
consumer packaged goods, 37, 135
Consumer Reports, 152
consumers, 17, 342–344, *see also* Customers
contingency approach, 186–187
contingency plans, 313–314
contingent valuation, 268–269
Continuum
    external ideation sources, 60
    investment, 76
    problem analysis, 65–66
    sleepy markets, 76
controlled introduction (CI), 311–312, 319
controlled sales, 276–277, 282
convenience, sustainable products, 370
conventionals, 371, 375
convergent approach, 282
Cooper, Robert G., 20, 24, 125
Coors Brewing Company, 279
core benefit proposition (CBP)
    defined, 116, 244
    market components testing, 264
    overview, 153, 172
    primary needs, 221

proactive new product development process, 218–219
product definition, 113
product designs, 226
product use testing, 258
quantitative studies, 102
single product evaluation, 263
testing advertising, 265
voice of the customer analysis, 99
core team, 329, 344
Cornell business school, 169
corporate cultures, 29–30
corporate strategies, 31–49
cosmetic ingredients, 335
cost reductions, 17, 23, 356
counterfeiting, 343–344
Counterfeiting Intelligence Bureau of the International Chamber of Commerce, 343
country-of-origin, 342
cradle-to-cradle life cycle
    applying to organization, 379
    defined, 376
    marketing sustainable products, 370
    overview, 364–368
    product life cycle management, 315
cradle-to-grave life cycle, 319, 364
Craftsman line of tools, 35
Creative Destruction, 50
creative destruction, 41
creativity approach, 239
creativity resources, 84–85
Crest products
    fluoride and parenting skills, 240, 242
    Tartar Control toothpaste, 17, 22
    toothpaste caps, 236
    Whitestrips, 310
criticisms, sequential processing, 13–15, *see also* Stage-gate innovation process
crowdsourcing
    defined, 81, 344
    global market, 333–334
    new product ideas, 66–67
    opportunity identification and idea generation, 66–67
CRT-based monitors, 364
Crystal beverage, 315
CT design, 235

Cuisinart, 40
culture, 29–30, 296
cumulative variance, 162, 177
cumulative % variance, 173
cumulative variance explained rule, 162, 174
curious disbelief, 267
customer cost, 370
customer/market analysis, 187, 206
customer/market research techniques, 186
customers, *see also* Consumers
    avoiding failures, 18
    categories, innovation strategy, 34
    external ideation sources, 60
    ideation, reactive processes, 40–41
    industry concerns, launch, 299
    input, determinant gap maps, 158
    priorities, perceptual maps, 169–172
    purchases, 153–155
customer solutions, 370
customization, development process, 16–17

D

data evaluation, 109–110
Datril, 40
DCF, *see* Discounted cash flow (DCF)
de Bono, Edward, 71, 82
decision trees, 184, 206
decline, product life cycle, 315
defensive strategy, 40, 50
degree of projectization, 296–297, 319
Dell
    aligning opportunities, 72
    growth potential, 74
    industrial setting, 37
    market identification, 73
    market selection, 78
    opportunity identification, 55–57
    outsourcing, 332
Deloitte & Touche, LLP., 331
Delphi method, 183–184, 206
Delta Airlines, 340
dematerialization, 369, 376
Democratizing Innovation, 42
deodorant, 154
descriptive names, 301
design
    eco-efficiency, 362, 363

firms, external ideation sources, 60
methods, 244
new product development process, 219–222
process, 244
product concerns, launch, 294–295
sustainability, 358–361, 364
DESIGNOR, 273
design phase
  benefit chains, 105–107
  building the business case, 111–115
  data evaluation, 109–110
  defined, 23, 116
  eavesdropping, 107–108
  elicitation techniques, 105
  empathic design, 103–104
  experiential interviews, 102–103
  glossary, 116–118
  go/no go decision making, 91–94
  idea screening gate, 94–96
  idea selection process, 94–95
  learning objectives, 89
  number of ideas, 95–96
  overview, 91
  product definition, 112, 113–114
  project justification, 112, 114
  project plan, 112, 114–115
  scoping, 96–110
  scoring models, 99
  Segway Personal Transporter, 89–91
  situational analysis, 111–113
  summary, 115
  user observation, 103–104
  voice of the customer, 99–102
  Web-based eavesdropping, 107–108
design thinking
  core attributes, 240
  creativity approach, 239
  critiques, 249
  defined, 244
  methods, 239–241
  overview, 235
  problem-solving approach/process, 236
  process and methods, 236–239
  process stages, 236, 238
  references, 248
  user-centered approach, 239
destruction, minimizing, 358, 361

determinant gap maps, 158–159, 174
determining goal, 127–128
development process customization, 16–17
development stage, 8, 10, 116
Dexter Corporation, 72
diagnostic information, 137–138
diagnostic questions, 136
Dial deodorant, 191
diapers, 17
DieHard battery, 114
Diet Coke, 34, 41, 167, 194
differentiation
  brand extensions, 306
  defined, 320
  sustainability, 357
diffusion models, 185, 206
diffusion of innovation
  analogy, forecasting by, 200–202
  estimating coefficients, 198–202
  historical estimates of coefficients, 199–200
  overview, 196–198
  sales history, 199
Di Giorno brand, 339
digital waiters, 131, 136, 137, 142
direct marketing, 276, 282
DIRECTTV, 201
disadvantages, perceptual maps, 172, 173
discounted cash flow (DCF), 44
discovery stage
  concept test, 125
  defined, 23
  diagnostic information, 138
  fuzzy front end, 9
  overview, 8
  through scoping, 9–10
dish soap, 153
Disney, Walt, 91
Disneyland, 307
displacement, 68
disruptive technology, 41, 50
distinctiveness, 230–231, 244
distributed (virtual) team, 330, 344
distribution channels, 61
distribution options, 271, 272
diversity, 331
division, 68
Dow Jones Sustainability Index, 358

Dr. Oetker's frozen pizza, 342
Dreamliner, 290–291, 333, *see also* Boeing
drifters, 371, 376
drinks industry, 379–380
Drucker, Peter, 353
drugs, *see* Pharmaceutical drugs
dry beer, 40
Dry Idea deodorant, 191
DuPont, 41, 60
durable consumer goods, 280–281, 312
Dyson vacuum cleaner
    additions to existing product lines, 17, 22
    building the business case stage, 10–12
    development stage, 10–12
    discovery stage, 9
    evolution of designs, 13
    individual creativity, 70
    launch, 12
    managing product lifecycles, 13
    overview, 8
    postlaunch review, 12–13
    prototyping, 11
    scoping stage, 9–10
    testing stage, 12

E

eavesdropping, *see* Web-based eavesdropping
ECG unit, 329, 336
Eclipse Aviation, 3, 18
EcoEasy Challenge (Staples), 333
eco-effectiveness, 376
eco-efficiency, 369, 376
eco labels, 371, 373
Ecomagination, 38
economic models, 44–45, 50
economies of scale, 75, 319
EDS, *see* Electronic Data Systems (EDS)
Edsel, 18
EGO, *see* Eigenvalue greater than one
    (EGO) rule
eigenvalue greater than one (EGO) rule,
    162–163, 174
eigenvalues, 162, 174
elbow of plot, 162–163, 175
electric vehicles (EV), *see also specific brand*
    determinant gap maps, 159
    perceptual maps, 157

scoping, 98
Electronic Data Systems (EDS), 69
elicitation techniques, 105
emerging markets, innovation, 335–337
emotional names, 301
empathic design, 103–104, 116
employees, 5, 357
end-of-life (EOL), 364
Energy Star label, 371
engineering, ideation source, 59
entrepreneurial optimism biasness, 95, 116
entrepreneurial strategy, 41–42, 50
EOL, *see* End-of-life (EOL)
Equator Principles, 358
Equinox Fuel Cell vehicle, 261
estimating coefficients, 198–202
estimating sales potential
    agent-based models, 186
    analogy, forecasting by, 200–202
    ATAR with cannibalization, 194–195
    casual/regression modeling, 185
    customer/market research techniques, 186
    diffusion of innovation, 196–202
    estimating coefficients, 198–202
    expert systems, 186
    forecasting techniques, 182–186
    forecast prediction, 196
    glossary, 205–208
    historical estimates of coefficients,
        199–200
    iPad success, forecasting, 179–181
    judgment techniques, 183–184
    learning objectives, 179
    managerial use of model, 204–205
    neural networks, 186
    new product forecasting strategy, 186–188
    overview, 181–182
    parameters estimation, 203–204
    probability scales, 195–196
    purchase intention, 188–195
    quantitative techniques, 184–186
    regression, purchase probability estimates,
        202–204
    repeat purchasing, 192–194
    sales history, 199
    summary, 205
ethnographic studies, 103–104
"Etsy" for services, 294

EU Flower label, 373
EV, *see* Electric vehicles (EV)
evoked set, 190, 206
Excedrin, 203
Excel spreadsheet, 40
exclusion, inventive templates, 68
executive opinion, 183, 206
"expense," 155–156, 158
experience curve, 75, 81
experiential interviews, 102–103, 116
experimental variations, 262, 263, 282
expert systems, 186, 207
explorer, 70, 81
exponential smoothing, 184–185, 207
exposure, 266, 282
extended team, 329, 345
external ideation sources, 60–61
Extraction of Sums of Squared Loadings, 177
extrapolation, 139, 185, 207
Extra Strength Tylenol, 40, 155
Exubera, 3, 18

F

Facebook, 353, 357
Face of the Customer (FOC), 103
factor loadings, 163–164, 174
factors
    analysis, 160–161, 166–167, 174
    defined, 174
    perceptual maps, 155
    scores, 164–165, 174
    SPSS, 177
factors, dimensions
    identification, 163–164, 167–168
    number of, 161–163
    summary, 167
failures
    avoiding, 18–19
    new product development risks, 3–4
family brand, 319
farming, organic/sustainable, 360
Fartfull workbench, 303
fast follower, 40, 50
fast-moving consumer goods (FMCGs)
    defined, 282
    launch timing, 312–313

probability scales, 196
    simulated test markets, 273
fax machines, 16, 24
features, 126, 144, 174
Febreeze
    failure turned around, 317
    meaningfulness, 268
    new-to-the-world products, 16
    overall product diagnostics, 136
    relaunch, 316
FedEx (Federal Express), 91
feedback, advertising copy, 266
field testing, 258, 282–283
field trials, 262
Fiesta vehicle, 299
final specifications, 221, 244
finite resources, 354–355, 358–359, 376
Firefox browser, 60
First Energy Pvt. Ltd., 337
1st principle of sustainability, 375
fit, 219, 244
five-step process, 57–58
"fly on the wall," 103
FMCG, *see* Fast-moving consumer goods (FMCGs)
FOC, *see* Face of the Customer (FOC)
"Follow-Me-Home" program, 104
Forbes, 296
Ford Motor Company
    failures, 18
    Fiesta vehicle, 299
    psychosocial cues, 264
forecasting, *see also* Sales forecasts
    agent-based models, 186
    by analogy, 184, 200–202, 207
    casual/regression modeling, 185
    conducting concept tests, 128
    customer/market research techniques, 186
    estimating sales potential, 196
    expert systems, 186
    form, 182
    judgment techniques, 183–184
    level, 182
    neural networks, 186
    overview, 182–183
    quantitative techniques, 184–186
    references, 210
    time horizon, 182

time interval, 182
form, 219, 244
formal review process, 49
four paradigms, 362
4th principle of sustainability, 375
Franklin Templeton Investments, 154
freemium, 309, 319
frequency of purchase, 135
frequency questions, 135
Frito-Lay/SunChips
    minimizing physical destruction, 361
    redesign, 363
    solar panels, 359
    sustainability, 352–353
frontline personnel, 5
FTSE4Good, 358
Fuji camera, 229–230
Fujitsu, 153
full proposition, 131, 144
Full Service Network, 217
functional, 297, 319
functional innovation, 364–368
functional matrix, 297, 319
functional names, 301
functional new product management, 23
function innovation, 376
functions, 174, 220, 244
Fusion razor, 17, 23
future conditioning, 133
future plans, 91–92, 116
fuzzy front end, 23

G

Galaxy Nexus, 340
Galaxy SII, 295
Galaxy S IV, 310
Gantt planning tool, 232
Gardner Technologies, *see* MetaCork
Gartner, Inc., 180
gates
    defined, 23
    review, 92–94, 116
    stage-gate innovation process, 6–7
GE, *see* General Electric (GE)
GE Healthcare, 327–329, 336
general availability, 310, 319
General Electric (GE), 38, 353

General Foods, 42, 80
General Mills
    brand extensions, 305
    country-of-orgin implication, 342
    product platform, 228
General Motors
    beta testing, 261
    innovation, emerging markets, 335
    inventive templates, 69
    preliminary business/financial assessment,
        98
    Volt vehicle, 124
generating product designs, 226–227
generating product ideas, 58–63
geographical segmentation, 279
Gerry Weber, 344
Gillette division, 17, 23, 341
Gimmer 5 program, 366–368
Ginger, see Segway Personal Transporter
Glad bags, 153
glass recycling, 367
Glazer, David, 30
global brands, top 10, 338, 339
Global Business Services (IBM), 5
global innovation grid, 61–62, 82
global new product development, *see also*
        Proactive new product development
        process
    adaptation, 341
    behaviors and actions, team leader, 331
    benefits, co-located/global teams, 330
    bottom-of-the-pyramid, 336–337
    branding, 338–341
    closed innovation, 332
    consumer perceptions, 342–344
    emerging markets, innovation, 335–337
    GE Healthcare, 327–329
    glossary, 344–345
    innovation, emerging markets, 335–337
    launch, 338
    learning objectives, 327
    open innovation, 332–334
    organizing for, 329–331
    packaging, 341–342
    physical proximity of teams, 329–331
    protection of, 343–344
    reverse innovation, 335–336
    standardization, 341

summary, 344
global team, 330, 345
Global Technology Services (IBM), 5
"glocal" strategies, 112, 330
glossaries, *see end of each chapter*
Golden Chicken & Noodles soup, 73
Golden Eye, 340
Gompertz Curve, 185
Gonan, Ron, 361, 379
go/no go decision making
    design phase, 91–94
    launch management, 313
    stage-gate innovation process, 6–7
Google
    Android smartphone, 160, 299
    beta testing and versions, 30, 31, 261
    brand extensions, 307
    competitive attractiveness, 76
    corporate culture, 30
    crowdsourcing, 333
    degree of projectization, 297
    focus, 33
    industrial setting, 37
    innovation, 2, 33
    looping, 15
    market/customers/product categories, 34
    mission statement, 31–32
    portfolio mix, 43
    proactive processes, 41
    reactive processes, 40
    reactive *vs.* proactive strategies, 39
    resources, 35
    risk *vs.* reward, 36
    technology, 34
Go to Development, *see* Go/no go decision
    making
green marketing activities, 370, 372
greenwashing, 371, 373–374
Griffin, Abbie, 103
group creativity, 71
group sessions, 64
growth, 73–74, 315
Gulfstream G150, 154
Gurel-Atay, Eda, 107

H

Haagen Dazs, 307, 342
Hallmark, 296
Hamilton Beah, 40
Hanes Corporation, 35
hardtop, retractable, 167
Harvard Business Review, 354–355, 357
Harvard business school, 168–169
Hasbro, 332
Hasso-Plattner-Institut (HPI) D-School,
    236, 239
Hastings, Reed, 296
Hauser, John, 103
HCI, *see* Human-computer interface (HCI)
HDTV, 17, 23, 308, *see also* Television tubes
heavyweight team, 329, 345
hedonic scale, 263
Hewlett-Packard (HP)
    physical proximity of teams, 331
    product definition, 113
    reactive processes, 40
    regulation compliance, 357
    service economy, 6
    TouchPad discontinuation, 77
Hills Brothers, 80
Hindustan Lever, 336
historical estimates of coefficients, 199–200
holistic approach, 337, 363
home delivery measures, 276
Home Depot, 342
Honda, 296
Honey Nut Cheerios, 228
Honey Nut Clusters, 228
hospital, 114
House of Quality, 222–226, 244
Howe, Jeffery, 67
HP, *see* Hewlett-Packard (HP)
HP Bulmers Ltd. case study, 379–380
HPI, *see* Hasso-Plattner-Institut (HPI)
    D-School
Hughes Electronics Corporation, 69
human-computer interaction (HCI), 236,
    244
human-computer interface (HCI), 239
humorous names, 302
Hurricane Alex, 299
Hwang, Davy, 327–328

I

IA, *see* Information acceleration (IA)
IBM
    Dell computers, 55
    external ideation sources, 61
    open innovation, 61
    service economy, 5
idea generation, 57, 82
IdeaPad, 294
idea screening
    defined, 116
    gate and gate review, 92, 94–96
    overview, 8
ideate stage, 238
ideation methods, 63–64
identification
    factor dimensions, 163–164, 167–168
    new product development process, 16
IDEO
    external ideation sources, 60
    problem analysis, 65–66
    problem-solving approach/process, 236
Ikea products, 230, 303
imitative innovation strategy, 50
improvements/revisions
    defined, 23
    overview, 17
    sustainability, 362
inclusion, inventive templates, 68
incremental innovation, 16, 23
incremental redesign, 363
index of attractiveness, 94–95, 116
individual branding, 339, 345
individual creativity, 69–71, 82
industrial products/services
    launch timing, 312
    market testing, 280–281
industrial setting, 37–38
industry concerns
    competitors, 298–299
    customers, 299
    defined, 319
    logistics, 299
    overview, 298
    partners/suppliers, 299
    technical environment, 298
industry segmentation, 279

informal selling, 276, 282
information acceleration (IA)
    concept statement preparation, 133–134
    conducting concept tests, 133–134
    defined, 144, 282, 319
    market testing, 280–281
    pricing, 310
Ingeo, 352
inhalable insulin, 3
initial eigenvalues, 177
Innocentive, 67, 82, 334
innovation
    arena, 33
    corporate cultures, 29–30
    defined, 23
    emerging markets, 335–337
    OECD viewpoint, 16
    proactive processes, 41
    requirements, 2
    sources of, 58
    strategy, 50
    type importance, 17–18
    waves, 354
innovation strategy
    channels, 34–35
    competition, 34
    customer categories, 34
    defined, 50
    industrial setting, 37–38
    innovation arena, 33
    market categories, 34
    new product development strategy, 38–39
    overview, 22, 33
    product categories, 34
    resources, 35–36
    risk *vs.* reward, 36
    selection of, 42
    technology, 34
integration planning, 234, 244
Intel, 76
intent to purchase, 128, 144
intent transition model, 207
intent translation model, 188
interactive media, 132
internal ideation sources, 59–60
internal rate of return (IRR), 43–45
Internet
    simulated test markets, 274

virtual reality modeling language, 132
introduction, product life cycle, 315
Intuit, 104
invented names, 301
inventive templates, 68–69, 82
investment, 76–77
iPad
    failure turned around, 317
    forecasting success, 179–181
    innovation, 2
    Kindle comparison, 180
iPhone
    attribute rating perceptual maps, 160
    co-branding, 340
    cost reductions, 17, 23
    estimating position, 165
    innovation, 2
    innovation strategy, 42
    innovation type importance, 17
    Pre positioning, 151–152
    value maps, 155
    web-based eavesdropping, 108
iPod
    additions to existing product lines, 17, 22
    innovation, 2, 16
    innovation type importance, 17
    new-to-the-firm products, 17
    product architecture, 231
    reactive processes, 40
Ipsos, 273
iRobot, 77
IRR, *see* Internal rate of return (IRR)
Ishrak, Omar, 328
issues
    beta testing, 262
    conducting product use tests, 263–264
    launch, 291
    preuse reactions, 259–260
iTouch, 231
iTunes, 39
ITV, 6

J

jackknife, 144, 174
Jaguar, 301
James Bond movie, 340
jeans, 17, 24

Jobs, Steve, 41, 90–91, 296
Johnson & Johnson, 40
joining, inventive templates, 68
Journal of Advertising Research, 107
judge, 70, 82
judgmentally unaided recall, 19, 207
judgment and past product experience, 283
judgment techniques, 183–184, 207
Juster's probability scale, 195–196
justification, project, 112, 114, 117

K

Kalypso, 334
Kamen, Dean, 89, 94, 95
Kate Spade, 343
Kelley, Tom, 66
Kellogg's, 338–339
Khurana, Girish, 328
kill decisions, *see* Go/no go decision making
Kin, 3
Kindle, 180
Kirin, 40
Klenert, Josh, 179
Klopfer, Eric, 274
knowledge, product platform, 228
knowledge asset planning, 234, 244
Koch, Franz, 364
Kodak camera, 229–230
Kraft, 339
Kyoto Protocol, 356

L

labels, 371, 373
laboratory measurement, 276
laddering, 107, 117
lateral thinking
    defined, 82
    idea generation, 64
    individual creativity, 71
launch, *see also* Postlaunch
    branding, 304–307
    business strategy, 294
    checklists, product name, 302, 303
    company structure, 295–296
    competitors, 298–299
    considerations, 293

culture, 296
customers, 299
defined, 319
design, 294–295
Dreamliner, 290–291
failures, 259
freemium strategy, 309
global new product development, 338
glossary, 318–320
industry concerns, 298–299
issues, 291
launch timing, 295, 310–313
learning objectives, 289
life-cycle management, 23, 315–318
logistics, 299
management, 313–314
market orientation, 297–298
market segments, 294
monitoring launch, 313–314
naming pitfalls, 302–303
new-to-the-world products, 309–310
NPD team structure, 296–297
organizational concerns, 295–298
overview, 8, 291–292
partners/suppliers, 299
penetration, 309
postlaunch, 314
prelaunch, 292
pricing, 307–310
processes, 292, 319
product concerns, 293–295
product failure, 315–318
production, 295
product names, 301–307
skimming, 308–309
speed to market, 296
stage-gate process, 12–13
strategic launch, 293–299
summary, 318
tactical launch planning, 299–303
technical environment, 298
timing, 295, 310–313
what to avoid, 302, 304
Launch eValuate, 273
laundry detergent, 305, 337, 339
LCD monitors, 364
lead users, 62–63, 82
LEAF vehicles, 98, 124

leapfrogging strategy, 41
learning objectives, *see beginning of each chapter*
leasing, 369
legal teams, 60
L'eggs panty hose, 35
Lego toy bricks, 62
Lenovo, 294
Levi Jeans, 17, 24
LG's Optimus G Pro, 295
life-cycle management, launch, 23, 315–318
Lifestyles of Health and Sustainability (LOHAS), 370–371, 376
lightweight team, 329, 345
likability
    advertising copy, 266–268
    defined, 283
    single product evaluation, 263
likelihood of purchase ratings, 140–141
Likert scale, 135–136, 159
limited availability, 310–311, 319
linear regression, 185, 207
linking, inventive templates, 68
Local Motors Inc., 334
Logistic Curve, 185
logistic regression, 185, 207
logistics, industry concerns, 299
LOHAS, *see* Lifestyles of Health and Sustainability (LOHAS)
"long alcoholic drinks," 380
long-range planning, 234, 245
Look Smart, 40
looping, 14–15, 23
Lopp, Michael, 29
Lotus123 spreadsheet, 40
Louis Vuitton, 343
Lovins, Amory, 379
low-calorie dimension, 167
Lumia 710 (cellphone), 340
Lycos, 40

M

MAC 400, 328–329
MacBook, 103
Macy's, 280
"Made in America," 342
mainframe data storage, 364

MakerBot Industries, 299
management, launch, 313–314
manager, new products, 21
managerial input, 158
managerial use of model, 204–205
maps, 45–47
M/A/R/C, 273
market, *see* Customer/market analysis
market analysis, 112
market assessment, preliminary, 117
market categories, 34
market components testing, 186, 206–207, 283
market/customer innovation strategy, 42, 50
market identification, 57, 73–74, 82
marketing
    internal ideation sources, 59
    involvement, stage-gate process, 20
    role in design, 240, 242
    sustainability, 370–374
Marketing Science Institute, 3
Marketing Vision Research, 266
market orientation, 297–298, 320
market planning, 320
market profile analysis, 72, 82
market segments, 294
market selection
    defined, 82
    opportunity identification, 57
    opportunity identification and idea
        generation, 78
market size, 18
market testing
    customer/market research, 186
    defined, 207, 283
    durable consumer goods, 280–281
    industrial products/services, 280–281
    information acceleration, 280–281
    overview, 256, 257, 277
    rollouts, 279–280
    summary, 281
    test markets, 277–279
Marks and Spencer (M&S), 357
mass customization, 229, 245
maturity, product life cycle, 315
Maxwell House brand, 339
Mayer, Marissa, 30
Mazda, 264

MBMW Z3 Roadster, 340
McDonald's
    Arch Deluxe failure, 315
    corporate name protection, 302
    proactive processes, 42
    risk *vs.* reward, 36
    standardization or adaptation, 341
McDonough, William, 364, 379
McKnight, William L., 63
McMahon, Greg, 274
meaningfulness, 266–268, 283
means-end chain, 106, 117
mechanistic firms, 41, 295–296
MeeGo, 76
Mennen deodorant, 191
mental flexibility, 70
Mercedes Benz, 3
Merck, 37
MetaCork
    alpha testing, 261
    beta testing, 261
    product/market testing, 253–255
    product use testing, 258–259
    risk reduction, 257
metals recycling, 367
methods, design thinking, 236–240, *see also*
        Processes
MICHELIN X One tires, 17, 23
Michelob Dry, 40
Michigan business school, 169
Microsoft
    customer perceptions, 153–154
    Excel spreadsheet, 40
    new product development strategies, 3
    risk, new product development, 3
microwave oven, 133
Mini Cooper EV, 98, 231
minimarkets, 276–277, 282–283
mission critical items, 313
Mitchum deodorant, 191
MITRE, 41
M&M's candy, 231, 270–271
Mobile World Congress, 295
model mock-ups, 132
monitoring launch, 313–314
morphological analysis, 64
Motorola, 180, 332
mountain bike riders, 42

Mountain View, 36
moving average, 184, 207
MRI design, 235
Mrs. Budd's Foods, 61
M&S, *see* Marks and Spencer (M&S)
MSN TV, 316
multiple branding, 345
Mustang vehicle, 75
My Starbucks Idea Web site, 334

N

naming products, *see* Product names
Narita International Airport, 290
NASCAR racers, 62
National Football League (NFL), 340
naturalites, 371, 376
Natural Marketing Institute (NMI), 370
NatureWorks, 352
NBC Universal, 38
Needham & Co., 180
Neiman Marcus, 298
nested segmentation approach, 279, 283
Nestlé, 80
Nest (thermostat), 294
Netflix, 296
net present value (NPV), 43–45
neural networks, 186, 207
Newby, Paul, 218
New Coke, 18, 255, 316, 318
new product development strategy, 50, *see also*
        Proactive new product development
        process
new product forecasting strategy, 186–188
new product innovation strategy
    Apple, 29, 32
    channels, 34–35
    competition, 34
    corporate cultures, innovative, 29–30
    corporate strategies, 31–49
    customer categories, 34
    economic models, 44–45
    formal review process *vs.* reality, 49
    glossary, 50–51
    Google, 30, 32
    industrial setting, 37–38
    innovation, 33–38
    innovative corporate cultures, 29–30

learning objectives, 29
maps, 45–47
market categories, 34
new product development strategy, 38–39
new product portfolio management,
        43–44
overview, 30–32
portfolio management tools, 44–45
portfolio maps, 45–47
portfolio review process, 48–49
proactive processes, 41–42
product categories, 34
reactive processes, 39–41
resources, 35–36
risk *vs.* reward, 36
selection of, 42
strategic buckets, 47–48
summary, 50
technology, 34
new product portfolio management
    maps, 45–47
    overview, 43–44
    strategic buckets, 47–48
    tools, 44–45
new product position estimation, 165–166
new product program management, 23–24
Newton
    failures, 18
    ignoring problems, 317
    risk, new product development, 3
    risk *vs.* reward, 36
new-to-the-firm products, 17, 24
new-to-the-world products, 16, 24, 309–310
NFL, *see* National Football League (NFL)
Nielsen Company, 188, 273
Nieman Marcus Holiday Catalog, 340
NIH, *see* Not-invented-here (NIH)
Nike
    brand extensions, 307
    protection of brand, 343
    sustainability, 353
Nine Sigma, 334
Nintendo Wii, 279
Nirma, 337
Nissan LEAF vehicles, 98, 124
NIVEA Calm and Care deodorant, 191
NMI, *see* Natural Marketing Institute (NMI)
Nokia

co-branding, 340
competitive attractiveness, 76
psychosocial cues, 264
risk, new product development, 3–4
nondurable goods, 135
nonlikability, *see* Likability
nonlinear regression, 185, 207
Nook, 180
Nordstroms, 298
Northeastern University in Boston, 168–169
Northwestern business school, 169
not-invented-here (NIH), 61, 344
NPV, *see* Net present value (NPV)
number of factor dimensions, 161–163
number of ideas, 95–96
Nuprin pain reliever, 59, 170
Nutrasweet, 167
nutrient cycles, 366

O

observe stage, 238, 245
offensive strategy, 41, 50
OK Soda, 255, 304–305, 315
Old Spice deodorant, 191
Old World Vegetable soup, 73
online banking, 369
online video presentations, 378–379
OnStar, 69, 201–202
Oorjastove, 337
open/democratizing innovation, 64
open-ended questions, 135–136
open innovation
    defined, 82, 345
    external ideation sources, 61
    global markets, 332
    global new product development, 332–334
    physical proximity of teams, 321
Open Innovation Marketplace, 67
Opera browser, 60
operant *vs.* operand resources, 368
operators, inventive templates, 68
opportunities
    analysis, 57
    situational analysis, 113
opportunities, identification and idea
        generation

aligning opportunities with NPD strategy,
    71–72
best opportunities selection, 80–81
brainstorming, 67–68
competitive attractiveness, 76
crowdsourcing, 66–67
defined, 24
Dell, 55–56
economies of scale, 75
external ideation sources, 60–61
five-step process, 57–58
generating product ideas, 58–63
global innovation grid, 61–62
glossary, 81–83
group creativity, 71
growth potential, 73–74
ideation methods, 63–64
individual creativity, 69–71
internal ideation sources, 59–60
inventive templates, 68–69
investment, 76–77
lead users, 62–63
learning objectives, 55
market identification, 73–74
market selection, 78
overview, 57
portfolio alignment, 73
problem analysis, 65–66
reward, 77
risk, 77–78
scenario generation, 64–65
substitution, 78–80
summary, 81
Optimus G Pro, 295
options theory, 45
Orange (Europe), 340
Oreo cookie, 68
organic cider beverage, 380
organic firms, 295–296
"Organic Monitor," 360
organizational concerns, launch
    company structure, 295–296
    culture, 296
    defined, 320
    market orientation, 297–298
    NPD team structure, 296–297
    overview, 295
    speed to market, 296

Organization for Economic Cooperation and innovation, 16
organizing, 329–331
OS, *see* Overall similarities (OS) gap maps
outcomes axis, 169
outsourcing, 332–333, 345
Overall Product Diagnostic Questions, 144
overall similarities (OS) gap maps, 168–169, 174
overestimating purchase intentions, 132

P

packaged goods, 135
packaging, *see also* Frito-Lay/SunChips
    global new product development, 341–342
    petroleum-free plastic bottle, 359
    value maps, 172
Page, Larry, 41
pain relievers, 156, *see also* Aspirin; *specific brand*
Palmolive dish soap, 153
Palm's smartphone (Pre)
    estimating position, 165
    perceptual maps, 151–152
    value maps, 155
Panasonic tablet introduction, 180
panty hose, 35
paper recycling, 367–368
paradigms, sustainability, 362
parameters estimation, 203–204
parity plus product, 222, 245
partners/suppliers, 299
past product experience, *see* Judgment and past product experience
payback analysis, 45
pay-per-services, 369
paywalls, 309, 320
PDA, *see* Personal digital assistants (PDAs)
PDMA, *see* Product Development Management Association (PDMA)
penetration pricing, 309, 320, *see also* Pricing
people, product platform, 228
people, sustainability, 353, 377
PepsiCo
    Crystal beverage, 315
    customer ideation, 41
    "fridge" packs, 191

    Kona beverage, 278
    petroleum-free plastic bottle, 359
perception models, 102, 117
perceptions, consumers, 342–344
perceptual maps
    advantages and disadvantages, 172, 173
    air conditioning needs, 158
    attribute rating perceptual maps, 159–161, 164–165
    benefits and value, 154–155
    customer priorities, 169–172
    customer purchases, 153–155
    defined, 174
    determinant gap maps, 158–159
    factor analysis summary, 166–167
    factor dimensions, 161–164, 167–168
    glossary, 173–175
    identification, factor dimensions, 163–164, 167–168
    learning objectives, 151
    new product position estimation, 165–166
    number of factor dimensions, 161–163
    overall similarity gap maps, 168–169
    overview, 153, 155–157
    pain relievers, 156
    perceptual dimensions, 170–171
    Pre (Palm's smartphone), 151–152
    pricing, 170–171
    product factors, 156
    summary, 172
    transport options, 157
    types, 158–166
    value maps, 169–172
Perrier bottle, 341
personal-care hospital, 114
personal computers, 133, 143
personal digital assistants (PDAs), 152
personal names, 301
perspectives, varied, 64
PERT planning tool, 232
petroleum-free plastic bottle, 359
Pfizer, 3, 18
P&G, *see* Proctor & Gamble (P&G)
pharmaceutical drugs, 3, 7
Phillips, 332
physical destruction, minimizing, 358, 361
physical evidence, 219, 245
physical product testing, 258, 283

physical proximity of teams, 329–331
Pier One Imports, 230
Pillsbury company, 342
pirate radio stations, 5
PLA, *see* Polylactic acid (PLA)
Plan A, 357
planet, sustainability, 353, 377
Planter's brand, 339
plastics, 366–368
platform, product, 228–230
PLC, *see* Product life cycle (PLC)
Pocari Sweat, 341
pointing devices, 103
point of view stage, 238, 245
Polaroid
    risk, new product development, 3
    Zink mobile printer, 125–126, 131,
        138–139
Polo Harlequin vehicle, 68–69
polylactic acid (PLA), 352
pooling, 369
pop-up stores, 271–272
Porsche, 307
portfolio alignment, 73, 82
portfolio management, 50
portfolio management tools, 44–45
portfolio maps, 45–47, 50
portfolio review process, 48–49
positioning
    conducting concept tests, 128
    estimating, 165–166
    statement, 141–142, 144
Post-it notes, 63
postlaunch
    analysis, 314
    life-cycle management, 23
    monitoring, 292
    overview, 8
    stage-gate process, 12–13
Pre (Palm's smartphone)
    estimating position, 165
    perceptual maps, 151–152
    value maps, 155
preference, 266, 283
preference models, 102, 117
prelaunch strategizing and tactics, 292
preliminary business/financial assessment,
        98, 117

preliminary market assessment, 97, 117
preliminary technical assessment, 97
premarket testing
    controlled sales, 276–277
    customer/market research, 186
    defined, 207, 283
    laboratory measurement, 276
    overview, 256, 257, 272
    pseudosale, 272–275
    summary, 277
    trial/repeat measurement, 275–276
presentation impact, 134
presentations, online video, 378–379
Preserve, 366–368
preuse reactions, 259–260, 283
price testing
    auctions, 270
    conjoint study, 268
    contingent valuation, 268–269
    overview, 268
    testing distribution options, 270–271
    transaction data, 269–270
pricing
    conducting concept tests, 137
    freemium strategy, 309
    new-to-the-world products, 309–310
    overview, 307–308
    perceptual maps, 170–171
    resources, 322
    skimming, 308–309
    value maps, 172
primary level, 220–221, 245
principle of commonality, 306, 320
principle of differentiation, 306, 320
principle of prominence, 306, 320
principle of relevance, 306, 320
principle of simplicity, 306, 320
principles, sustainability product design,
        358–361
Prius vehicle, 16, 24
proactive new product development process,
        see also Global new product
        development
    architecture, 230–231
    building the business case, 10–12
    classification, 16
    creativity approach, 239
    design, 219–222

design thinking, 235–240
development process customization, 16–17
discovery through scoping, 9–10
failures, avoiding, 18–19
fit, 219
form, 219
function, 220
generating product designs, 226–227
glossaries, 22–25, 244–246
identification, 16
innovation type importance, 17–18
launch, 12–13
learning objectives, 1, 217
manager, new products, 21
marketing involvement, 20
methods, 239–240
overview, 2–3, 218–219
platform, 228–230
postlaunch review, 12–13
problem-solving approach/process, 236
process and methods, 236–239
process stages, 236, 238
product architecture and platform,
        227–231
risk, 3–4
role of marketing in design, 240, 242
sequential processing criticisms, 13–15
service economy, 5–6
stage-gate process, 6–13, 20
summary, 22, 243–244
technology roadmapping, 231–235
testing, 12
TiVo, 217–218
types of new products, 16–17
user-centered approach, 239
validation, 12
voice of the engineer/customer, 222–226
Whirlpool, 1–2
proactive processes, strategies, 41–42
probability of purchase, 126, 144
probability scales, 195–196, 208
problem analysis, 65–66, 113
problem-solving approach/process, 236
processes, *see also* Methods
    design thinking, 236–239
    launch, 292, 319
    product platform, 228
process innovation, 59, 82

process planning, 234, 245
process stages, 236, 238
Proctor & Gamble (P&G)
    aligning opportunities, 71–72
    branding strategies, 339
    external ideation sources, 61
    failure turned around, 317
    global markets, 332
    improvement/revisions to existing
            products, 17, 23
    industrial setting, 37
    innovation, emerging markets, 335
    investment, 76–77
    launch monitoring, 314
    meaningfulness, 268
    new product development strategies, 3
    open innovation, 61
    positioning, 154
    pricing, 310
    proactive processes, 42
    problem analysis, 65
    relaunch, 316
    reward, 77
    sleepy markets, 76
    standardization or adaptation, 341
product architecture and platform
    architecture, 230–231
    defined, 245
    overview, 227
    platform, 228–230
    technology roadmapping, 231–235
product categories, 34
product concerns, launch
    business strategy, 294
    defined, 320
    design, 294–295
    launch timing, 295
    market segments, 294
    overview, 293
    production, 295
    timing, 295
product definition, 112, 113–114, 117
Product Development Management
        Association (PDMA)
    innovation strategy selection, 42
    new product development risk, 4
    new product managers, 21
    new product portfolio management, 44

product diagnosis, overall, 135–136
product factors, 156, *see also* Factors
product improvement, 362, 376
production, ideation sources, 59
product life cycle (PLC)
  defined, 208
  forecasting, 187
  management, 315–318
  system innovation, 364–365
product manager viewpoints, 159
product/market testing
  advertising copy evaluation, 265–268
  alpha testing, 260–261
  auctions, 270
  beta testing, 261–262
  blind tests, 263
  channel segmentation, 280
  conducting product use tests, 262–263
  conjoint study, 268
  contingent valuation, 268–269
  controlled sales, 276–277
  direct marketing, 276
  durable consumer goods, 280–281
  experimental variations, 263
  geographical segmentation, 279
  glossary, 281–284
  home delivery measures, 276
  industrial products/services, 280–281
  industry segmentation, 279
  informal selling, 276
  information acceleration, 280–281
  issues, 263–264
  laboratory measurement, 276
  learning objectives, 253
  market components testing, 264–272
  market testing, 277–281
  MetaCork, 253–255
  minimarkets, 276–277
  overview, 255–257, 272
  premarket testing, 272–277
  preuse reactions, 259–260
  price testing, 268–270
  product use testing, 258–264
  pseudosale, 272–275
  risk reduction, 257–258
  rollouts, 279–280
  scanner market testing, 277
  simulated test market, 273–274
single product evaluation, 263
  speculative sale, 274–275
  summary, 281
  testing advertising, 265–268
  testing distribution options, 270–271
  test markets, 277–279
  transaction data, 269–270
  trial/repeat measurement, 275–276
product names
  checklists, product name, 302, 303
  naming pitfalls, 302–303
  overview, 301–303
  what to avoid, 302, 304
product-oriented services, 368–369, 376
product planning, 232, 245
product platform, *see* product architecture
      and platform
product redesign, 376
product use testing
  customer/market research, 186
  defined, 208, 283
  product/market testing, 256, 258–264
professional wet cleaning, 360
profiling variables, 137, 144
profit, sustainability, 353, 377
program planning, 234, 245
Project Glass, 297
projectization, degree of, 296–297, 319
project justification, 112, 114, 117
project management, *see* Team leader
project matrix, 297, 320
project plan, 112, 114–115, 117
project team, 297, 320
prominence, principle of, 306
  defined, 320
protection, new product development,
      343–344
prototype stage, 132, 238, 245
pseudosale, 272–275
psychological comfort, 155–156, 263–264
public transportation, 157, 369
Puma, 364
purchase intent basis, 139
purchase intention
  ATAR with cannibalization, 194–195
  overestimating, 132
  overview, 188–192
  questions, 134–135, 138–139, 145, 147

repeat purchasing, 192–194
purchase probability estimates, 202–204

Q

QFD, *see* Quality function deployment (QFD)
quadrant analysis, 141, 145
qualitative studies and techniques
concept test, 125
new product forecasting strategy, 186
scoping, 101–102
quality function deployment (QFD), 222–226, 246
quality of execution, 91–92, 117
quantitative studies and techniques
concept test, 125
defined, 208
forecasting techniques, 184–186
new product forecasting strategy, 186
scoping, 102
quasistrategic greening, 372
questionnaire development, 134–137
questions
concept diagnostics, 127–128
forecasting, 128
industry concerns, 298
innovative strategy, 32
interface, innovation/new product strategy, 38
launch timing, 312
name and branding strategies, 301–302
overall product diagnostics, 135
positioning, 128
pricing, 308
product concerns, 293
product use testing, 258–259
purchase intention, 134–135, 138
scenario generation, 64
specific attributes, 136
quick and dirty scoping, 96
Quicken, 104
QuickSnap camera, 229–230
Qwikster, 296

R

radical innovations, 16, 24

radio frequency identity (RFID) chips, 344
Rally Fighter vehicle, 334
Ramesh, S.S., 328
Ramsay, Mike, 217
reactive innovation process, 50
reactive *vs.* proactive strategies, 39–42
realism axis, 169
reality *vs.* formal review process, 49
really new innovations, 16, 24
really new products, 309–310, 320
rebound effect, 363, 376
recyclable products, 367, 377
Recycle Bank, 361, 379
redesign, sustainability, 362–363
regression, 164, 202–204, *see also* Causal/regression modeling
regulation compliance, 356–357
relationships, product platform, 228
relevance, principle of, 306, 320
relevant set, 190, 208
renting, 369
repeat, *see* ATAR model; Trial and repeat
repeat purchasing, 192–194
repeat shares, 193
replacement, 68
repositioning, 17, 24
reputation, 357
research & development, 4–5, 59
researcher bias, 163
resources, innovation strategy, 35–36
restaurant digital waiters, 131, 136, 137, 142
results, interpreting and reporting, 137–138
results-focused services, 369, 377
retractable hardtop, 167
revealed preference, 269–270, 283
reverse innovation, 335–336, 345
review questions, *see end of each chapter*
revisions, 17, *see also* Improvements/revisions
revitalization stage, 317
reward, 77
RFID, *see* Radio frequency identity (RFID) chips
Right Guard deodorant, 191
risks
culture, 296
new product development process, 3–4
opportunity identification and idea generation, 77–78

reduction, 257–258

-reward bubble diagram, 43–44

vs. reward, 36

roadmapping, technology, 231–235

road noise, 224

Roberts, Karl-Henrik, 358

rock slide, *see* Scree plot

Rocky Mountain Institute, 379

Roddick, Anita, 167

Rollerblades, 301

rollouts, 279–280, 283

Rolls Royce Silver Mist, 302–303

"roof" matrix, 223–224, 246

roofs, retractable on vehicles, 167

rotated factor loading matrix, 163–164, 174

Rotation Sums of Squared Loadings, 177

rotation sums of squared loadings, 174

Ryanair airline, 76

S

saccharin, 167

Safari browser, 60

Saks Fifth Avenue, 298

sales

    analysis, 187, 208

    force composite, 208, 183208

    history, 199

    internal ideation sources, 59

    potential, 181, 208

sales forecasts, *see also* Forecasting

    conducting concept tests, 139–141

    defined, 208

    estimating sales potential, 181

    likelihood of purchase ratings, 140–141

    purchase intent basis, 139

sample size determination, 129

sampling guidelines, 128–129

Sam's Club, 280

Samsung

    co-branding, 340

    estimating position, 165

    launch timing, 295, 310

    tablet introduction, 180

"sanity check," 98

Sanka, 80

scale of liking, 263, *see also* Likability

scanner market testing, 277, 282–283

scenario analysis, 183, 208

scenario analysis (what if), 187–188, 208

scenario generation, 64–65, 82

Schumpeter, Joseph, 50

scoping

    benefit chains, 105–107

    data evaluation, 109–110

    elicitation techniques, 105

    empathic design, 103–104

    experiential interviews, 102–103

    overview, 96–99

    scoring models, 99

    user observation, 103–104

    voice of the customer, 99–102

    web-based eavesdropping, 107–108

scoping stage

    benefit chains, 105–107

    concept test, 124

    data evaluation, 109–110

    defined, 24, 117

    elicitation techniques, 105

    empathic design, 103–104

    experiential interviews, 102–103

    fuzzy front end, 9

    overview, 8, 96–99

    scoring models, 99

    stage-gate process, 9–10

    user observation, 103–104

    voice of the customer, 99–102

    web-based eavesdropping, 107–108

scoring models, 99, 117

scree plot, 162, 175

scree rule, 162, 175

Scrumpy Jack beverages, 380

Seahorse Power, 301–302

sealed bid auctions, 270, 284

Sears products, 35, 114

secondary level, 220–221, 246

second but better, 40, 50, 60

Second Life, 108, 274

2nd principle of sustainability, 375

Secret deodorant, 191

segmentation, product concerns, 294

Segway Personal Transporter

    data evaluation, 109–110

    design phase, 89–91

    idea selection process, 94

    ignoring problems, 317

index of attractiveness, 95
value to consumer, 107
voice of the customer analysis, 100
selection, innovation strategy, 42
sequential processing criticisms, 13–15, *see also* Stage-gate innovation process
service/capability planning, 232, 246
service personnel, ideation sources, 59
services
    economy, 5–6
    "Etsy" for, 294
    industrial community, 280–281, 312
    results-focused, 369, 377
    role of, 369
    sustainability, 368–369
    user-oriented, 369
    voice-over-the-Internet telephone, 105
servicescape, 219, 246
sexiness, virtual reality concept statement, 133
Shakti entrepreneurs, 361
shopping vouchers, 357
SIC, *see* Standard Industrial Classification (SIC)
Silkience Self-Adjusting Shampoo, 113
similarity scaling, 168
simple moving average (SMA), 207
simplicity, principle of, 306, 320
simulated test markets (STMs), 269, 273–274
single product evaluation, 262, 263, 284
situational analysis, 111–113, 118
SK-II products, 335
skimming, 308–309, 320
skunkworks, 297, 320
Skype, 229, 331
sleepy markets, 76, 82
SMA, *see* Simple moving average (SMA)
Smart car, 17, 24
smartphones, *see also specific brand*
    attribute rating perceptual maps, 159–166
    perceptual map, 153
    Pre, 151–152
SmartTV, 76
snake plot, 159–160, 174
social media, 334, 357
social objectives, sustainability, 353
social problems, solutions, 358, 361
Soft & Dri deodorant, 191

solar panels, 359
solar-powered air conditioners, 157–158
Solyndra, 3
Sony
    Walkman, 16, 24
    Xperia Z smartphone, 295
Soup for One, 73
Sourcemap, 355
Southwest Airlines, 76
Spanx, 297–298, 301
specialized M&M's candy, 231, 270–271
specific attribute questions, 136–137, 145
speculative sale, 274–275
"speed and convenience," 155–156, 158
speed to market, 296
splitting, inventive templates, 68
Spring Tide deodorant, 191
stage-gate innovation process, 24
stage-gate process
    building the business case, 10–12
    criticisms, 13–15
    discovery through scoping, 9–10
    launch, 12–13
    marketing involvement, 20
    overview, 6–8
    postlaunch review, 12–13
    sequential processing criticisms, 13–15
    testing, 12
    validation, 12
stakeholders, sustainability, 354–355
Standard Industrial Classification (SIC), 74
standardization, 341
Stanford business school, 168–169
Staples, 333
Starbucks, 67, 80, 334
stated intentions, 139, 145
stated preference, 269, 284, *see also* Conjoint analysis and studies
statistical software packages, 161
STM, *see* Simulated test markets (STMs)
Storm cellphone, 151
strategic buckets, 44, 47–48, 50
strategic considerations, 293–299
strategic greening, 372
strategic launch, 338, 339
strategic planning, 234, 246
strategy development, sustainability, 358–362
Strongbow beverage, 380

subproblems, 226–227, 246
substitution, 78–80, 83
summaries
  concept test, 143–144
  conducting product use tests, 264
  design phase, 115
  estimating sales potential, 205
  factor dimensions, 167
  global new product development, 344
  launch, 318
  market components testing, 272
  market testing, 281
  new product development process, 22, 243–244
  new product innovation strategy, 50
  opportunity identification and idea generation, 81
  perceptual maps, 172
  premarket testing, 277
  product/market testing, 281
  sustainability, 374
Sunbeam, 40
SunChips/Frito-Lay
  minimizing physical destruction, 361
  redesign, 363
  solar panels, 359
  sustainability, 352–353
Sunday Times, 379
Suntory, 40
Super Bowl game, 340
suppliers, ideation sources, 60
Sure deodorant, 154, 191
Surge beverage, 255
surveys
  format selection, 129–130
  population selection, 128–129
  sample size determination, 129
  sampling guidelines, 128–129
sustainability in innovation
  brand value, 357
  building the business case, 355–358
  capital investment, 357–358
  costs reduction, 356
  differentiation, 357
  eco labels, 371, 373
  employees, attract and retain, 357
  finite resources, 355, 358–359
  four paradigms, 362

Frito-Lay/SunChips, 352–353
functional innovation, 364–368
glossary, 375–377
green marketing activities, 372
greenwashing sins, 373–374
HP Bulmers Ltd. case study, 379–380
learning objectives, 351
marketing products, 370–374
online video presentations, 378–379
overview, 353–355
paradigms, 362
physical destruction, minimizing, 358, 361
product design principles, 358–361
product improvement, 362
product-oriented services, 368–369
redesign, 362–363
regulation compliance, 356–357
reputation, 357
results-focused services, 369
role of services, 369
social problems, solutions, 358, 361
stakeholder pressure, 355
strategy development, 358–362
summary, 374
system innovation, 364–369
toxic materials, elimination, 358, 359–361
transparency, 355
user-oriented services, 369
Sustainability Marketing, 370
Swaddlers Sensitive New Baby diapers, 17, 23
Swatch's Smart car, 17, 24
Swiffer product line
  aligning opportunities, 71–72
  industrial setting, 37
  investment, 76
  problem analysis, 65
  problem-solving approach/process, 236
  reward, 77
  scoping, 98
  sleepy markets, 76
Swiss Army knife, 144, 174
Synergy, 152
synetics, 64
Synovate, 274
system innovation, 364–369, 377

T

Tab beverage, 167
tactical greening, 372
tactical launch
    branding, 304–307
    checklists, product name, 302, 303
    freemium strategy, 309
    global new product launches, 338, 339
    launch timing, 310–313
    naming pitfalls, 302–303
    new-to-the-world products, 309–310
    overview, 299–301
    penetration, 309
    pricing, 307–310
    product names, 301–307
    skimming, 308–309
    what to avoid, 302, 304
tailgate appliance, 5, 15
target specifications, 221, 246
Tata Nano, 335–336
team leader, 24, 331
team structure, 296–297
technical environment, 113, 298
technical nutrients, 366–368, 377
technology
    innovation strategy, 34
    roadmapping, 231–235
Teflon cookware, 60
television tubes, 364, *see also* HDTV
"Tell Award," 357
Terrachoice, 371
tertiary level, 220–221, 246
Tesla vehicles, 76
testing and validation stage
    defined, 284
    overview, 8
    purpose, 256
test markets, 277–279, 284
tests and testing, *see also* Product/market
        testing
    advertising, 265–268
    alpha, 260–261
    beta, 261–262
    blind tests, 263
    conducting product use tests, 262–263
    defined, 24, 246
    distribution, price testing, 270–271

    market components, 264–272
    market testing, 277–281
    premarket testing, 272–277
    pricing, 268–270
    product use, 258–264
    stage, 238, 246
    stage-gate process, 12
The Art of Innovation, 66
"The Best a Man Can Get," 341
The Body Shop, 60, 167
The Mountain View, 36
The Natural Step, 358
The New York Times, 309
theory of inventive problem solving, 64, 68
ThinkPad, 294
3rd principle of sustainability, 375
TH!NK vehicle
    concept statement, 130–131
    concept test, 123–124, 127, 146–147
    positioning statement, 143
    sampling guidelines, 129
    specific attribute questions, 137
3M, 3, 42, 63
Thunderbird vehicle, 75
Tide laundry detergent, 305, 339
Tiffany & Co., 353
Time, 90
Timeberland boots, 343
time series model, 184–185
timing, avoiding failures, 18
TiVo, 217–218, 226–227
T-Mobile, 340
TNS-Research International, 188, 273
TNT, *see* Turner Network Television (TNT)
tools, 44–45
toothpaste, 10, 17, 236, 314
top management, 60
top-two box score, 138
Toshiba, 153, 180
total, SPSS, 177
Toubia, Olivier, 65
TouchPad, 77
touchpoints, 219, 246
toxic materials, elimination, 358, 359–361
Toxic Use Reduction Act, 360
Toyota, 305–306, *see also specific vehicle*
tracking system, design, 314, 316–317
trade partners, 60

transaction data, 269–270, 284
transparency, 355
transport options, 157
trial, *see* ATAR model
trial and repeat, 284
trial probabilities, 193
trial/repeat measurement, 275–276
triple-bottom line, 353–354, 377
TRIZ, 64, 68
Tropical Storm Bonnie, 299
Truvia, 189, 195, 196
Turner Network Television (TNT), 301–302
Twitter, 357
two-stage market segmentation, 279, 284
Ty-D-Bol, 301
Tylenol, 40, 155, 170, 202–203

U

UberSense, 298
UCD, *see* User-centered design (UCD)
Ugg boots, 343
umbrella branding, 338, 345
U2 (music group), 16
Unattended Box idea, 2
unbelievability of product, 143
unconcerned consumer segment, 371, 377
underlying factors, 160, 175
understand stage, 238, 246
Unilever PLC
    bottom-of-the-pyramid, 336
    social problems solutions, 361
    sustainability, 353
unlinking, inventive templates, 68
unmarketable products, 367, 377
user-centered approach, 239
user-centered design (UCD), 236, 239, 246
user observation, scoping, 103–104
user-oriented services, sustainability, 369
user testing, 258, 283–284
use service, 377
US News, 65

V

validation, 12, *see also* Testing and Validation
    stage
valuation, contingent, 268–269

value
    customer judgments, 143
    customer purchases, 154–155
    defined, 175
value maps
    customer priorities, 169–172
    defined, 175
    overview, 155
Vanilla Coke, 41
vaporware, 275
variations, experimental, 263
varied perspectives, 64
Verizon, 340
Veuve Clicquot champagne, 153
Vickery auctions, 270, 284, 308
video presentations, 132, 378–379
virtual reality, 132–133, 274
virtual reality modeling language (VRML),
    132
virtual team, 330, *see also* Distributed
    (virtual) team
Visa, 340
Visions, 21
visuals only, 131
VOC, *see* Voice of the customer (VOC)
Vodaphone, 340
VOE, *see* Voice of the engineer (VOE)
voice of the customer (VOC)
    concept diagnostics, 127
    concept test, 126–127
    data evaluation, 109
    defined, 25, 118, 145, 246
    design stage, 11
    elicitation techniques, 105
    market orientation, 297
    new product development process,
        222–226
    proactive new product development
        process, 218–219
    scoping, 96, 99–102
    situational analysis, 112–113
voice of the engineer (VOE), 222–226, 246
voice-over-the-Internet telephone services,
    105
Volkswagen, 68–69, 264
Volt vehicle, 98, 124
Von Hippel, Eric, 42, 62–63
von Oech, Roger, 70, 83

vouchers, shopping, 357
VRML, *see* Virtual reality modeling language (VRML)

W

Walkman (Sony), 16, 24
Walmart
    channel segmentation, 280
    costs reduction, 356
    external ideation sources, 61
    internal ideation sources, 59
    pop-up stores, 271
    sustainability, 353
Walt Disney, 91
warrior, 70, 83
waves of innovation, 354
Wayne, Ron, 41
Web-based eavesdropping, 107–108, 118
WebCrawler, 40
WebEx, 331
WebTV, 316
weighted moving average, 184, 207
Weilbacher, William, 69
Westinghouse, 15
wet cleaning, 360
Whack on the Side of the Head, 70, 83
Wharton business school, 169
Whirlpool, 1–2
Whitwam, David R., 1
willingness to pay (WTP), 268–270
Wilson, Fred, 309
Wired, 62, 67, 90, 153
Wirtschaftswoche, 364
Wolf, Charlie, 180
Woodchuck beverages, 380
"Woody" station wagon, 75

words and visuals, 131–132
words only, 130–131
Wozniak, Steve, 41
WTP, *see* Willingness to pay (WTP)

X

Xerox, 296, 307
Xperia Z smartphone, 295

Y

Yahoo! Directory, 40
Yoplait yogurt, 342
Yugo vehicle, 342

Z

Zaarly web site, 294
Zaltman's metaphor elicitation technique (ZMET), 105, 118
Zastava Corporation, 342
"zero point," 172
Zima beverage, 279
Zink mobile printer
    concept statement, 131
    concept test, 125–126
    purchase intention questions, 138
    sales forecasts, 139
Zipcar service
    determinant gap maps, 158
    Th!nk vehicles, 124
    user-oriented services, 369
ZMET, *see* Zaltman's metaphor elicitation technique (ZMET)
Zogby International, 180